Manfred M. Fischer · Peter Nijkamp (Eds.)

# Geographic Information Systems, Spatial Modelling and Policy Evaluation

With 66 Figures

Springer-Verlag

Berlin Heidelberg New York
London Paris Tokyo
Hong Kong Barcelona
Budapest

Professor Dr. MANFRED M. FISCHER
Department of Economic and Social Geography
Vienna University of Economics and Business Administration
Augasse 2−6
A-1090 Vienna, Austria

Professor Dr. PETER NIJKAMP
Department of Regional, Urban and Environmental Economics
. Free University Amsterdam
De Boelelaan 1105
NL-1081 HV Amsterdam, The Netherlands

ISBN-13:978-3-642-77502-4        e-ISBN-13:978-3-642-77500-0
DOI: 10.1007/978-3-642-77500-0

Library of Congress Cataloging-in-Publication Data. Geographic information systems: spatial modelling and policy evaluation / Manfred M. Fischer, Peter Nijkamp (eds.). p. cm.
ISBN-13:978-3-642-77502-4
1. Geographic information systems. I. Fischer, Manfred M., 1947−  . II. Nijkamp, Peter.
G70.2.G459 1992 910'.285−dc20 92-36358 CIP

© Springer-Verlag Berlin Heidelberg 1993
Softcover reprint of the hardcover 1st edition  1993

Typesetting: K+V Fotosatz GmbH, Beerfelden

2142/7130-5 4 3 2 1 0 − Printed on acid-free paper

# Preface

In recent years we have witnessed an increasing interest in Geographic Information Systems (GIS). Despite the wave of popularity for these systems it is noteworthy that they have been around for almost thirty years. It is the development of universally available computers and new types of software which has made these systems so popular and widespread at the present time. Geographic Information Systems have moved from a concern with geo-processing in the 1960s, to management of geographic information in the 1970s and to first steps of spatial decision support systems in the 1980s. Today, geographic information systems incorporate many state-of-the-art principles such as relational database management, powerful graphics algorithms, elementary spatial operations such as calculating the accessibility or nearness of geographical objects to one another, buffering in vector maps, polygon overlay with logical operations, interpolation, zoning and simplified network analysis, as well as a high degree of user-friendliness through windows, icons, menus and pointers interfaces. However, the lack of analytical and modelling functionality is widely recognized as a major deficiency of current systems.

There is a wide agreement in both the GIS community and the modelling community that the future success of GIS technology will depend to a large extent on incorporating more powerful analytical and modelling capabilities. There is an increasing need to link spatial depictions in a GIS framework to spatial analysis techniques (statistics and modelling). Such a cross-fertilisation will also be of great value for policy analysis and evaluation in space. The various contributions in this volume bring together reflective, technical and applied research work at the interface of GIS, modelling and planning. A limited subset of these contributions has also been published as a special issue of the Annals of Regional Science (Volume 1, 1992). The volume originated from special sessions on GIS and spatial analysis organised under the auspices of the IGU Commission on Mathematical Models and in the framework of the 31rst European Congress of the Regional Science Association in Lisbon in August 1991.

We acknowledge the support rendered to this publication by the Netherlands Institute for Advanced Study in the Humanities (NIAS) in Wassenaar and the Department of Economic and Social Geography of the Vienna University of Economics and Business Administration.

Vienna/Wassenaar, June 1992                     MANFRED M. FISCHER
                                                PETER NIJKAMP

# Contents

## Part C   Applications

# Contributors

L. ANSELIN, National Center for Geographic Information and Analysis, Department of Geography and Department of Economics University of California, Santa Barbara, CA 93106, USA

M. BATTY, National Center for Geographic Information and Analysis, State University of New York at Buffalo, 301 F Wilkeson Quad, Buffalo, NY 14261-0001, USA

P. J. B. BROWN, The Urban Research and Policy Evaluation Regional Research Laboratory, Department of Civic Design, University of Liverpool, P.O. Box 147, GB-Liverpool L 69 3 BX, Great Britain

CHR. BRUNSDON, North East Regional Research Laboratory, CURDS, Newcastle University, UK-Newcastle Upon Tyne NE1 7RU, United Kingdom

A. CAN, Department of Geography, Syracuse University, Syracuse, NY 13244, USA

S. CARVER, North East Regional Research Laboratory, CURDS, Newcastle University, UK-Newcastle Upon Tyne NE1 7RU, United Kingdom

V. K. DESPOTAKIS, Department of Geography, National Technical University, Zographon Campus, GR-Athens, Greece

S. EBERHARDT, Institution for Ecosystems and Environmental Studies, Austrian Academy of Sciences, Kegelgasse 27, A-1030 Vienna, Austria

M. M. FISCHER, Department of Economic and Social Geography, Vienna University of Economics and Business Administration, Augasse 2–6, A-1090 Vienna, Austria

R. FLOWERDEW, North West Regional Research Laboratory, Lancaster University, UK-Lancaster LA1 4YB, United Kingdom

A. GETIS, Department of Geography, San Diego State University, San Diego, CA 92182, USA

M. GIAOUTZI, Department of Geography, National Technical University, Zographon Campus, GR-Athens, Greece

M. GREEN, North West Regional Research Laboratory, Centre for Applied Statistics, Lancaster University, UK-Lancaster LA1 4YB, United Kingdom

D. A. GRIFFITH, Department of Geography and Interdisciplinary Statistics Program, Syracuse University, H. B. Crouse Building, Syracuse, NY 13244-1160, USA

W. D. GROSSMANN, Institution for Ecosystems and Environmental Studies, Austrian Academy of Sciences, Kegelgasse 27, A-1030 Vienna, Austria

M. GROTHE, Department of Economics, Free University, De Boelelaan 1105, NL-1081 HV Amsterdam, The Netherlands

A. F. G. HIRSCHFIELD, The Urban Research and Policy Evaluation Regional Research Laboratory, Department of Civic Design, University of Liverpool, P.O. Box 147, GB-Liverpool L69 3BX, Great Britain

Y. LEUNG, Department of Geography, The Chinese University of Hong Kong, Shatin NT, Hong Kong

J. MARSDEN, The Urban Research and Policy Evaluation Regional Research Laboratory, Department of Civic Design, University of Liverpool, P.O. Box 147, GB-Liverpool L69 3BX, Great Britain

P. NIJKAMP, Department of Economics, Free University Amsterdam, De Boelelaan 1105, 1081 HV Amsterdam, The Netherlands

S. OPENSHAW, School of Geography, Leeds University, UK-Leeds LS2 9JT, United Kingdom

G. G. ROY, Department of Computer Science, University of Western Australia, Perth, Australia

K. RYBACZUK, Department of Geography, Trinity College, University of Dublin, Dublin 2, Ireland

H. J. SCHOLTEN, Department of Economics, Free University, De Boelelaan 1105, NL-1081 HV Amsterdam, The Netherlands

F. SNICKARS, Department of Regional Planning, Royal Institute of Technology, S-10044 Stockholm, Sweden

# Design and use of geographic information systems and spatial models

**Manfred M. Fischer[1] and Peter Nijkamp[2]**

[1] Department of Economic and Social Geography, Vienna University of Economics and Business Administration, Augasse 2–6, A-1090 Vienna, Austria
[2] Department of Regional, Urban and Environmental Economics, Free University Amsterdam, De Boelelaan 1105, NL-1081 HV Amsterdam, The Netherlands

**Abstract.** Geographical Information Systems (GIS) are capable of acquiring spatially indexed data from a variety of sources, changing the data into useful formats, storing the data, retrieving and manupulating the data for analysis, and then generating the output required by a given user. Their great strength is based on the abilly to handle large, multilayered, heterogeneous databases and to query about the existence, location and properties of a wide range of spatial objects in an interactive way. The lack of analytical and modelling functionality is, however, widely recognised as a major deficiency of current systems. There is a wide agreement in both the GIS community and the modelling community that the future success of GIS technology will depend to a large extent on incorporating more powerful analytical and modelling capabilities. This paper discusses some major directions and strategies to increase both the analytical and modelling capabilities and the level of intelligence of geographic information systems.

## 1. Introduction

The analytical tools used in planning have changed drastically in the past years. Since the mid 1980s, major breakthroughs in cost, speed and data-storage capacity of computer hardware in general and the advent of stand-alone workstations and PCs capable of running Geographic Information Systems (GIS) applications in particular, as well as sufficient progress in GIS software have made GIS technology nowadays available, affordable and accessible to many users. Such clients range from local, regional and federal agencies to small and medium sized companies in a variety of fields, such as, for example, environmental management, economic development, marketing and public service delivery. The increasing interest in GIS is also reflected in government funding of academic research, with the establishment of the National Center for Geographic Information and Analysis (NCGIA) in the USA (see Abler 1987) and the Regional Research Laboratory in the framework of the Economic and Social Research Council (ESRC-RRL) in the UK (see Masser and Blackmore 1988), as well as in the establishment of large scale professional and academic conferences dedicated to GIS (among them the IGIS Symposium in 1987, the GIS/LIS Conferences which started in 1988 and the EGIS Conferences which started in 1990). Every geographic department with some degree of self esteem has now its own GIS experts; every national physical planning agency is establishing a GIS, and even middle and large sized towns use increasingly GIS for urban planning purposes (van Teefelen 1991 and Worrall 1991). There is a wide agreement in both the GIS and the modelling community that the future success of GIS technology will depend to a large extent on incorporating more powerful analytical and modelling capabilities (see Fischer and Nijkamp 1992).

Birkin et al. (1987) speak in this context of a necessary marriage between the model-based methods and the techniques from GIS to provide adequate tools to assist decision makers. They distinguish techniques for: transformation of data, synthesis and integration of data, updating information, forecasting, impact analysis, and optimisation. They conclude that the two approaches have barely come together because of different historic traditions and research foci. A number of examples, however, can be quoted in which the power of this integration is beginning to be used (Cliff and Haggett 1988; de la Barra et al. 1984; Fedra 1986; Fedra et al. 1987).

It seems meaningful to start an endeavour for a coherent integration by a methodological reflection. Both GIS and spatial modelling aim at providing operational insight into a complex spatial system. The behaviour of actors cannot be 'explained' by GIS, but GIS may provide the tools for a meaningful collection of information as building blocks for a model, while it may also represent the output of a model in a visually attractive form, thus allowing a more appropriate interpretation of results. The design and implementation of both a GIS system and a spatial model may be based on similar methodological steps. Here we follow Hafkamp (1987) who makes a distinction between three different levels of designing an integrated systems model, based on the so-called triple-layer principle. A triple layer approach provides a set of systematic design principles for integrated systems modelling and policy analysis by making a distinction between three analysis levels:

- conceptual reflection,
- prototype experimentation,
- operational software development.

At the *conceptual* level the main components of the spatial system are identified in general terms, taking into consideration the interlinks as well as the external driving forces. At this reflective level variables are not yet necessarily included as quantified units, but may be represented as latent variables.

Next, at the level of *prototype experiments* a first (usually simplified) model structure is designed, which contains the key variables of the spatial system in such a way that its properties can be analyzed by means of simulation experiments on simple observed (manifest) variables or indicators.

Finally, the level of *operational software* refers to the integration of all data sources, information systems and calibrated models as ingredients for an applicable information systems model for conditional prediction or strategic policy analysis in a spatial system.

Methodologically, GIS and spatial models are not mutually contrasting but rather complementary modes of analysis, each having its own scope, limitations and merits. Therefore, in Sections 2 and 3 respectively, both GIS and spatial models will be given due attention.

## 2. Geographic information systems: some essential characteristics

In general a geographic information system may be defined as a computer-based information system which attempts to capture, store, manipulate, analyze and display spatially referenced and associated tabular attribute data, for solving complex research, planning and management problems. Such a system will normally embody:

- a database of spatially referenced data consisting of location and associated tabular attribute data,
- appropriate software components encompassing procedures for the interrelated transactions from input via storage and retrieval, and the adhering manipulation and spatial analysis facilities to output (including specialised algorithms for spatial analysis and specialised computer languages for making spatial queries), and
- associated hardware components including high-resolution graphic displays, large-capacity electronic storage devices which are organized and interfaced in an efficient and effective manner to allow rapid data storage, retrieval and management capabilities and facilitate the analysis.

A GIS may be considered as a subsystem of an information system which itself has five major component subsystems (see Burrough 1983, Smith et al. 1987, Guptill 1988, Linden 1990, Star and Estes 1990) including:

- data input processing,
- data storage, retrieval and database management,
- data manipulation and analysis,
- display and product generation, and
- a user interface.

These main components which are summarized in figure 1 will be described in some more detail in the sequel. Data input covers all aspects of transforming spatial and non-spatial (textual or feature attribute) information from both printing and digit files into a GIS data base. To capture spatially referenced data effectively, a geographic information system should be able to provide alternative methods of data entry. These usually include digitising (both manual and automatic), satellite images, scanning and keyboard entry. The data may come from many sources such as existing analogue maps, air photography, remote sensing (data from satellites and from sensors from other platforms), existing digit data sets (for example, from other geographic information systems), (field) surveys and other information systems. Often the data require operations of manual or automated processing prior to encoding, including format conversion, data reduction and generalisation of data, error detection and editing, merging of points into lines, edge matching, rectification and registration, interpolation (see the component of data manipulation). The level of measurement of these data may vary and range from categorical to ratio data, and from fuzzy and stochastic information to precisely measured data.

Database management functions control the creation of, and access to, the database itself. For the storage, integration and manipulation of large volumes of different data types at a variety of spatial scales and levels of resolution, a GIS has to provide the facilities available within a database management system (DBMS). Most commercial GISs (such as, for example, ARC/INFO) have a dual architecture. The non-spatial attribute information is stored in a relational database management system and the spatial information in a separate subsystem which enables to deal with spatial data and spatial queries. Such an architecture, however, reduces the performance because objects have to be retrieved and compiled from components stored in the two subsystems (Oosterom and Vijlbrief 1991). This problem is not easy to solve. Spatial data processing is performed with vector, raster or a combination of these geometric data formats (see Spencer and Menard 1989).

The most important distinguishing feature which a GIS has over a mere computer mapping system or CAD is the ability to manipulate and analyze spatial data. The manipulation and analysis procedures which are usually integrated in a GIS are often limited, however, to simple spatial operations such as:

- geometric calculation operators such as distance, length, perimeter, area, closest intersection and union,
- topological operators such as neighbourhood, next link in a polyline network, left and right polygons of a polyline, start and end nodes of polylines,
- spatial comparison operators such as intersects, inside, larger than, outside, neighbour of, etc.,
- multilayer spatial overlay involving the integration of nodal, linear and polygone layers, and to
- restricted forms of network analysis.

Product generation is the phase where final products from the geographic information system are created. The displays and products may take various forms such as statistic reports, maps and graphics of various kinds, depending upon the characteristics of the media chosen. These include video screens for an animated time-sequence of displays similar to a movie, laser printers, ink jet and electrostatic plotters, colour film recorders, micro film devices and photographic media.

The final module of a geographic information system (see Fig. 1) consists of software capabilities which simplify and organize the interaction between the user and the GIS software via, for example, menu-driven command systems.

It is noteworthy that most current geographic information systems are strong in the domains of data storage and retrieval, and graphic display. Their capabilities for more sophisticated forms of spatial analysis and decision making, however, are rather limited. This lack of analytic and modelling functionality is widely recognized as a major deficiency (see, for example, Goodchild 1991, Clarke 1990, Openshaw 1990 a, b, Worrall 1990, Fischer and Nijkamp 1992). As mentioned above, the analytical possibilities basically and usually refer to polygon overlay

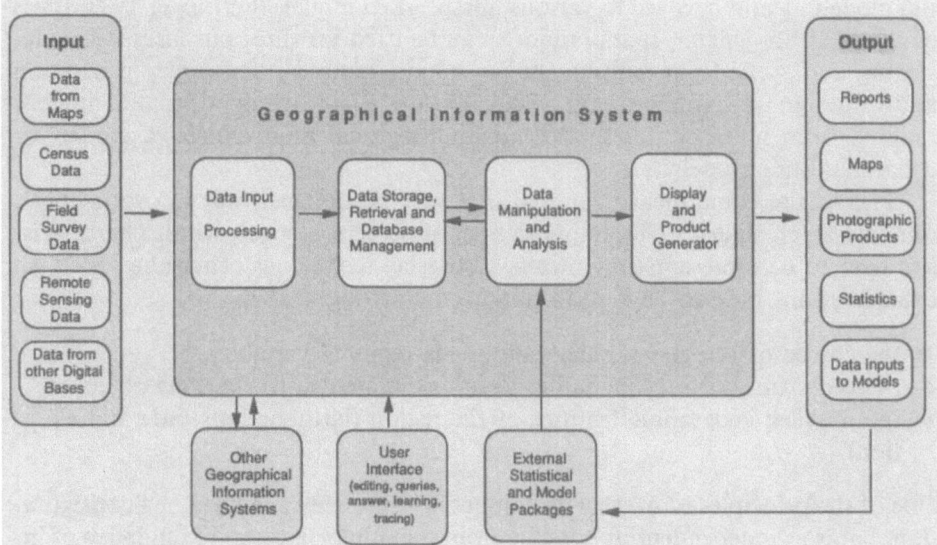

**Fig. 1.** Component subsystems of geographical information system

with logical operations, buffering in vector maps, interpolations, zoning and simplified network analysis. GIS capabilities for location- allocation problems, optimal land use allocation, and management routing vehicles for delivery of goods
and services, for example, are currently limited to simplistic types of analysis.

## 3. Spatial models

In recent years the size of models describing complex systems has dramatically
increased in size (see Burridge et al. 1991); models of more than 500 equations
are no longer spectacular. As a result, the presentation of results and their interpretation has become fraught with difficulties, especially if they are presented in
convention tabular forms. Such models tend to become unmanageable, caused by
the increase in computer speed, availability of library programmes for econometric routines and computerised data banks. There the use of modern information
systems may facilitate the task of the researcher. This also applies to the field of
spatial models, where especially GIS seems to offer a great potential.

Spatial models have already a long history, starting already in the last century
with Von Thünen's well-known land use model and Ravenstein's migration model
based on gravity principles. The tradition of models in regional science and quantitative geography has shown a strong orientation towards location-allocation
problems in space, witness the wide variety of transportation and location types
of models (see e.g. Haynes and Fortheringham 1988). Most of these models were
static in nature and regarded space as a set of discrete points (area, grids) rather
than as a continuum. The problems of ecological fallacy and spatial auto-correla-
tion are essentially caused by the sometimes arbitrary spatial demarcation in spa-

tial modelling and have led to various misspecified models (cf. Baxter 1987; Press et al. 1986). In general, spatial models can be used for three purposes, viz. forecasting and scenario generation, policy impact analysis, and policy generation and/or design (see Nijkamp et al. 1986). There is a wide variety of spatial models, ranging from input-output models to multiregional multi-objective models or urban land-use models.

The spatial component is present in a very pronounced way in overlay tools of analysis (cf. Bailey 1988). Suppose a set of objects in space, whose characteristics have to be represented by means of overlay techniques. Then the following considerations have to be kept in mind (cf. Burrough 1983):

- the choice of the geographic scale of the regional variables,
- the types of land use (including location) related to the various objects,
- the physical geographic features of the region (latitude, longitude and elevation).

One of the principles of overlay techniques is to present each layer in a configuration that is as independent as possible from the others (i.e. non-redundance of information) and as complete as possible (i.e., no loss of information). Clearly, all layers should have the same origin and refer to the same coordinate system.

This is essentially similar to a spatial model that encompasses various phenomena (or variables) and which also is based on the principle of decomposition of reality. Each layer may refer to a specific attribute (e.g. land use, transportation, thematic information, population distribution, public utilities), to the same extent as a model equation may refer to a specific variable. The level of information or the measurement level may be given in a vector or raster format; this is more or less analogous to cardinal data or interval data in modelling techniques. Similarly, also ranked and categorical data can be dealt with in overlay techniques.

Now the main question is whether it is possible to ensure a one-to-one mapping between GIS overlay information and model input or output (cf. Lewis 1977). This would require editing, updating, performing algebraic operations, using statistical methods, performing simultaneous queries and geographically displaying of all layer information. In this context, a relational data base system is a necessity. For instance, the well-known SPANS, in which the user can enter geographic data from digitizers, manually or by transforming existing types of data, allows to transform various layers of information from point, line or areal data in vector or raster form, or in a quadtree. The spatial resolution of the data depends then on the a priori established scale of the area and the data precision. Digital data sources such as digital elevation data, digital base maps and remote sensing data are necessary for a proper treatment of data input and output functions (by using e.g. scanning and manual digitizing), while both vector and raster technologies can be used for GIS colour monitor displays.

Finally it may be worthwhile to call attention for error analysis in this context (s. Heuvelink et al. 1989). The error sources may be subdivided into 3 categories, viz:

- numerical integration errors,
- locational and attribute errors in overlaying procedures,
- errors due to mathematical models.

Such error analyses are useful to identify their propagation in GIS models, their expected orders of magnitude and their impact on statistical inferences. Such approaches concern inter alia positional accuracy, attribute accuracy, logical consistency, resolution or completeness.

A different departure has been chosen in the history of (multi-)regional modelling. Here the aim has been much more to develop statistically and econometrically valid models with a rigorous application of data analysis techniques, but with much less concern for geographical dimensions and visual presentation. In general, spatial information systems were only regarded as a meaningful vehicle without too much indigenous merit.

Frequently, information systems for regional planning have been developed in close connection with multiregional models. Multiregional models − as an extension of traditional econometric modelling − aim at providing consistent and coherent information on a complex spatial world, so as to identify the main driving forces and the mechanism of a complicated multiregional system (see also Issaev et al. 1982). The aim of coherence and consistency will, in general, lead to a rejection of economic models that do not take into account the openness of a region. Thus, without a consideration of interregional and national-regional links, there is no consistency guarantee for the spatial system as a whole. Usually, there are various kinds of direct and indirect cross-regional linkages caused by spatio-temporal feedback and contiguity effects, so that regional developments may affect the competitive power of regions in a spatial system. For instance, a general national innovation policy may favour especially areas with large agglomerations. The diversity in an open spatial economic system requires coordination of planning activities on the national and regional level, leading to the necessity of using multiregional economic models in attempts to include regional profiles in national-regional development planning.

In principle, GIS and spatial modelling can be complementary approaches in two respects:

- in the area of spatial pattern and flow recognition, where spatial differences − in multiple dimensions − can be shown either by statistical representation or by GIS computer graphics and maps; the choice for one of the two representations depends on the complexity of the pattern to be represented (in various cases colour maps give a more direct visual image of main features of a data set);
- in the area of explanatory and predictive analysis, spatial models are usually more powerful than GIS in carrying out precisely numerical experiments (unless geographic information is very precisely digitized), but the final results can again be used by a GIS as an input for a user-friendly computer presentation.

The main problem is that current GISs are less suitable for explanatory or predictive modelling compared to conventional spatial analysis. Nevertheless, it is noteworthy that some progress has recently been made, e.g., in the field of spatial location-allocation modelling, where traditional spatial interaction tools have been linked to a GIS representation of the resulting patterns and flows. To some extent one may claim that GIS seems to be a more proper tool for perception/visualisation methods for impact analysis rather than a direct impact tool in itself.

Also in dynamic modelling the same remarks hold essentially. GIS is so able to produce dynamic maps in connection with the result of a dynamic descriptive explanatory or predictive dynamic model. In this context GIS is a meaningful vehicle for dynamic scenario analysis. A good example can be found in Despotakis (1991), who used a dynamic system model in a GIS environment in order to identify and evaluate various strategic development scenarios for a sustainable evolution of the Greek Sporades islands.

In recent years, much attention has been given to the coherent and joint use of GIS and spatial models in spatial analysis and planning. An integration of both modes seems for the time being a too ambitious task. However, there is a field where GIS and spatial modelling may provide immediately and consistently analytical support, viz. in the area of policy analysis (evaluation and planning).

## 4. GIS and spatial models for evaluation and planning

It is noteworthy that both GIS and spatial models are only an intermediate tool in order to generate a strategic integrated information system for spatial analysis and planning. Such a system may also comprise decision support systems and expert systems, evaluation methods, information from a referendum etc. This is sketched in Fig. 2.

Decision support and expert systems may be viewed as systems which achieve expert-level performance, utilising artificial intelligence (AI) concepts and programming techniques such as connectionist models (for example, the concept of neural networks), symbolic representation, inference and heuristic search. A fully fledged decision support and expert system may be considered to have four essential components (see e.g., Benfer et al. 1991, Forsyth 1989, Fischer and Nijkamp 1992); (see also Fig. 3):

- a knowledge base consisting of spatial and non-spatial knowledge about some substantive domain,
- an inference engine consisting of a set of general (search and) inference procedures to reason from knowledge in the knowledge base and to infer additional conclusion,
- a knowledge-acquisition module to assist in expressing knowledge in a form suitable for inclusion in the knowledge base,
- an user interface which assists the user to consult the spatial expert system.

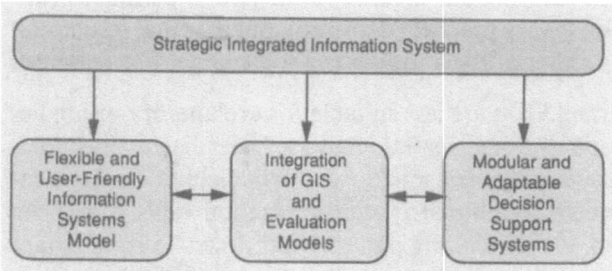

**Fig. 2.** Components of strategic integrated information systems

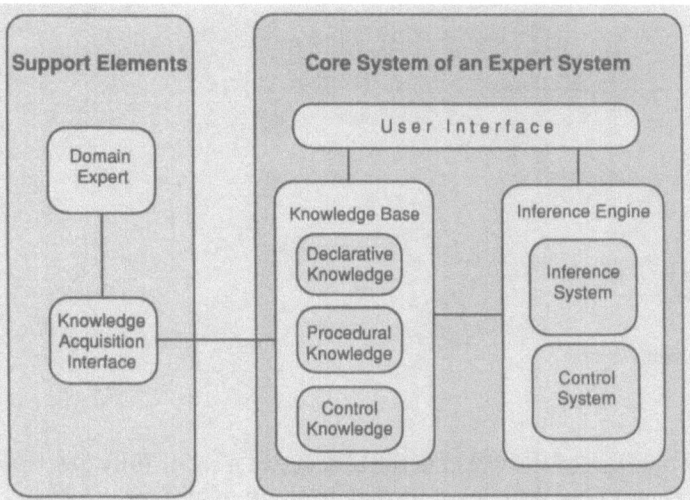

**Fig. 3.** System architecture of a knowledge-based geographic information system

The knowledge base is elicitated from a domain expert and reformulated as a collection of rules, a network of facts or a frame-based structure. The relative maturity of decision support and expert systems is reflected in the variety of alternative, and sometimes complementary knowledge representation techniques. The mainstream approach to knowledge representation has been inspired by the insights of mathematical logic and is symbolic in nature. The production rules format, simple condition-conclusion or condition-action statements (IF condition THEN action, or IF condition THEN action 1 ELSE action 2), is the most widely used way of symbolic encoding of knowledge. More complex representations are mostly frame or object-oriented systems. In contrast to a symbolic representation of knowledge the connectionist architectures, strongly inspired by the concept of neural networks, are more committed to the principle of inter-connectivity (Shadbolt 1989). One major difference between symbolic and connectionist types of knowledge refers to whether knowledge has to be explicitly or implicitly represented (see Forsyth 1989, Fischer 1992).

In general, the knowledge in a knowledge base may consist of three types:

- Declarative or factual knowledge (spatial object knowledge and useful spatial and non-spatial feature information about the objects),
- Procedural or strategic knowledge (the core of a knowledge base) including concepts and various relationships (definitional, taxonomic, associational and empirical relationships) among them,
- Control knowledge (usually based on rules of thumb) for a variety of control strategies which allow to rule out obviously wrong solutions and to focus on the mere promising ones.

Such knowledge-based systems are particularly useful in case of 'qualitative reasoning', one of the scientific fields that forms the basis for artificial intelligence (see Weld and De Kleer, 1990). Qualitative reasoning aims to make good quantiza-

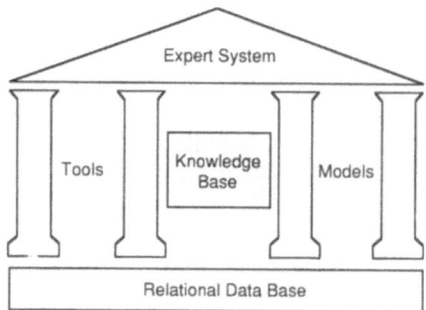

**Fig. 4.** The information systems temple

tions of continuous properties of the world in the absence of a set of fully descriptive equations (see Forbus 1988). Applications may relate to quality of life analysis in urban areas, neighbourhood quality analysis, search analysis on the housing market, marketing etc. All such issues are multidimensional in nature and a spatial mapping of such items via GIS based systems offers a great potential for a comprehensive and solid analysis.

The previous observations can also be summarized in the so-called information system temple sketched in Figure 4, where a relational data base is regarded as the foundation of a knowledge base, which − in conjunction with mathematical, statistical and software tools as well as a rule-based inference engine − form the ingredients for an expert system. It is evident, however, that so far the operationalisation of such analytical approaches is very rare, and therefore there is a need for a further investigation of experiences in this field.

### 4. Scope and structure of the book

The aim of the present volume is to bring together interesting experiences at the interface of GIS and spatial modelling with a particular view on the practical use or operational character of these approaches. Various results from different countries will be presented. The book is organized according to the following main subdivision into: *theoretical and conceptual issues* (part A), *technical issues* (part B), and *applications* (part C).

In part A a state of the art overview of recent developments at both the theoretical and conceptual level is given. After this introductory chapter by Fischer and Nijkamp on the design and use of GIS models, Openshaw calls attention for the potential of artificial intelligence methods in relation to spatial modelling and GIS. Next, Anselin and Getis focus attention on the connection between GIS and spatial statistical analysis. The final contribution in part A is offered by Batty, who positions GIS in the framework of urban planning and policy analysis.

Part B is primarily concerned with technical issues in GIS modelling. The first contribution is given by Flowerdew and Green, who discuss areal interpolation methods. Then Rybaczuk deals with the issue of information based rules for sliver polygon removal in GIS, followed by a paper by Griffith on the conversion of spatial statistics tools to GIS functions. Next, Brunsdon and Carver treat the accuracy of digital representations of two-and-three-dimensional objects by means of simulation methods, and Leung discusses conceptual and technical issues in developing a software environment for decision-making using GIS, fuzzy logic and expert systems technologies. Finally, Openshaw summarizes some basic features of artificial neural networks and applies neurocomputing to spatial interaction modelling.

In the final part of this volume a broad spectrum of applications is dealt with. Grossmann and Eberhardt operationalize GIS in the framework of dynamic modelling; Roy and Snickars provide an empirical illustration of computer-aided regional planning; Can gives an extensive treatment of residential quality assessment based on GIS; Hirschfield, Brown and Marsden discuss a prototype GIS for Urban Programme Impact Applications in the UK; Despotakis, Giaoutzi and Nijkamp present an extensive empirical dynamic GIS model for policy simulation and evaluation on the Greek islands; and Grothe and Scholten aim at modelling catchment areas including applications to residential zoning for the elderly and a spatial interaction model based on GIS for elderly care. All these illustrations demonstrate convincingly that GIS provides a great potential for spatial modelling and applied policy analysis.

## References

Abler RF (1987) The National Science Foundation Center for Geographic Information and Analysis. Int J Geogr Inf Syst 1:303–326

Aronoff S (1989) Geographic Information Systems: A Management Perspective, WDL Publications, Ottawa

Bailey RG (1988) Problems with Using Overlay Mapping for Planning and Their Implications for Geographic Information Systems. Environ Manag 12:11–17

Barra T de la, Perez B, Vera N, Tranus J (1984) Putting Large Models into Small Computers. Environ Plann B 11:87–102

Batty M (1988) Informative Planning: The Intelligent Use of Information Systems in the Policy-making Process. Technical Reports in Geo-Information Systems, Computing and Cartography, no 10.

Baxter MJ (1987) Testing for Misspecification in Models of Spatial Flows. Environ Plann 19:1153–1160

Benfer RA, Brent EE Jr, Furbee L (1991) Expert Systems. Sage University Paper 77

Birkin M, Clarke GP, Clarke M, Wilson AG (1987) Geographical Information Systems and Model-based Locational Analysis; Ships in the Night of the Beginnings of a Relationship? WP-498, School of Geography, University of Leeds, Leeds

Burridge MS, Dhar S, Mayes D, Meen G, Neal E, Tyrell N, Walker J (1991) Oxford Economic Forecasting's System of Models. Econ Modelling 8:227–413

Burrough PA (1983) Principles of Geographical Information Systems for Land Resources Assessment, Clarendon Press, Oxford

Clarke M (1990) Geographical Information Systems and Model Based Analysis: Towards Effective Decision Support Systems. In: Scholten HJ and Stillwell JCM (eds) Geographical Information Systems for Urban and Regional Planning, Kluwer, Dordrecht pp 165–175

Cliff AO, Haggett P (1988) Atlas of Disease Distributions; Analytic Approaches to Epidemiological Data, Basil Blackwell, Oxford

Despotakis V (1991) Sustainable Development Planning Using Geographic Information Systems, PhD Diss, Dept of Economics, Free University, Amsterdam

Fedra K (1986) Decision Making in Resources Planning; Models and Computer Graphics, Paper presented at the UNESCO symposium on „Decision Making in Water Resources Planning", Oslo, 5–7 May

Fedra K, Li Z, Wang Z, Zhao C (1987) Expert Systems for Integrated Development; A Case Study of the Shanxi Province, The People's Republic of China, SR–87–1. International Institute for Applied Systems Analysis, Laxenburg, Austria

Fischer M, Nijkamp P (1992) GIS and Spatial Analysis. Ann Reg Sci 26:3–17

Fischer MM (1992) Expert Systems and Artificial Neural Networks for Spatial Analysis and Modelling. Essential Components for Knowledge-Based Geographical Information Systems. NCGIA Specialist Meeting, San Diego, April 16–18, 1992

Forbus K (1988) Qualitative Physics: Past, Present and Future. In: Shrobe H and American Association for Artificial Intelligence (eds) Exploring Artificial Intelligence. M. Kaufmann, Palo Alto, Cal, pp 239–296

Forsyth R (ed) (1989) Expert Systems. Principles and Case Studies. Chapman and Hall, London

Goodchild MF (1991) Progress on the GIS research agenda. In: Harts J, Ottens HFL and Scholten MJ (eds), EGIS '91. Proceedings of the Second European Conference on Geographical Information Systems, vol 1. EGIS Foundation, Utrecht, pp 342–350

Guptill SC (ed) (1988) A Process for Evaluating Geographic Information Systems. Technology Exchange Working Group – Technical Report 1, Federal Interagency Coordinating Committee on Digital Cartography

Hafkamp WA (1987) Economic-Environmental Modelling in National Regional Systems, North-Holland, Amsterdam

Haynes KE, Fortheringham AS (1988) Gravity and Spatial Interaction Models, Sage Publ., Beverly Hills

Heikkila EJ, Kim JE, Moore II (1989) Incorporating Expert Geographic Information Systems into Urban Land Use and Transportation Planning Models, Sistemi Urbani 2:161–176

Heuvelink GBM, Burrough PA, Stein A (1989) Propagation of Errors in Spatial Modelling with GIS. Int J Geogr Inf Syst 3:303–322

Issaev B, Nijkamp P, Rietveld P, Snickars F (eds) (1982) Multiregional Economic Modelling: Practice and Prospect, North-Holland, Amsterdam

Laurini R, Thompson D (1991) Fundamentals of Spatial Information Systems, Academic Press, London

Leary M (1987) Development Control; The Role of Expert Systems, Town Planning Review 58:331–342

Lewis P (1977) Maps and Statistics. Wiley, New York

Linden G (1990) Education in geographical information systems. In: Scholten HJ and Stillwell JCM (eds), Geographical Information Systems for Urban and Regional Planning, Kluwer, Dordrecht, pp 191–201

Lo CP, Shipman RL (1990) A GIS Approach to Land-Use Change Dynamics Detection. Photogram Eng Remote Sensing 56:1483–1491

Maguire DJ, Goodchild MF, Rhind DW (eds) (1991) Geographic Information Systems: Principles and Applications. Longman, Harlow, Essex

Masser I (1991) Promoting GIS Awareness: The European Dimension. In: Harts J, Ottens HFL and Scholten HJ (eds) EGIS '91, Proceedings of the Second European Conference on Geographical Information Systems, vol 2. EGIS Foundation, Utrecht, pp 700–706

Masser I, Blakemore M (1988) The Regional Research Laboratory Initiative: The Experience of the Trial Phase. ESRC, Swindon

Nijkamp P, Rietveld P, Snickars F (1986) Regional and Multiregional Economic Models: A Survey, Handbook of Regional and Urban Economics. Elsevier, Amsterdam

Oosterom P van, Vijlbrief T (1991) Building GIS on the top of the open DBMS „Postgres". In: Harts J, Ottens HFL, Scholten HJ (eds) EGIS '91. Proceedings of the Second European Conference on Geographical Information Systems, vol 2. EGIS Foundation, Utrecht, pp 775–787

Openshaw S (1990a) Spatial analysis and geographical information systems: A review of progress and possibilities. In: Scholten HJ, Stillwell JCM (eds): Geographical Information Systems for Urban and Regional Planning. Kluwer, Dordrecht, pp 153–163

Openshaw S (1990b) A spatial analysis research strategy for the Regional Research Laboratory Initiative, Regional Research Laboratory Initiative Discussion Paper No. 3, ESRC, Swindon

Press WH, Flannery BP, Tenkolsky SA, Vetterling WT (1986) Numerical Recipes: The Art of Scientific Computing, Cambridge University Press, Cambridge

Sagalowicz D (1984) Development of an Expert system. Expert Systems 1:137–141

Scholten HJ, Stillwell JCH (eds) (1990) Geographical Information Systems for Urban and Regional Planning, Kluwer, Dordrecht

Shadbolt N (1989) Knowledge representation in man and machine. In: Forsyth, R (ed): Expert Systems. Principles and Case Studies. Chapman and Hall, London, pp 142–70

Smith TR, Menon S, Star JL, Estes JE (1987) Requirements and principles for the inplementation and construction or large-scale geographic information systems. Int J Geogr Inf Syst 1:3–31

Spencer R, Menard RD (1989) Integrating vector/raster technology in a geographical information system. In: American Congress on Surveying and Mapping, American Society for Photogrammetry and Remote Sensing, Association of American Geographers, Urban and Regional Information Systems and AM/FM International (eds): GIS/LIS '89. Proceedings, vol 1, pp 1–8

Star J, Estes J (1990) Geographic Information Systems. An Introduction. Prentice-Hall, Englewood Cliffs

Teeffelen PBM. van (1991) Analyzing interaction data with GIS techniques in developing countries: The use of „flowmap" in Mali and Mexico. In: Harts J, Ottens HFL, Scholten HJ (eds) EGIS '91. Proceedings of the Second European Conference on Geographical Information Systems, vol 2, EGIS Foundation, Utrecht, pp 1058–1066

Weld D, De Kleer J (eds) (1990) Qualitative Reasoning about Physical Systems, M Kaufmann, Palo Alto, Cal

Worall L (ed) (1990) Geographic Information Systems, Belhaven, London

Worall L (ed) (1991) Spatial Analysis and Spatial Policy Using Geographic Information Systems, Belhaven, London

**Part A**

**Theoretical and conceptual issues**

# Some suggestions concerning the development of artificial intelligence tools for spatial modelling and analysis in GIS

**Stan Openshaw**

School of Geography, Leeds University, Leeds, LS2 9JT, UK

**Abstract.** The last 5 years have seen the development of artificial intelligence (AI) methods that are capable of being applied to many practical problems. The paper looks at some of the actual and potential applications for AI in the area of space analysis and modelling relevant to GIS. Of particular importance here is the development of data exploration tools for pattern description, relationship seeking, and modelling. The paper reviews some of the applicable heuristic search, artificial life, genetic optimization, and neurocomputing methods.

## 1. Introduction

The last five years have witnessed significant advances in both the speed and price performance of computer hardware and by the emergence of practical and useful artificial intelligence (AI) technologies. Many of these AI developments are in principle generally applicable to geography and regional science. The purpose of this paper is to make brief suggestions as to how some of these AI methods might be used to solve some of the outstanding analysis and modelling problems relevant to GIS.

It is not a matter of having to start from scratch but of looking for GIS relevant applications of existing AI procedures. Indeed some of the existing AI tools are of such general applicability that it is probably not too soon to start talking about the prospects for a third quantitative and computational revolution in geography based on AI rather than statistical techniques or mathematical modelling. In this context, the decline of interest in traditional quantitative methods and mathematical modelling procedures during the 1970's and 1980's may well have been extremely beneficial to this impending AI revolution. The weakening of the hold of orthodox quantitative methods in the face of an onslaught of qualitative methodologists may ensure that geography will be more receptive to the new methods and more advanced approaches than otherwise might have been the case. At the same time, GIS provides a degree of cybernetic cohesion that may reverse the ongoing process of geographic disintegration (Openshaw 1991). Another view of history would be that the 1960's quantitative revolution in geography and regional science had run out of steam, run out of new ideas, had become too dependent on old-fashioned statistical methods, and in places had become so mathematically complex and overly theoretical at the expense of realistic empirical applications, that the field is ready for a new era of advanced computing based tools that are essentially data rather than theory driven, and applied rather than theoretical in their motivation.

Another essential component is the Geographical Information Systems (GIS) revolution which above all else emphasises geographical data analysis in the context of data led computer systems. Suddenly, and increasing, the world is awash with more geographically referenced data than ever before in human history. The emergence across a broad front of new, computerised, digital map data sources and map related information systems are providing a wealth of analysable geographical information in a large number of diverse areas. Gradually, but at an ever increasing pace, most of the world's mappable information is becoming available in computer databases for display, manipulation, decision support, and analysis. This is happening at all scales, from local, to regional, to national, and increasingly global. The flood of data causes a number of major crises concerned with its management, its storage, and soon its modelling and analysis. Some of these problems will be resolved by further advances in computer hardware. Some of the needs for analysis and modelling will be met by existing technologies. However, it is also apparent that many of the necessary modelling and analysis tools needed to cope with this new data rich but theory poor state do not yet exist in sufficient numbers to meet the likely demands from applications.

The GIS toolkits that exist today have mainly focused on the data capture, storage, and cartographic display processes. What is often classed as sophisticated spatial analysis and modelling is often no more that map data manipulations such as polygon overlay and buffering. By contrast the inventory of existing models and model-making methodologies relate to an era when data was often rare, and the challenge was to develop high quality, theoretically plausible, models mainly for conceptual rather than empirical objectives (see for example, Wilson and Bennett 1985). There is no denying that in the short-term the GIS revolution will result in a re-birth of many of these models in applied contexts, whereas previously they were mainly academic curiosities of no great empirical relevancy because of the absence of adequate data (Openshaw 1989). But there are limits to what they can achieve. Quite often they are too complex, too sophisticated in a theoretical sense, inherently poor performers because they are optimised for theoretical representation rather than data representation, and they are available in too few application areas. Many of these models also reflect a normal science paradigm. A similar situation characterises the existing inventory of spatial analysis methods (see Openshaw 1991). Most reflect a strongly inferential approach based on attitudes and methods that many statisticians would probably now regard as inappropriate. The irony is that not only are many of the more widely used statistical methods inherently inapplicable but they make assumptions about the nature of spatial data that are known to be wrong.

The argument is expressed that the GIS revolution is causing a major crisis that affects both conventional mathematical models and many spatial analysis techniques. These problems are being highlighted at a time when it seems to be increasingly feasible to start again using AI technologies. GIS emphasises the need for a data driven and more exploratory approach but there are few relevant tools able to operate in that way. The new models of the GIS era will have to be created in the first instance from data rather than from theory and they will be increasingly computationally dependent rather than analytical in nature. New spatial analysis methods are also needed that are able to spot patterns and relationships without being told either *where* to look or *what* to look for or *when* to look. AI offers some prospect of being able to meet some of these needs and this paper provides a speculative and broadly based review of some of potentially promising new developments.

## 2. Why do we need AI?

The answer reflects perceived difficulties with the existing and established analysis and modelling methods when faced with the explosion of information being created by the GIS revolution. The deficiencies associated with conventional approaches and models are discussed in Openshaw (1990a). The problems include the following:

(1) The insubstantial nature of much of the 'established' theory that has been traditionally used to specify models.
(2) The systems and processes being modelled are very complex and contain a large stochastic component which makes model specification hard and patterns and processes difficult to specify.
(3) The modelling problems constitute Weaver type 3 systems. The available analytical mathematical model building methods find these difficult to handle in any realistic manner.
(4) There is seemingly a shortage of new analogies and potentially applicable ideas that remain to be borrowed from other disciplines that might be relevant to the GIS world. Historically, this ideas 'borrowing' process has been an important source of new modelling concepts but the lode seems to have been mined out. The time has come to rely on indigenous thinking about the geography of problems. The general systems paradigm has broken down: if indeed, it was ever of any great use.
(5) GIS is actively stimulating the emergence of a data rich but theory poor environment in which there are often no longer any good prior ideas of what patterns and relationships might exist or what to expect. Additionally, the available data tends to be: (a) highly serendipitous in nature, quality, and quantity whilst being largely uncontrollable in terms of content and detail in that it is available as a byproduct of computerisations developed for essentially bureaucratic rather than research or modelling purposes; (b) there are often large quantities of data with whole population data sets being available; (c) there are increasing numbers of databases to be analysed; (d) much data now exists in areas where there is little, if any, prior theory; (e) the spatial details are greatly enhanced but the existing methods tend not to be particularly good at handling such details; (f) the tem-

poral detail are similarly improved but there are hardly any relevant space-time modelling methodologies that can be applied; and (g) the data is often noisy either because of error propagation in GIS and, or, due to the nature of the data source, yet few methods can handle spatial data uncertainty in a reasonable way. (6) The geographical nature of the information means that considerable reliance has to be placed on the analysis of surrogates as substitutes for the non-availability of more direct measurements of process or causal factors or important variables. This both complicates the analysis, sets limits to what can be achieved, but is increasingly a fact of GIS-life that cannot readily be changed. Modelling and analysis have to be performed within the constraints of available data. This has always been the case, but seemingly the GIS modeller has little control over the available information. On the other hand, there is now no shortage of data. There is often a plethora of possibly relevant, semi-relevant, and seemingly totally irrelevant variables available. Data selection now becomes a major task.
(7) Geographical data also possess special characteristics that have traditionally been ignored in modelling work. These include: (a) spatial dependencies, (b) space-time dependencies, (c) non-stationarities, (d) varying degrees of data reliability that are geographically structured, (e) sensitivity to scale and aggregation effects; (f) complex measurement errors; (g) multivariate non-normality; (h) nonlinearities and discontinuities; (i) mixtures of measurement types; and (j) data that are noisey. The task is often to achieve the best that is possible given difficult, sometimes poor, data. The GIS's complement to exploratory data analysis (EDA) might well be termed poor-data analysis (PDA). Yet it is often important to analyse data that are known to be wrong; for instance, the analysis of a rate variable based on the wrong denominator because it is out of date but no better variable exists.
(8) Finally, there is an increasing imperative to do something with the available data. The research design process is outside the spatial analyst's or modeller's control but there is often a strong demand for analysis and modelling. It will be difficult to ignore these requests and there is, therefore, an urgent need to develop computer technologies that can cope with them without making too many dangerous assumptions or running the risk of producing results which can only be described as rubbish. The great danger is that many existing methods which were not conceived of as applied tools will be rediscovered and run in this new context and in the process they will implicitly malfunction and produce results which may be wrong, harmful, and sometimes totally misleading.

The GIS data revolution needs to be accompanied by an appropriate methods revolution. The new challenge is to discover how best to model and analyse what are essentially non-ideal data (from a scientific research design point of view), in areas where there is little previous work to provide guidance, little if any relevant theory, and an increasingly strong imperative for the work to be performed. It is also important to avoid making basic assumptions that are unsustainable in order to model and analyse data. The secret is to develop methods that can be used to model and analyse GIS data in a manner that is sympathetic to its nature and foibles rather than grossly abuse if by forcing upon it palpably unrealistic and unreasonable assumptions in order to do anything at all. The challenge is also to discover how best to use the virtual absence of any meaningful constraints on computational power by devising more appropriate modelling and analysis procedures, based on substituting machine time for intellectual effort in those areas where such a substitution is possible. AI increases the scope and changes the role of the computer from a calculation engine to a model designer and builder, from a statistical calculator to an automated analysis tool that should be able to perform a much greater proportion of the hard work itself. Indeed, the aim must be to develop computer based methods that can perform the hard analyses, leaving the analyst free to concentrate on the computer impossible task of providing a substantive interpretation and validation of the results.

The other major change is the increased emphasis on application. GIS emphasises problem solving in applied and highly empirical contexts. Discovering how best to move modelling and analysis into these rather different, and essentially less academic environments, is now becoming increasingly urgent. There is certainly no shortage of data sets waiting to be analysed and evidence of an increasing urgency for the work to be performed quickly and efficiently. Modelling and analysis is less likely than ever before to be a leisurely, harmless, and relatively arcane academic activity but one which has to support decision making in real-world contexts and, soon, also in real-time computing environments.

From a modelling point of view, there is a need to both maximise the utility of the available tool-kit and also an urgent need to create new models that can utilise the new data-based opportunities under difficult and non-ideal conditions. Creating new models has always been difficult outside the statistical field. Indeed, it is likely the mathematical modelling era of geography saw the creation of only a very small number of genuinely "new" models as distinct from literally thousands of either re-inventions or refinements of previously existing models. Suddenly, in the GIS era, there is a requirement for hundreds, if not thousands, of fundamentally new models in all sorts of application areas, many of which are different from those previously studied by geographers and regional scientists. New model building tools are urgently needed which can meet the necessary productivity and creativity targets and do something useful with the flood of geographically referenced data. It is argued that many current methods are inadequate for this new situation and that AI offers the basis for new approaches, some of which may well establish a new style and a new era of modelling and analysis relevant to GIS. At the very least, it is important to try and discover what these new tools might be and to investigate how they might best be used. No small part of this task is purely demonstrating the need to look.

AI covers a wide range of subjects, not all of which have any immediate or obvious or relevant applications to GIS. Three keys area of interest here are: expert systems, heuristic search procedures, and neurocomputing. The common task involved in modelling and analysis relevant to GIS is the need to essentially exploit the power of AI and fast computer processors to ease the problems involved in developing appropriate, applicable and better performing models and analysis methods. This is a particularly focused view of how AI may be used, there are certainly other viewpoints which would stress other aspects of the technology and may seek to tackle a different set of problems.

Section 3 briefly reviews expert systems approaches. Sections 4 and 5 are the main sections of the paper. Section 4 considers the use of heuristic search procedures as analysis and modelling tools, whilst Sect. 5 briefly outlines the utility of neurocomputing based tools. Finally Sect. 6 speculates about future trends.

## 3. Expert systems and intelligent knowledge based systems

There is an argument that expert systems have relatively little to offer to many spatial model building and analysis tasks, despite a widespread popularity and implicit assumption of utility. Moreover, expert systems, sometimes termed intelligent knowledge based systems (IKBS), are often mistaken assumed to be all that AI can offer in a GIS context. For example, both Chorley (1987) and the AGI yearbook (Shand and Moore 1989) give definitions of AI which imply that AI only consists of expert systems and that this is the only possible application of AI technology to GIS; see Rhind (1987) or Dangermond (1990). However, even here there is as yet little evidence of practical applications rather than theoretical speculation and design. Nevertheless, there is little doubt that IKBS have some use as intelligent assistants in areas where such skills are needed; for example, in cartography, in GIS interfaces and tutors. Otherwise IKBS might well be regarded as not particularly useful to spatial analysis and modelling for the following reasons: (1) the GIS inspired spatial modelling and analysis tasks are neither well defined nor easily represented in a rules database; (2) there are relatively few experts who could contribute to such a database; (3) most of the existing methods and models might well be considered inappropriate for the GIS era so that an expert systems approach may not be relevant since it would be representing obsolete knowledge; (4) the modelling and analysis aspects of GIS constitute a class of hard problems and it is uncertain as to how traditional analytical based approaches can be successfully used even when applied by an expert; and (5) at best the expert system will possess the ability of the expert(s) from whom the knowledge was obtained and this is unlikely to be good enough or sufficient in a GIS context. Surely, the real objective has to be to develop AI based tools that can improve on the performance levels that are achievable by the best experts, replicating them is not sufficient in areas where even the best might not be good enough.

Suitable problems for expert systems are those which are well defined with clearcut limits on the problem area to be tackled, there needs to be a considerable amount of high quality knowledge, and the availability of reasoning methods which work well for the tasks in hand. Unfortunately, there are few if any GIS relevant modelling and analysis problems which are even broadly equivalent to the class of problem that DENDRAL and MYCIN type expert systems can handle. The whole basis of an expert systems approach is to capture the problem solving expertise and knowledge of human beings in a specific problem domain and then to represent that knowledge in a computer so that non-experts can use it to solve useful problems. It is assumed here that the current area of interest is still in far too a primitive state of development for it to be worthwhile building such systems. What little is known about modelling and analysis in the GIS world suggests that it is too complex and too ill-defined for such a simplistic approach. Indeed, the argument is expressed that the class of problems faced in GIS are too complex and too hard for human beings and that the added intelligence needed to make new advances in this area may have to be machine rather than human-being based; or at least based on a combination of the two, each dealing with what they are best at.

## 4. Smart search procedures as a basis for analysis and modelling systems

Traditionally, a major interest in AI has been focused on problems which can only be approached by trial and error procedures involved a heuristic search process (see Shirai and Tsujii 1982). The aim in this section is to outline some of this heuristic search technology that can be used to aid analysis and modelling tasks relevant to GIS.

### 4.1 Brute force searches for good models and interesting patterns

Here the emphasis changes from a manual, human-being, orientated model design and analysis process, to one in which computers are able to perform most of the hardwork. This philosophy implies that the simplest way to utilise the increasing power of computers is to switch analysis and modelling styles, from one which assumes that computer time is extremely expensive and needs to be conserved which was sensible at a time when computer power was severely limited, to a new state in which computer power is in effect limitless and free. Although in practice there are still some limitations on what is computationally feasible, these limitations are no longer critical. This re-orientation is absolutely fundamental because it provides a means of avoiding many of the constraints imposed by an explicitly analytical approach on mathematical modelling and analysis, permitting progress to a new state in which virtually anything is possible via a computationally intensive process. Computational approaches to hard problems are not new but what is new is the lack of any meaningful computational barriers to the application of this less elegant but infinately more flexible style of modelling and analysis. Of course, some subjects such as theoretical physics have long used extensive computation as the basis for experimentation and knowledge creation. Unfortunately, this was not possible in geography because until fairly recently; even the fastest computers were not really fast enough to cope with the much harder computational problems that are envisaged here. This situation has now changed, there are no longer any severe computer power obstacles and even less are envisaged in the near future. Suddenly, the opportunity exists to take a giant step forward not by becoming clever in an analytical sense, but purely by becoming cruder and more computationally orientated in a way that allows the computer to do most of the work. It is almost as if computer power is to be used to make-up for our ignorance of the nature and behaviour of the complex systems under study.

   In spatial analysis this style of thinking has previously resulted in the development of automated zoning system designers (Openshaw 1978). In that system, the zoning system's design was optimised to maximise its sensitivity to either the pattern being described or the model being applied, whilst the configuration of the resulting zones was to be used as a pattern and process visualisation aid. The same philosophy re-appears later in the development of Geographical Analysis Machines (Openshaw et al. 1987; Openshaw 1990) and in some subsequent developments (for example, Besag and Newell 1991). Yet even a few years ago the idea of searching for spatial pattern without knowing either what to look for or where to look by considering all locations or particular types of location, would have been both computationally infeasible and totally alien. The latter difficulties still remain but the computational problems have disappeared. In less than three years, the early GAMs developed on a Cray X-MP supercomputer can now be run on a standard UNIX workstation. Subsequent developments of this search everywhere philosophy have resulted in a spatial relationship seeker known as a Geographical Correlates Exploration Machine (GCEM) (Openshaw et al. 1990) and in the development of a Space Time Attribute Machine (STAM) (Openshaw et al. 1991).

The STAM is worthy of a fuller description as it illustrates what might be achieved by the large scale automation of fairly simple analysis methods. By searching all locations it is possible to overcome our ignorance of not knowing where to look. The idea in a STAM is to seek pattern in data sets with measurements taken from at least three different spaces (geographical, temporal, and attribute). Previously, most work on spatial clustering has occurred in purely geographic space, any temporal patterns being removed by aggregation into time periods, and the effects of attributes either taken into account as covariates or ignored altogether. Yet, increasingly, geographic data bases stored in GISs contain high resolution details from all three spaces and it is a pity to throw away such potentially useful information. For example, a national UK cancer case database exists which is geographically referenced to about 100 m resolution, the temporal resolution is measured in days, and a number of case attributes are also available but are expressed in a variety of different measurement scales. Typically, a conventional space-only analysis of these data would involve an aggregation to census eds (a 100 fold coarsening of geographic space), aggregation into 20-year or 10-year time periods (a 365 to 730 fold reduction in temporal resolution), the selection of a single disease code and age sex group, and the use of census population at risk data which is updated only once every ten years (a major source of errors and uncertainty). Indeed, the analysis of important rare disease data may be severely affected by the forced use of incorrect census data and by fuzzyness in the disease categorisations.

The STAM attempts to look for pattern in all three spaces at the finest level of resolution available by looking for areas which are "similar" at the same geographic, temporal, and attribute level of similarity. The search strategy can either be based on a grid (as in the GAMs) or on observed cases, or by sampling space along coverages of interest; for example, only near incinerators or within a 1 km road buffer or a certain set of rock types. Approximate significance levels can be determined by Monte Carlo methods based on permuting the data. The first versions of STAM kept the geographic coordinates fixed in order to prevent purely spatial clusters from appearing; in the context in which it was used the search was for space-time or space-attribute or time-attribute associations. A typical analysis may allow for 10 different spatial similarities, 10 different temporal ones, and 5 different attribute match levels; note though the granularity imposed on the search process to keep it within the domain of computational feasibility. A STAM applied to 10000 cases would then involve 500×10000 searches each with 500 random simulations; it would not be difficult to reduce the search increments to yield 1 000 times more solution spaces. Nevertheless, in both cases the resulting program is very short and straightforward. Even with the coarse search increments it needs 4 h of CPU time on a Cray X-MP running at 65 megaflops, or a week on a workstation; but soon it will only take a few hours. Better still, run it on a parallel processor; the search is entirely a parallel process.

The space-time-attribute cluster analysis being performed by a STAM is new territory for spatial analysis. It is made feasible by a fast computer, a brute force algorithm, and a computational philosophy which states that such things are worthwhile. The knowledge and insights that may be generated are currently unobtainable by any other route. Although the intelligence is limited in the repeated application of an extremely simple search and analysis procedure, it may nevertheless be fairly powerful because it can be applied to a very large number of potential pattern locations under the control of an exhaustive search heuristic. The intelligence added by the computer results from being able to conduct this extensive search process; thereby answering the questions "where" might there be some evidence of pattern, and "what" characteristics do these patterns possess. By examining many different locations in a complex, multidimensional, and multivariable geographic-temporal-attribute space it may be possible to uncover new knowledge about pattern and process via mapping the results and applying human insight. The real end user intelligence is still provided by the human being but the evidence on which it is based is now computer generated.

## 4.2 Model design as a genetic optimisation process

Complete enumeration brute force searches are always optimal but not always possible or necessary. Quite often the same underlying problems can be tackled more efficiently and with greater elegance by using more intelligent search algorithms, although there is no longer any certainty of finding global optima in complex problems. A very promising approach is to code the search in a form suitable for a genetic optimisation procedure (see Holland 1975; Goldberg 1989). Genetic algorithms are search methods which are based on simulating the mechanics of selection and genetic operators found in living biological systems. The power of this approach is derived from a very simple heuristic assumption. It is assumed that the best solutions will be found in regions of the search space that contain relatively high proportions of good solutions and that these regions can be identified by a judicious and robust sampling of the search space. Holland (1975) shows that simple mathematical models of population genetics using search operators such as crossover, mutation, and inversion, can efficiently and implicitly perform both tasks. Viewed in the context of a function minimisation problem, genetic algorithms offer a means of searching for the global or near global optimum of complex, non-convex, perhaps discontinuous, functions which may possess many local suboptima. Normal nonlinear optimisation procedures simply cannot cope with such complexity.

Openshaw (1988) outlines a deceptively simple model design process termed an automated modelling system (AMS) which can be based on a genetic optimisation approach. The idea is to define a model as a bit string which is decoded to yield a mathematical equation. It is noted that models usually consist of a number of pieces: observed variables, constants, unknown parameters, mathematical functions, binary operators, binary functions, and sequencing information needed to assemble a logically and syntaxically valid equation. The assembly of model pieces takes place under the constraint that the basic rules of equation writing syntax have to be obeyed; for example, parentheses should balance, the number of binary operators should be one less than the sum of the variables, constants, or parameters that are used, and the usual rules of the precedence of arithmetic operations apply. The task is to breed models that are well adapted, i.e. provided a good fit, to the prevailing data environment(s). In any realistic problem there are usually a vast number of alternative "models" that could be generated. The task is to use a genetic algorithm as a means of searching this nearly infinate universe of alternative models for "good" performance. Multiple data sets, bootstrapping, and related validation methods may need to be used to avoid wrapping the model around a given set of observations such that its ability to generalise is gravely impaired.

In AMS the user provides the basic set of model pieces that prior knowledge suggests might be relevant to a given modelling task. The system is then left with the harder problem of finding some relatively good performing models for the user to interpret. The intelligence partly comes from the initial selection of model pieces, mainly from the search process which is using data to create good performing models, and partly from the subsequent human being based interpretation of the results. Limited empirical observation suggests that it works and that better performing models can be readily found; see Openshaw (1988) for an example involving the design of spatial interaction models. Interpretation is, however, far less easy and causes a major model culture shock, since suddenly it is no longer possible to attach a plausible story to either the model's equation or its parameters. Viewed from a traditional, pre-AI perspective, this is a fundamental criticism. However, if the resulting model can be shown to be robust and empirically recurrent then clearly it is saying something about the patterns and processes contained in the data sets used to generate it. The challenge is now to re-express this information in plain english. However, this approach is still in its infancy and there is much more research needed before it can be legitimately claimed that here is a practical, generally applicable, model design tool.

The principal remaining technical problem is identifying suitable representational schemes for mapping model structures on to a bit string in such a highly general fashion as to provide an adequate reflection of the range of possibilities and still provide sufficient gist for the genetic mill to operate on. Other technical issues concern the genetic algorithm itself, and the extent to which practical problems such as the loss of diversity can be remedied by changes to the basic algorithms. Another model question concerns whether or not parameter estimation should be left to the genetic algorithm to perform at the same time as it is dealing with model design, or else embed a parameter estimation subproblem within the genetic algorithm to take care of the model calibration process. Alternatively, the genetic algorithm could be replaced by another type of focused random search procedure; such as simulated annealing. The feasibility of this AMS approach is totally dependent on the nature of the available computing hardware. It is expected that this approach to model building will become more attractive as it becomes more computationally feasible and that useful fully automated and human-being assisted AMS procedures will become commonplace during the second half of the 1990's.

## 4.3 Using artificial life to search out patterns in data bases

So far the focus has been on developing a whole system model that applies to complete data sets. This is useful but there are dangers in being forced into making various strong assumptions; for instance, that one or a small number of global model exist or that spatial pattern and regularity if it occurs at all, occurs more or less everywhere. These are strong assumptions which may not always be appropriate. There is no reason to assume that such large scale spatial homogeneities exist. Geographic data may well be characterised by chaos, apart from in a small number of localised areas. Under these circumstances, the global modelling and analysis task will fail or else the results will be heavily dependent on the selection of study region boundaries. In a spatial epidemiological context, the raison d'etre for the GAM was the need to escape from global tests of clustering by being able to search out and find a small number of highly localised areas of pattern (or data anomalies) against a background of noise. If there are clusters in a large number of locations then this would be obvious from the GAM results, as it would if clusters only occurred in one or two locations. In an AI context, it seems important to try and develop tools that can adapt to the data environments they face. Instead of imposing possibly inappropriate structure on the data, why not let the data themselves define what structures are relevant and where they may exist. This requires analysis and modelling tools that are both highly flexible and data context adaptive with a minimum of preconditioning as to what is and what might not be important.

The logic underlying the artificial life (AL) literature is particularly appealing here. Beer (1990) emphasises this perspective when he writes "To me, this penchant for adaptive behaviour is the essence of intelligence: the ability of an autonomous agent to flexibly adjust its behavioral repertoire to the movement-to-movement contingencies which arise in its interaction with its environment" (p. xvi). The AL view of AI emphasises that intelligence is not necessarily achieved by imposing global top-down rules of logic and reasoning (for instance, as provided by an expert system) which then dictate the behaviour of what is now no longer an autonomous agent. Rather it is adaptive behavior, defined as a much broader and flexible ability to cope with a complex, dynamic, unpredictable world as a means of survival. Artificial life offers an opportunity to synthesise pattern and relationship seeking lifeforms. Such life-forms have no biological relevancy but maybe they have much to offer as intelligent pattern and relationship modellers. The flexibility on offer here may be extremely important. Langton (1989) explains "Natural life emerges out of the organised interactions of a great number of nonliving molecules, with no global controller responsible for the behavior of every part" (p. 2). The emphasis is on bottom-up rather than top-down modelling, on local rather than global control, simple rather than complex specifications, emergent rather than prespecified behaviour, and on population rather than individual simulation.

The analogy borrowed here from AL is simple enough. A few deterministic rules (cf. cellular automata) or even a genetic optimisation procedure can be used to breed AL forms that feed and multiply only if they discover patterns or relationships (or other forms of statistical anomalies) in the data environment in which they live. In this context pattern or relationships could be defined in several ways; for instance, as results which are unusual within a set of $k$ random data permutations. The most able forms survive, the weaker and less successful ones die out. Mapping and following, perhaps by computer movie, the life and times of these artificial life forms would provide fascinating insights into pattern and process; combining animation with locally adaptive autonomous intelligent creatures.

Once you start thinking along these lines, it is only a small step towards adopting AL analogies as the basis for a new generation of intelligent spatial modelling and analysis creatures. It should be possible to create various types of artificial life forms that can roam around databases at night (using spare CPU time), looking for pattern, "foraging" for relationships as "food", evolving as a result of interaction with their data environment, and by tracking their behaviour it may be possible to gain deep insights about structure and process, unobtainable by a more constrained or traditional approach. There is only a need to impose an external measure of success as a means of ensuring a minimum of direction towards the problems that are of interest. AL should not exist purely for itself, but then maybe there are also arguments that it, or they, should! Nevertheless, there is no prior knowledge of where or what to look for. The hope is expressed that artificial life forms will evolve over a number of generations that are best able to find and highlight any major patterns or relationships that are sufficiently important to be of potential interest.

An example may help. Suppose a space-time-attribute database is to be searched for evidence of localised pattern. Pattern in this context would be defined as an unusually strong concentration of cases with similar but not necessary identical characteristics. Suppose that the AL form consists of a multidimensional jelly fish, the idea being that its shape will engulf a region of the database, based around a particular location, where something unusual is going on. Its radius in geographic space, its temporal dimension, and its attribute similarity envelope indicate where and what the unusualness consists of. The degree of unusualness can be defined by Monte Carlo simulation. Multiple AL forms can be created to roamed around the database and those that prosper most will be grabbing locations where it may well be worth further investigation. This is similar to the STAM problem discussed previously but the search spaces are now infinately elastic, they need not be continuous, and could contain holes. Additionally, the creatures are searching many different locations in an intrinsically parallel manner. Because they are using local intelligence rather than a global search heuristic, they need far less computer time and can, more significantly, indulge in a far more detailed search process than with a brute force heuristic. Experiments with demons designed to search out pattern in a crime analysis database, have already yielded interesting results, although whether the resulting data anomalies are real or reflect data errors is a separate issue. The fact that such things can be found by any of several different AL routes suggests that there is the basis here for a number of highly novel analysis and modelling procedures.

## 5. Neurocomputing

### 5.1 The neurocomputing revolution

Another major recent development is the emergence of neurocomputing as a subject that has captivated the interest of large numbers of technologists, scientists, and mathematicians. It is a highly infectious technology that has seemingly much to offer now and even more in the near future. Suddenly it has become possible to apply computation to problems previously restricted to human intelligence, with the appearance of procedures that enable machines to learn and remember in ways that bear some resemblance to human mental processes (Anderson 1988). Neurocomputing or artificial neural nets (ANN) are also biologically inspired but this time are based on a loose analogy with the supposed workings of the human brain. They are composed of elements that perform in a manner analogous to the most simple type of neuron and they are organised in a way that appears to be related to the anatomy of the brain. Despite only a superficial resemblance, ANNs exhibit a surprising number of the brain's characteristics. They have the ability to spontaneously learn a desired function from training samples. They learn from experience and can often generalise from learned examples to new ones and they can abstract essential characteristics from inputs containing noisey or irrelevant data. Introductory texts are provided by Wasserman (1989), Khanna (1990), and Aleksander and Morton (1990).

Neural nets can in principle be used to represent and model almost any complex function or system; no matter how dynamic or difficult the task may appear to more conventional approaches. They can cope well with fuzzy and qualitative reasoning. They can be taught to recognise patterns and are particularly effective when faced with noisey data. Wasserman (1989) summarises the current state of ANN as follows: "We are presented with a field having demonstrated performance, unique potential, many limitations, and a host of unanswered question" (p. 7). However, it is also apparent that in many application areas, ANN are already sufficient good to be a practicable and useful technology for certain classes of problem, often where there are hard problems to be solved. There are two major types of neural net with many different architectures; supervised and unsupervised.

### 5.1 Supervised nets for analysis and modelling

Supervised nets have to be trained to reproduce a desired function based on data. The development of practical training methods since the mid-1980's has occasioned a surge of interest in this technology (Rumelhart et al. 1986). Back propagation training methods have turned a fairly limited utility net architecture, the multi-layer feed forward semi-linear perceptron, into a useful and widely applicable modelling tool. In a multi-layer net the inputs to any neuron K is a nonlinear function of the weighted sum of the outputs received from all neurons in the preceeding layer. The value is then squashed on to the range 0 to 1 or $-1$ to 1. This process repeats from input layers, through one or more intermediate layers, to an output layer. Scaling of the inputs and outputs allows the net to handle any size of number ranges. Training is the process of assigning values to the weights such that the sum of the squares of the errors between the net's outputs and the desired result is minimised. This might be regarded as a parameter estimation process but usually with many thousands of parameters to estimate.

This type of net is probably the ultimate in black box modelling technology. Within broad limits, it can be trained to represent the structure and processes contained in data regardless of what it is and it can represent virtually any arbitrary function regardless of how complex, with a level of descriptive accuracy that often exceeds conventional models (where there are alternatives). A few examples may help. A spatial regression model that takes into account spatially autocorrelated errors is a fairly complex task (Upton and Fingleton 1985). The same problem modelled via a neural net is fairly trivial and may yield a far better level of performance. Imagine now that the dependent variable is an unordered categorical multi-state. There is no longer any viable statistical modelling procedure that could be used here which would also take into account spatial autocorrelation. To the neutral net there are merely a few more output neurons. It learns about spatial autocorrelation from the training data, if it is relevant. ANNs applied to spatial modelling problems offer a high degree of flexibility; for example, spatial interaction nets can be engineered in which the interaction structure of entire regions can be captured. Openshaw and Wymer (1990) give some other examples. Time series and space-time modelling constitute other classes of application. Neural nets can also be trained to classify data. In some ways they can act as a general purpose analysis tool that is broadly applicable to an extremely wide range of problems for which reasonably sized training data sets exist. In theory, and also in practice, it would be possible to develop ANN equivalents to virtually all the existing spatial and non-spatial models of interest to geographers and regional scientists. Proof of this assertion will be self-evident in the impending flood of neural net based modelling applications.

There are also some problems: (1) neural nets are not easily described or understood in any traditional way, indeed it only by numerical experimentation is it possible to understand what is going on; (2) training times can be excessive, there could be several thousands of weights to estimate, and convergence of the nonlinear optimisation procedures tend to be very slow; (3) validation can be problematic in that the training data should be broadly representative of the range of situations the net is supposed to represent; (4) nets can be over-trained and learn how to reproduce random noise as well as structure; and (5) net architectures are still an art characterised by a high level of subjectivity; for example, how many layers and how many neurons in each layer. Nevertheless, there is considerable practical potential.

## 5.2 Unsupervised nets for pattern identification

A second major type of net is relevant to situations where there is no training data. Unsupervised or competative learning nets seek to discover patterns and relationships without being trained on anything. Such a net is supposed to discover for itself the features of the data that matter most, by developing its own featural representation that captures the salient characteristics of the input data

(Grossberg 1976; Kohonen 1984, 1988; Rumelhart et al. 1986). The only assumption is that some structure exists.

In a competative net the neurons compete amongst themselves for the right to represent a given input. In a single layer version, each input neuron is fully connected to each output and the neuron with the highest score wins an input case and updates its weights. Repeated scans through samples of the input data allows data structure to gradually emerge in a bottom-up fashion. These methods are essentially pattern detectors and classifiers but unlike conventional classifiers there is no global objective function being explicitly optimised and there is no knowledge of statistical theory, instead a process of self-organisation allows any structure to emerge from the bottom-up. Kohonen (1984)'s self-organising feature map is one example of this procedure and it offers the additional advantage of trying to retain topology.

Most of the research on unsupervised nets seems to have been focused on speech recognition, signal and image processing, and robotic control systems. In a GIS context, they mainly provide the basis for immensely flexible classifiers. Limited empirical comparison with more conventional cluster analysis methods in Openshaw and Wymer (1991) suggests that they perform better, contain no assumptions about the nature of the structure contained in the data, are relatively unaffected by spatial dependencies, inter-correlation, and data errors, and are useful platforms for retaining fuzzyness. They can also take in account spatially varying data reliability; something that existing methods ignore. In principle, the same procedures can also be applied to classification problems in more complex spaces; for instance, the space-time-attribute-net (STAN) of Openshaw et al. (1991) developed to handle similar problems to the STAM.

## 6. AI futures in GIS, geography, and regional science

The early 1990's are witnessing the beginnings of a new quantitative revolution that contains technology very relevant to GIS. This might also be termed the AI revolution. The stimulous is threefold: faster computers, practical and applicable AI tools, and a new emphasis on exploratory data analysis and modelling caused by the GIS revolution. They present an opportunity to start again, to re-assess the aims, objectives, and philosophy of spatial modelling and analysis relevant to GIS, and in the process seek to create a new geo-modelling and analysis tool-kit of general relevance to geography, planning, and regional science. There certainly seems to be an opportunity to solve some of the outstanding hard problems by throwing computer power at them, in a manner that has previously characterised areas of physics, chemistry, and biology. It is argued, therefore, that it is quite possible that the modelling and analysis activities in GIS related fields will become dependent on AI technologies. It has been assumed previously that computers cannot be creative, yet seemingly AI methods will increasingly be able to assist the human user to generate new knowledge and concepts from data, insights that may well be unobtainable by any other route.

It is perhaps useful if a number of key landmark in this development process could be recognised. They would be: (1) demonstrations of the potential of the new technologies using padagogic and other illustrative data examples; (2) empirical evidence that the new methods out perform traditional alternatives when run on historic data for which there are existing results so that comparisons can be made; (3) illustrations that the new methods work in the data rich environments being created by GIS where there are no benchmarks based on traditional methods; and (4) the incorporation of AI procedures, perhaps hidden away, in general purpose analysis and modelling tools. In general, the current state is somewhere around stages (1) and (2). Yet it might confidently be predicted that in certain specialised areas, stages (3) and (4) will be reached fairly soon. However, consolidation across a broader range of examples, applications areas, topics, and the development of general purpose tools may well take at least another decade. At least a start has been made and it is argued that the eventual goal is not in doubt; the only uncertainty concerns when the scale of these developments might be sufficiently so that others may recognise that an AI revolution has in fact taken place.

*Acknowledgement.* Thanks are due to the ESRC for funding the North East Regional Research Lab where many of the developments briefly described have been applied and tested on spatial data.

# References

Aleksander I, Morton H (1990) An introduction to neural computing. Chapman and Hall, London

Anderson ZA (1988) Neural information processing systems. American Institute of Physics, New York

Beer RD (1990) Intelligence as adaptive behavior: an experiment in computational neuroenthology. Academic Press, Boston

Besag J, Newell J (1991) The detection of clusters in rare diseases. J R Statist Soc Ser A 154:143–156

Chorley R (1987) Handling of geographic information. HMSO, London

Dangermond J (1990) The future of GIS technology. GIS Source Book, GIS World, Colerado, pp 7–11

Goldberg DE (1989) Genetic algorithms in search, optimisation, and machine learning. Addison-Wesley, Reading, MA

Grossberg S (1976) Adaptive pattern recognition and universal recoding: part 1: parallel development and coding of neural feature detectors. Biol Cyber 23:121–134

Holland JH (1975) Adaptation in natural and artificial systems. University of Michigan Press, Ann Arbor

Khanna T (1990) Foundations of neural networks. Addison-Wesley, Reading, MA

Kohonen T (1984) Self-organisation and associative memory. Springer, Berlin Heidelberg New York

Langton CG (1989) Artificial life. In: Langton CG (ed) Artificial life. Addison-Wesley, Reading, MA, pp 1–48

Openshaw S (1978) An empirical study of some zone design criteria. Environ Plann A 10:781–794

Openshaw S (1988) Building an automated modelling system to explore the universe of spatial interaction models. Geogr Anal 20:31–46

Openshaw S (1989) Computer modelling in human geography. In: Macmillan B (ed) Remodelling geography. Blackwell, Oxford, pp 70–89

Openshaw S (1990a) Spatial analysis and GIS: a review of progress and possibilities. In: Scholten H, Stillwell JCH (eds) Geographic information systems for urban and regional planning. Kluwer, Dordrecht, pp 153–163

Openshaw S (1990) Automating the search for cancer clusters: a review of problems, progress, opportunities. In: Thomas RW (ed) Spatial epidemiology. Pion, London, pp 48–78

Openshaw S (1991) A view on the GIS cris in geography, or, using GIS to put Humpty-Dumpty back together again. Environ Plann A 23:621–628

Openshaw S (1991) Developing appropriate spatial analysis methods for GIS. In: Maguire D, Goodchild MF, Rhind DW (eds) Geographical information systems: principles and applications, vol 1. Longman, London, pp 389–402

Openshaw S, Charlton M, Wymer C, Craft A (1987) A mark I geographical analysis machine for the automated analysis of point data sets. Int J GIS 1:335–358

Openshaw S, Cross A, Charlton M (1990) Building a prototype geographical correlates exploration machine. Int J GIS 3:297–312

Openshaw S, Wymer C (1990) Some applications of neural nets in urban and regional modelling. NE.RRL Discussion Paper, CURDS, Newcastle University

Openshaw S, Wymer C (1991) A neural net classifier for handling census data. In: Murtagh F (ed) Neural networks for statistical and economic data. Munotec Systems, Dublin

Openshaw S, Wymer C, Cross A (1991) Using neural nets to solve some hard problems in GIS. Proceedings of EGIS '91, vol 1–2. EGIS Foundation, Utrecht, pp 797–807

Rhind DW (1987) Recent developments in GIS in the UK. Int J GIS 1:229–235

Rumelhart DE, Hinton GE, Williams RJ (1986) Learning representations by back propagating errors. Nature 323:533–536

Rumelhart DE, McCelland JL (1986) Parallel distributed processing, vol 1 and 2. MIT Press, Cambridge, MA

Shand PJ, Moore RV (1989) The association for geographic information yearbook. Taylor and Francis, London

Shirai Y, Tsujii J (1982) Artificial intelligence. Wiley, Chichester

Upton G, Fingleton B (1985) Spatial data analysis by example, vol 1. Wiley, Chichester

Wasserman PD (1989) Neural computing: theory and practice. Van Nostrand Reinhold, New York

Wilson AG, Bennett RJ (1985) Mathematical methods in human geography and planning. Wiley, New York

# Spatial statistical analysis and geographic information systems

**Luc Anselin[1] and Arthur Getis[2]**

[1] National Center for Geographic Information and Analysis (NCGIA), Department of Geography and Department of Economics, University of California, Santa Barbara, CA 93106, USA
[2] Department of Geography, San Diego State University, San Diego, CA 92182, USA

**Abstract.** In this paper, we discuss a number of general issues that pertain to the interface between GIS and spatial analysis. In particular, we focus on the various paradigms for spatial data analysis that follow from the existence of this interface. We outline a series of questions that need to be confronted in the analysis of spatial data, and the extent to which a GIS can facilitate their resolution. We also review a number of exploratory and confirmatory techniques that we feel should form the core of a spatial analysis module for a GIS.

## 1. Introduction

Space plays a central role as an organizing concept in regional science. It is therefore to be expected that the analysis of spatial data and the specialized techniques that this requires have received considerable attention in the research literature. The emphasis of this research has been on theoretical and methodological aspects, such as the role of spatial dependence and spatial heterogeneity, the effect of spatial scale, and the development of estimation methods for spatial process models (for a review, see Anselin 1988). However, as pointed out by Anselin and Griffith (1988), the dissemination of these results to the practice of data analysis in empirical work has been rather limited. In part, this has undoubtedly been due to the lack of an easy and effective way to explicitly incorporate the spatial aspects of data. This problem is now largely eliminated, due to recent advances in the technology of geographic information systems (GIS). In spite of this, the effectiveness and importance of GIS and spatial information systems as an enabling technology is only slowly becoming recognized in regional science (Nijkamp and Rietveld 1983; Nijkamp 1988, 1990; Anselin and Madden 1990). In particular, some fundamental questions related to the role played by GIS technology in defining the research agenda for spatial data analysis have not received the attention they deserve. We thought it timely to identify a number of important issues that pertain to the interface between spatial statistical analysis, GIS and regional science. Our objective is not so much to review the recent literature, but to outline and discuss alternative viewpoints, to summarize the current state of the art in spatial statistical analysis and how it relates to GIS, and to suggest directions for future research.

The focus of attention in GIS tends to be on the display, organization and simple manipulation of information in spatial data bases. As a result, most commercial GIS implementations are rather limited in what they offer in terms of statistical tools for the analysis of spatial data. This lack of analytical capacities of a GIS is by now a familiar complaint in the research literature (e.g., Goodchild 1987; Burrough 1990; Couclelis 1991) and several efforts have been initiated to alleviate this situation (Abler 1987). In those, spatial data analysis and spatial statistics are often perceived as playing a central role among the components of the analysis function in a GIS (HMSO 1987; Gatrell 1987; Goodchild and Brusegard 1989; Bailey 1990; Openshaw 1990; Csillag 1991; Goodchild et al. 1992). There are two aspects to this. First, there is the incorporation of spatial statistical techniques as part of the toolbox provided with a GIS, by adding statistical functions to the menu of GIS capacities, or by providing an easy link between a GIS and a statistical package. A second, and potentially more interesting aspect is the extent to which statistical and even spatial statistical techniques are appropriate for use with a GIS, and the resulting need to develop new "spatial" analysis tools. In this paper, we focus on both aspects.

We start by discussing the interface between GIS and spatial analysis, and the various paradigms for spatial data analysis that follow from this. Included is a section on the special qualities of spatial data. We next outline a series of questions that need to be confronted in the analysis of spatial data, and the extent to which a GIS can facilitate their resolution. We also briefly review a number of exploratory as well as confirmatory techniques that in our opinion should form the core of a spatial analysis module for a GIS. We close with some remarks about the role of GIS and spatial analysis in regional science research in general.

## 2. Interfacing spatial analysis and GIS

Traditionally, geographic information systems are considered to perform four basic functions on spatial data: input, storage, analysis, and output (Goodchild 1987). Of these, analysis has received least attention in commercial systems. Typically, a variety of map description and manipulation functions are defined by commercial vendors as being "spatial analysis," but they have little bearing on the use of this concept in the academic community (Couclelis 1991). In the GIS literature spatial analysis has become narrowly defined. For example, Gatrell (1987) defines it as "the application of statistical methods to the solution of geographical research questions" (see similar uses of the concept in Openshaw 1990; Openshaw et al. 1991; Ding and Fotheringham 1991; Goodchild et al. 1992). Clearly, to support research in regional science (as a spatial information system or a spatial decision support system) a large set of techniques should be included under the rubric of spatial analysis, such as location-allocation and other operations research methods, urban and regional modelling, and spatial demographics.

We give a highly simplified schematic representation of the interaction between the four basic functions of GIS in Fig. 1. At one end of the graph is "reali-

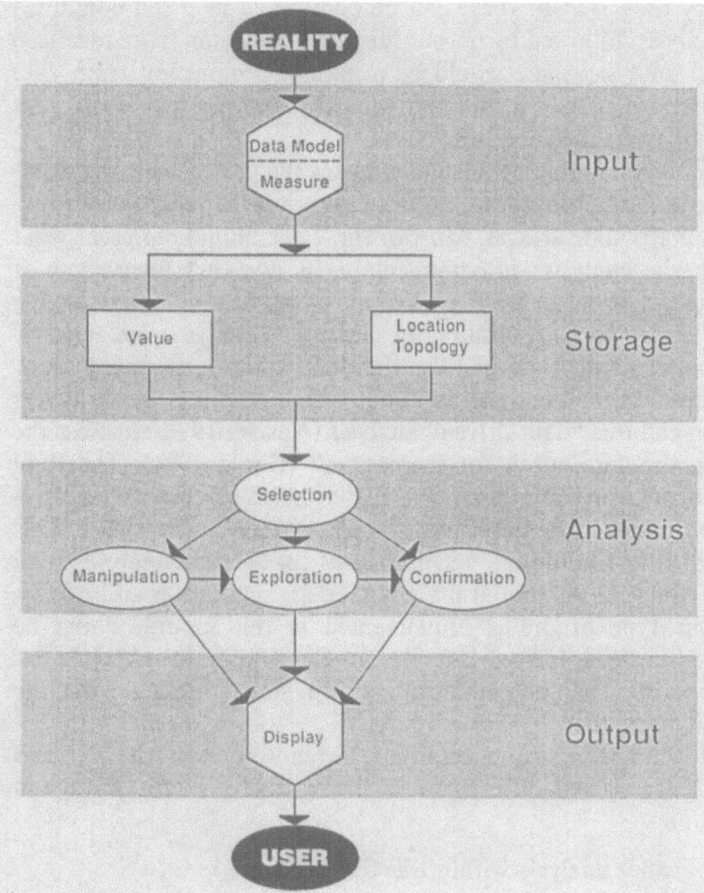

**Fig. 1.** Functions of a GIS

ty", at the other the "user", concerned with policy or theory development. In between are the four functions, input (data model and measurement), storage (of data values and their location and topology), analysis (data selection, manipulation, exploration and confirmation), and output (display). Our focus is on the analysis functions and their interface with the storage function. The latter is typically associated with a relational database. In a GIS, this database not only contains information on value, but also on the location and spatial arrangement (topology) of observational units. The way in which reality is measured and structured into a spatial data base is determined by the data model, which has important implications for the types of spatial analyses that can most effectively be carried out (Peuquet 1984, 1988; Goodchild 1992). We return to this point below.

In the analysis module we distinguish between four important functions: one is the selection or sampling of observational units from the data base and the choice of the proper scale of analysis. The other three functions consist of increasing degrees of abstraction from the data. We call them data manipulation, exploration and confirmation. In Fig. 1 they are represented on the same level to illustrate the property that each of these can be considered as a self-contained module of spatial analysis, followed by output (display). However, in an idealized framework of spatial analysis, there would be a natural progression from data manipulation to exploratory analysis and confirmatory analysis, obviously with multiple feed-backs between the modules.

In our framework, data manipulation encompasses the partitioning, aggregation, overlay and interpolation procedures needed to convert the selected information into meaningful maps and surfaces. Most of these techniques represent what is understood by "spatial analysis" in commercial GIS, and they form some of the more powerful aspects of the technology in terms of the flexibility in changing observational units. Under data exploration we classify inductive approaches to elicit insight about pattern and relations from the data, without necessarily having a firm pre-conceived theoretical notion about which relations are to be expected. We could also call this "data-driven" analysis (Anselin 1990) to stress the emphasis on "letting the data speak for themselves" (Gould 1981). The final module is then confirmatory analysis, where the point of departure is a theoretical notion or model ("model-driven" analysis in Anselin 1990). This would include most of the "traditional" techniques of spatial data analysis, such as hypothesis tests, estimation of spatial process models, simulation and prediction. In principle, the type of model implemented in this module could be anything where space (location, region) is a relevant element, and thus could encompass a wide range of urban, regional and multiregional models from the toolbox of the regional scientist.

The way in which the analysis function in a GIS (as we perceive it) is linked to the database has been forged differently. Logically, there are three ways in which this can be done:

1. fully integrate all spatial analysis within the GIS software;
2. construct modules of spatial analysis that efficiently link with the GIS and effectively exploit the "spatial" information in the database;
3. leave the GIS and spatial analysis as two separate entities and simply import and export data in a common format between the two.

The third approach is really a non-solution, since it ignores the distinctive characteristics of a spatial data base for use in spatial data analysis. Nevertheless, it seems to be the approach most taken in practice, due to the problems with proprietary data formats in commercial GIS and the limited facilities of often awkward macro languages (for an extensive discussion, see Kehris 1990a; Bivand 1990). Examples of this strategy are the joint use of GRASS and S for exploratory data analysis in Farley et al. (1990) and Williams et al. (1990); the combination of SPANS and SYSTAT to carry out stepwise regression in Bonham-Carter et al. (1988); and the use of ARC/INFO and BMDP for logistic regression in Warren (1990).

The second approach is similar to the so-called "modular" design in integrated regional modelling and consists of developing self-contained modules for various types of spatial analyses. These modules are then linked to the specific data structures used in a commercial GIS. They are thus not "generic," but limited to a particular combination of GIS and technique. Among many examples are the work of Walker and Moore (1988) on combining GIS and other statistical packages and the linkage of ARC/INFO and GLIM in Flowerdew et al. (1991). Most of these modules are written and compiled separately and access the data structure of the GIS by means of proprietary library functions. In general, the use of the GIS macro facilities is avoided, given its poor performance in terms of speed (Bivand 1990; but see Ding and Fotheringham 1991, for an example of the use of the ARC/INFO AML macro language to construct a measure of spatial association). Even though this second approach links a statistical package to a GIS, it is generally limited to simple descriptive measures, such as univariate measures of spatial association (e.g., Kehris 1990b; Ding and Fotheringham 1990; for an exception, see Bivand 1990).

Finally, the first strategy ("encompassing" in the terminology of integrated modelling) is basically non-existing, due to the lack of analytical capabilities in most commercial GIS (partial exceptions are SPANS, IDRISI and GIS-PLUS/TRANSCAD). It is most closely approximated by the idea behind the "spatial analysis toolkit" of Openshaw and associates, if it were not that spatial analysis is limited to a small number of generic functions in their approach (Openshaw 1990; Openshaw et al. 1991). In contrast, our vision of the spatial analysis function in a GIS is much wider and would include at least all of the "traditional" techniques. The determination of an unambiguous set of "generic" functions of spatial analysis is an important and still largely unresolved question.

## The nature of spatial data

The implementation of a generally useful spatial analytical capability within a GIS can be achieved once it is recognized that the solution to spatial problems inevitably must be based on the special character of spatial data. From an analytical point of view, it is more than the fact that data are spatially referenced that differentiates spatial data from other types of data. When the data are spatially referenced one must go beyond Tukey for spatial exploration and beyond standard statistics and econometrics for confirmatory analysis.

What is it that makes spatial data special? Anselin (1990) explains in considerable detail why one must treat spatial data differently than other types of data. In essence, the point is that spatial effects complicate any straightforward understanding of spatial data. "Spatial effects" has two interrelated meanings. The first is that embodied in Tobler's First Law of Geography (Tobler 1979a), where "everything is related to everything else, but near things are more related than distant things." This simply implies that we should expect stronger relationships within and among variables that are sampled at places that are spatially near to one another rather than far from one another. The more troublesome second meaning, however, is that because of the size and configuration of spatial units we find relationships within or among variables that are due as much to the nature of the spatial units as to the nature of the variables being studied. The first type of spatial effect can be handled, for the most part, with conventional data analytical procedures, but not the second. Since all spatial data are subject to the second effect, one must take it into account when devising systems for analysis.

Spatial effects can be divided into two types: dependence and heterogeneity. Spatial dependence refers to the relationship between spatially referenced data due to the nature of the variable(s) under study and the size, shape, and configuration of the spatial units. The smaller the spatial units, the greater the probability that nearby units will be spatially dependent. If units are spatially long and narrow, the chances of spatial dependence with nearby units will be greater than if the units are more compact. Spatial heterogeneity occurs when there is a lack of spatial uniformity of the effects of spatial dependence and/or of the relationships between the variables under study. A dependence structure that is inconsistent across the study area lacks homogeneity. In a sense, then, spatial heterogeneity can be thought of as a special case of spatial dependence. It represents a complex realization of the nature of the variable(s) under study and the effects of the size, shape, and configuration of spatial units.

## 3. GIS and perspectives on spatial data analysis

Spatial analysis ranges from simple description to full-blown model-driven statistical inference. As outlined in Anselin (1990) many different perspectives can be taken towards spatial data analysis. For the purposes of our discussion, we will classify techniques into exploratory (or data-driven), and confirmatory (or model-driven), although in practice many techniques incorporate aspects of both (Haining 1990b).

In an exploratory data analysis (EDA, Tukey 1977) the data are used in an inductive fashion to gain new insights. To date this is by far the most common approach towards data analysis using GIS, although it has taken at least three different forms.

In one, represented by the work of Openshaw and associates (Openshaw 1990; Openshaw et al. 1987, 1988, 1990, 1991), the role of the analysis is limited to pure description and indication of potentially interesting spatial patterns. In fact, Openshaw goes so far as to reject most of the traditional spatial analytical methods of the type outlined in Berry and Marble (1968). Furthermore, the role of the "spatial analysis" function in GIS is restricted to pure description of map pattern, without explanation, since "what is causing the pattern is not a subject matter for the geographer" (Openshaw 1990, p. 158). His geographical analysis machine (GAM) and geographical correlates exploration machine (GCEM) are computation intensive approaches to elicit patterns in the data. In that sense, they can be considered to be examples of exploratory data analysis. However, the lack of indication of "significance" and the admitted possibility that the patterns could be spurious are a far cry from the usual interpretation of EDA (Tukey 1977; Mosteller and Tukey 1977). In addition, while most EDA exploits the high dimensionality of data (using various clustering and graphical cross tabulation methods), Openshaw's examples so far pertain to univariate and bivariate situations. The extent to which these "machines" can be made operational and cost-effective to address more complex research questions remains to be resolved.

In a second approach to combining GIS with EDA, data are exported from a GIS into a standard statistical package for analysis (typically, S; Becker et al.

1988). This stands in sharp contrast to Openshaw's rejection of such a linkage as worthless and "an irrelevant distraction" (Openshaw et al. 1991, p. 788). The types of analyses that are carried out use standard EDA tools, such as box plots, Chernoff faces, Tukey stars, scatterplot matrices and hierarchical clustering (e.g., Farley et al. 1990; Williams et al. 1990). Although such techniques are very useful in generating insight into patterns and potential associations, they are a-spatial. Moreover, to the extent that measures of fit and tests of significance are included (e.g., as in added variable plots; Haining 1990a) the presence of spatial dependence (and/or spatial heterogeneity) can easily lead to spurious conclusions. One comes close to "spatial" analysis in the work of Kehris (1990b) and Ding and Fotheringham (1991), who provide a link between a GIS and specialized routines to compute measures of spatial association. However, this is still fairly rudimentary, and a true "spatial" EDA, or ESDA (exploratory spatial data analysis) does not yet exist (see also Anselin 1990).

A third approach consists of some recent developments in the use of statistical graphics, where the "map" is included as one of a series of dynamically linked graphs (for an extensive discussion of this concept, see Cleveland and McGill 1988). This is typified by the work of Haslett and associates (Haslett et al. 1990, 1991; Stringer and Haslett 1991; see also MacDougall 1991) on interactive graphic environments (SPIDER and REGARD) that combine a map, histogram and scatterplot view of the data, as well as various lists. Selection of any subset of the observations in one of the dynamically linked windows affects the representation in all other windows. This allows for an intuitive and visual impression of the correspondence between value association (scatterplot) and locational association (map), although no quantification of the latter is provided. In a sense, the focus is on spatial heterogeneity (regional differentiation) rather than on spatial dependence. Ideally, of course, both should be included in a framework for ESDA.

In contrast to the recent flurry of research activity in exploratory data analysis and GIS, very little has been achieved in terms of model-driven or confirmatory analysis. Most applications are non-spatial applications of regression analysis and fail to exploit the information on the topology of the observations that is contained in a GIS. As pointed out in Anselin and Griffith (1988) and Haining (1990b), the general problem is one of lack of software to carry out the complex and nonlinear estimation and inference for spatial process models. A number of recent advances have been made in software development for spatial data analysis in the form of libraries of macro routines for commercial statistical packages (e.g., Griffith 1988; Griffith et al. 1990; Bivand 1990, 1991). A self-contained spatial data analysis software package, SpaceStat, is introduced in Anselin (1991b). So far, however, the linkage between this software and a GIS is very limited (e.g., the work of Bivand 1990, which exports data from ARC/INFO into a SYSTAT module).

## 4. The GIS data model and spatial statistics

As defined by Goodchild (1992), the data model implicit in a GIS is the "discretization" of geographical reality necessitated by the nature of computing devices. Commercial GIS can be classified as following either a raster (or grid) or vector data model, i.e., a regular or irregular tesselation of the plane (see also Peuquet 1984, 1988). The raster or vector structure defines the spatial unit of observation that can be used in spatial analysis. In the former, the unit is the grid (or other regular tesselation) and all points within the grid are assumed to take on the same value. This is an implicit form of spatial sampling. Clearly, if the grid does not exactly correspond to the spatial arrangement of values in the underlying process there will be an inherent tendency for spatial dependence. Similarly, if the scale of the grid cell has an imperfect match with the scale of the process studied, various types of misspecification may result, often called ecological fallacy or the modifiable areal unit problem (MAUP).

When a vector structure is used, the choice of the points, lines and polygons that will be represented, their spatial resolution and spatial arrangement are also an implicit form of spatial sampling. Similar to the raster approach, homogeneity is assumed within the point, line or areal unit of observation. For the latter in particular, this may only be a crude approximation and spatial dependence as well as scale problems are likely to be present.

The implied spatial sampling is also a component of alternative conceptualizations of data models, such as the distinction between a so-called "field view" ("infinite sets of tuples approximated by regions and segments", Goodchild 1992), and an "object view" ("planes littered with objects," Goodchild 1992). For the former it necessitates a choice of the size of the region and their relative spatial arrangement (the so-called container view of GIS, Couclelis 1991), for the latter the selection of which "objects" will be included in the database.

It is important to keep this in mind, since this sampling process structures the database and precedes any sampling the analyst may want to carry out (the data selection module in Fig. 1). It is often dictated by administrative or policy (or political) concerns which may or may not be founded on "accepted" theoretical concepts of the time. Examples are the delineation of administrative regions which pre-determine the collection of many socio-economic data. In a sense then, even though spatial analysis may be exploratory, the data that are available and the way in which they are collected and arranged are often constrained by the accepted theoretical knowledge of the time (which variables are important, etc.) and its implications for spatial resolution (or, rather, the lack of interest in spatial resolution).

Obviously, this sampling will lead to a sampling error and the resulting problem of accuracy in spatial data bases (see Goodchild and Gopal 1989, for an overview). The error in spatial databases pertains both to value (the usual problem) as well as to location and spatial arrangement (topology). As a result, what we perceive as "observations" can be conceptualized as a mixture of signal (truth) and noise (error), or, more precisely, as either a sample from an unknown population or a realization of a stochastic process. The objective of spatial analysis is to elicit information about the signal, taking into account the fact that noise is present. The presence of this "error" does not preclude a statistical methodology (and its associated inference) as argued by Openshaw (1990), but is in fact the very essence for its need.

## 5. Implementation issues

In the implementation of a framework for spatial analysis within a GIS many issues can be addressed by means of familiar techniques. These techniques do not necessarily fit neatly within our classification of spatial analysis into the four modules of data selection, data manipulation, exploratory and confirmatory analysis (Fig. 1), but many methods are important in more than one module. In order to make our discussions less abstract, we next review a number of ways in which specific techniques would be incorporated into our framework. It is important to keep in mind that this will only give a general flavor of what we envisage as a general purpose spatial analysis system, since a detailed inventory of techniques is beyond the scope of this paper. Also, much remains to be addressed, and many tricky methodological problems have not yet found a satisfactory solution.

Most of the decisions made about the selection, manipulation, and analysis of spatial data can be thought of as strategies designed to avoid, specify, or account for the effects of spatial dependence. The data available in a GIS are rarely referenced in spatial units that are appropriate for final analysis. For example, pixel data, which are highly spatially dependent, must be aggregated for land use studies. In the data selection process (the first analysis module in Fig. 1), the nature of the data dependence should be evaluated before a representative sample can be designated.

In a well-known study, Openshaw and Taylor (1979) summarize the results of extensive experimentation in which scale changes radically altered the correlative and autocorrelative relationships among variables. Arbia (1989) claims that it is the spatial autocorrelation, or the dependence of nearby spatial units on one another, that is responsible for changes in summary measures as scale is changed. If units are summed into larger units, the mean increases, the covariance increases, and the correlation decreases in absolute value in proportion to the change in the size of the units. In all but a few circumstances, however, the variance increases in relation to the changed size of units and to the correlation between specified neighboring units. Immediately it becomes clear that statistical tests will be affected by the chosen scale (see also Haining 1991). This being the case, the selection of an appropriate sample is a crucial decision to which a great deal of attention must be given. This is particularly important, since it often is not clear whether the so-called modifiable areal unit problem is indeed an artifact of a particular data set, as is typically assumed, or instead should be attributed to the use of an improper model and/or technique, as argued by Tobler (1989).

If is difficult to predict how the moments of a spatial sample will change with changing scale in all but the simplest circumstances, that is, when the specification of the relationship between spatial units is simple, and therefore, not particularly interesting. In addition, when spatial units are of unequal size, weighting schemes to "equalize" them must be arbitrary and, as a consequence, one must settle for a range of test results rather than a specific value. It is clear that any multi-purpose GIS must be capable of assisting the data selection process by containing flexible clustering and aggregation algorithms.

The manipulation of spatial data (the second spatial analysis module in Fig. 1) may result in the creation or smoothing of a surface or the partition of data units into polygons. These types of operations rest to a large extent on the evaluation of the degree of spatial dependence present in the data. The creation of a surface by interpolation is based on the nature of trends or regularities in the data. Filtering a complex surface into a smooth one is essentially an exercise in specifying a structure for spatial dependence. In order to carry out these operations, a GIS might contain a number of measuring devices that evaluate dependence. Various cross product statistics (Hubert et al. 1981) such as Moran's I, Geary's c, the variogram, and Getis and Ord's G are all helpful in this regard (Cliff and Ord 1981; Haining 1990b; Cressie 1985; Getis and Ord 1992). In addition, smoothing techniques can be based on spectra (Rayner 1971), trend surfaces, spatial adaptive filtering (Foster and Gorr 1986), and smooth pycnophylactic interpolation (Tobler 1979b), to name only a few commonly used methods.

For the creation of partitions, meaningful criteria should be based on the dependence structure of the spatial data under investigation. The techniques mentioned in the last paragraph can be used for this purpose as can clustering algorithms. Similarly, Thiessen polygons and associated tesselation techniques are often-used partitioning devices (Boots 1985).

Perhaps of greatest importance for the preparation and manipulation of spatial data for further analysis is the need to fill a surface with estimates of variable values when data are missing. For example, a GIS may contain data at points when the analytical interest is in areas. This problem of missing spatial data has received considerably attention (for a review, see Griffith et al. 1989) and many techniques have been implemented in operational GIS, e.g., based on kriging (Cressie 1986; Davis 1986; Oliver and Webster 1990).

As pointed out before, the precise allocation of techniques to the exploratory spatial data analysis (ESDA) and confirmatory spatial data analysis (CSDA) modules is not always clear (the third and fourth spatial analysis modules in Fig. 1), although there are some major differentiating characteristics between the viewpoints taken in each (see Anselin 1988; Haining 1990b). Suffice it to say here that ESDA is that phase of analysis in which spatial patterns and structures are revealed, hypotheses proposed and models suggested. In contrast, CSDA includes the entire roster of techniques and methodologies for hypothesis testing, the determination of confidence intervals, estimation, simulation, prediction and the assessment of model fit. In ESDA one searches for structure and association, while in CSDA one evaluates the evidence. As Haining (1990b) points out, one alternates in the application of the two aspects of spatial data analysis, similar in spirit to the idea behind EDA advanced by Tukey (1977).

The various elements of ESDA include those which aid in the identification and description of patterns and variables, elicit the characteristics of variables and patterns, help determine the extent of data dependence and heterogeneity. In addition, ESDA should also allow for simple modeling, especially so that residuals can be evaluated and the selection of a "best" subset of explanatory variables can be determined.

A wide array of techniques are available for ESDA. These include the standard tools of EDA and statistical graphics, such as box plots, star plots, Chernoff faces, etc., as well as many of the measures mentioned above. In addition, pattern recognition devices such as those discussed in the artificial intelligence and spatial statistics literatures are highly relevant here, e.g. as outlined in the work of Ahuja and Schachter (1983), Pielou (1977), Ripley (1981) and Boots and Getis (1988). However, the "spatial" aspects of ESDA have to date not been fully developed. In this respect, approaches that blend the analytics of the traditional techniques with the computing power and interactive graphics of some of the recent developments could show great promise.

In addition to the predominantly non-parametric approach taken in traditional EDA, one often also needs to know moments, errors, and other parametric characteristics of samples and surfaces at different scales. For example, the parameters of simple linear regression, trend surfaces, periodicities, semi-variograms and correlograms are often useful. Directional statistics and spatial ANOVA are tools that could be included in any exploratory analytical module. In addition, categorical variables are often mapped by GIS users. Thus, logit analyses of overlapping variables would prove useful in the exploratory stage of analysis.

It is here that the distinction between ESDA and CSDA becomes difficult. Indeed, the standard tools of CSDA consist of estimation algorithms for a wide range of specifications, both linear and nonlinear. The spatial aspects of such analysis are often identified with the field of spatial econometrics, i.e., "the collection of techniques that deal with the peculiarities caused by space in the statistical analysis of regional science models" (Anselin 1988, p. 7). In essence this boils down to four broad categories of methods: 1) diagnostics for the presence of spatial dependence and spatial heterogeneity in regression analysis (this includes ANOVA and trend surface models as special cases); 2) methods to estimate and obtain inference (e.g., based on maximum likelihood, instrumental variables or bootstrap estimators) for various types of regression models for cross-sectional and space-time data that explicitly take into account spatial effects (e.g., spatial process models); 3) methods to estimate and obtain inference that are robust to the presence of spatial effects; 4) spatial measures of model validity. Although much methodological progress has been made in these areas, a number of very tricky issues remain to be resolved, such as the issue of spatial dependence in models with limited dependent variables (e.g., logit, probit and Poisson regression models), the discrimination between spatial dependence and spatial heterogeneity, nonstationarity in models for space-time data, edge effects, etc. (Anselin 1990). To some extent then, the implementation of CSDA in a spatial analysis system is constrained by the state of the art, which to date is still unsatisfactory to be able to answer the range of questions faced by the users of a GIS.

## 6. GIS, spatial analysis and regional science

The fundamental issue in carrying out spatial analysis with a GIS is whether the "observations" (in the GIS) contain sufficient information to extract the signal and control for the noise. This is not necessarily satisfied, even though the technological sophistication of a GIS and the large size of many databases may give the opposite impression. The more one knows (or assumes) about the signal, the easier it will be to falsify preconceived notions and/or generate new hypotheses. Sometimes additional information can be obtained by combining observations on many different indicators (variables) and at many locations or spatial scales, but often even this is not sufficient. As is well known, in many instances different processes can generate observationally equivalent patterns of values. Failure to distinguish "significant" patterns or to gain insight into underlying causal relationships should not imply a rejection of the statistical methodology. Instead, more data (e.g., in the time dimension) and/or better theoretical notions may be needed. The statistical methodology provides one with a set of tools to assess the extent to which this is the case. This set of tools should not be used to the exclusion of others, but the GIS technology allows it to be complemented with powerful computation intensive approaches and innovative visualisation. A creative combination of the "old" spatial analysis with these new technologies, to form a "new" spatial analysis (similar to the change in perspective generated by the "new" urban economics of the 1970s) has not yet been achieved.

To suggest that the recent developments in GIS have already transformed the way in which spatial analysis is carried out in the field of regional science is clearly an overstatement. Although some embrace the new technology as an innovative means to look at the world in a different way, others tend to dismiss it as just another set of fancy color graphics. The opportunity in the use of GIS is that it indeed has made previously prohibitive computationally intensive and highly visual ways of spatial analysis accessible at reasonable cost. The challenge to the GIS field is that it has not yet been able to furnish or incorporate the types of analytical tools that are needed to answer the questions posed by regional scientists. Some would argue that those are the wrong questions and that using the existing GIS tools will lead to different and more interesting questions. Our position is that the technology should be led by theoretical and methodological developments in the field itself. Does this require an abandonment of the traditional spatial analysis? Clearly, approaches that were inspired by the lack of computational and graphical resources have now become redundant, but a considerable number of fundamental insights into the nature of spatial structure, spatial dependence and spatial processes remain relevant. An effective integration of these perspectives with the new technology may go a long way toward convincing researchers in regional science and other social sciences that the special role of space which underlies the essence of the field merits its own analytical toolbox. We suggest that spatial statistical analysis should play a central role in this toolbox.

*Acknowledgements.* Anselin's research was supported in part by Grant SES-8721875 from the National Science Foundation and by the National Center for Geographic Information and Analysis (NSF Grant SES-8810917).

## References

Abler RF (1987) The national science foundation national center for geographic information and analysis. Int J Geogr Inf Syst 1:303–326
Ahuja N, Schachter BJ (1983) Pattern models. Wiley, New York
Anselin L (1988) Spatial econometrics, methods and models. Kluwer, Dordrecht

Anselin L (1990) What is special about spatial data? Alternative perspectives on spatial data analysis. In: Griffith DA (ed) Spatial statistics, past, present and future. Institute of Mathematical Geography, Ann Arbor, MI, pp 63–77

Anselin L (1991 a) Quantitative methods in regional science: perspectives on research directions. In: Boyce D, Nijkamp P, Schefer D (eds) Regional science, retrospect and prospect. Springer, Berlin Heidelberg New York, pp 403–424

Anselin L (1991 b) SpaceStat: A program for the analysis of spatial data. Department of Geography, University of California, Santa Barbara, CA

Anselin L, Griffith DA (1988) Do spatial effects really matter in regression analysis? Papers Reg Sci Assoc 65:11–34

Anselin L, Madden M (1990) Integrated and multiregional approaches in regional analysis. In: Anselin L, Madden M (eds) New directions in regional analysis. Belhaven Press, London, pp 1–23

Arbia G (1989) Spatial data configuration in statistical analysis of regional economic and related problems. Kluwer, Dordrecht

Bailey TC (1990) GIS and simple systems for visual, interactive, spatial analysis. Cartogr J 27:79–84

Becker RA, Chambers JM, Wilks AR (1988) The new S language, a programming environment for data analysis and graphics. Wadsworth, Pacific Grove, CA

Berry BJL, Marble DF (1968) Spatial analysis. Prentice-Hall, Englewood Cliffs, NJ

Bivand R (1990) Spatial statistics: front-end interference support for GIS. Proceedings Third Scandinavian Research Conference on Geographical Information Systems. Helsingor, Denmark, pp 244–254

Bivand R (1991) SYSTAT-compatible software for modelling spatial dependence among observations. Paper Presented at the 7th European Colloquium on Theoretical and Quantitative Geography, Stockholm, Sweden

Bonham-Carter GF, Agterberg FP, Wright DF (1988) Integration of geological datasets for gold exploration in Nova Scotia. Photogram Eng Remote Sensing 54:1585–1592

Boots BN (1985) Voronoi (Thiessen) Polygons. Catmog No. 45. Geo Books, Norwich

Boots BN, Getis A (1988) Point pattern analysis. Sage, Newbury Park

Burrough PA (1990) Methods of spatial analysis in GIS. Int J Geogr Inf Syst 4:221–223

Cleveland WS, McGill ME (1988) Dynamic graphics for statistics. Wadsworth, Pacific Grove, CA

Cliff AD, Ord JK (1981) Spatial processes: models and applications. Pion, London

Couclelis H (1991) Requirements for planning-relevant GIS: a spatial perspective. Papers Reg Sci 70:9–19

Cressie N (1985) Fitting variogram models by weighted least squares. J Int Assoc Math Geol 17:563–586

Cressie N (1986) Kriging non-stationary data. J Am Stat Assoc 81:625–634

Csillag F (1991) Merging GIS and spatial statistics. Workbook for IGU-GIS Conference on Multiple Representations and Multiple Uses, Masaryk University, Brno, Czechoslovakia, April 22–25

Davis JC (1986) Statistics and data analysis in geology. Wiley, New York

Ding Y, Fotheringham AS (1991) The integration of spatial analysis and GIS. National Center for Geographic Information and Analysis, Buffalo, NY

Farley JA, Limp WF, Lockhart J (1990) The archaeologist's workbench: integrating GIS, remote sensing, EDA and database management. In: Allen KMS, Green FSW, Zubrow EBW (eds) Interpreting space: GIS and archaeology. Taylor and Francis, London, pp 141–164

Flowerdew R, Green M, Kehris E (1991) Using areal interpolation methods in geographical information systems. Papers Reg Sci 70:303–315

Foster SA, Gorr WL (1986) An adaptive filter for estimating spatially-varying parameters: Application to modeling police hours in response to calls for service. Manag Sci 32:878–889

Gatrell A (1987) On putting some statistical analysis into geographic information systems: with special reference to problems of map comparison and map overlay. Northern Regional Research Laboratory, Research Report No 5

Getis A, Ord JK (1992) The analysis of spatial association by use of distance statistics. Geogr Anal 24 (in press)

Goodchild MF (1987) A spatial analytical perspective on geographical information systems. Int J Geogr Inf Syst 1:327–334

Goodchild MF (1992) Geographical data modeling. Comput Geosci (in press)

Goodchild M, Brusegard D (1989) Spatial analysis using GIS: Seminar Workbook. National Center for Geographic Information and Analysis, University of California, Santa Barbara

Goodchild M, Gopal S (1989) Accuracy of spatial databases. Taylor and Francis, London

Goodchild M, Haining R, Wise S (1992) Integrating GIS and spatial data analysis: problems and possibilities. Int J Geogr Inf Syst 6 (in press)

Gould P (1981) Letting the data speak for themselves. Ann Assoc Am Geogr 71:166–176

Griffith DA (1988) Estimating spatial autoregressive model parameters with commercial statistical packages. Geogr Anal 20:176–186

Griffith DA, Bennett RJ, Haining RP (1989) Statistical analysis of spatial data in the presence of missing observations: a methodological guide and an application to urban census data. Environ Plann A 21:1511–1523

Griffith DA, Lewis R, Li B, Vasiliev I, Knight S, Yang X (1990) Developing Minitab software for spatial statistical analysis: a tool for education and research. Operat Geogr 8:28–33

Haining R (1990a) The use of added variable plots in regression modelling with spatial data. Profess Geogr 42:336–344

Haining R (1990b) Spatial data analysis in the social and environmental sciences. Cambridge University Press, Cambridge

Haining R (1991) Bivariate correlation with spatial data. Geogr Anal 23:210–227

Haslett J, Wills G, Unwin A (1990) SPIDER – an interactive statistical tool for the analysis of spatially distributed data. Int J Geogr Inf Syst 4:285–296

Haslett J, Bradley R, Craig PS, Wills G, Unwin AR (1991) Dynamic graphics for exploring spatial data, with application to locating global and local anomalies. Am Stat (in press)

HMSO (1987) Handling geographic information (The Chorley Report). London

Hubert LJ, Golledge RG, Costanzo CM (1981) Generalized procedures for evaluating spatial autocorrelation. Geogr Anal 13:224–233

Kehris E (1990a) A geographical modelling environment built around ARC/INFO. North West Regional Research Laboratory, Research Report No. 13, Lancaster University

Kehris E (1990b) Spatial autocorrelation statistics in ARC/INFO. North West Regional Research Laboratory, Research Report No. 16, Lancaster University

MacDougall EB (1991) A prototype interface for exploratory analysis of geographic data. Department of Landscape Architecture and Regional Planning, University of Massachusetts

Mosteller F, Tukey JW (1977) Data analysis and regression: a second course in statistics. Addison-Wesley, Reading, MA

Nijkamp P (1988) The use of information systems for regional planning. R Econ Reg Urb 15 (5):759–781

Nijkamp P (1990) Geographical information systems in perspective. In: Scholten HJ, Stillwell JCH (eds) Geographical information systems for urban and regional planning. Kluwer, Dordrecht, pp 241–252

Nijkamp P, Rietveld P (1983) Information systems for integrated regional planning. North Holland, Amsterdam

Oliver MA, Webster R (1990) Kriging: a method of interpolation for geographical information systems. Int J Geogr Inf Syst 4:313–332

Openshaw S (1990) Spatial analysis and geographical information systems: a review of progress and possibilities. In: Scholten HJ, Stillwell JCH (eds) Geographical information systems for urban and regional planning. Kluwer, Dordrecht, pp 153–163

Openshaw S, Taylor P (1979) A million or so correlation coefficients: Three experiments on the modifiable areal unit problem. In: Wrigley N, Bennett RJ (eds) Statistical applications in the spatial sciences. Pion, London, pp 127–144

Openshaw S, Charlton M, Wymer C, Craft A (1987) A Mark I geographical analysis machine for the automated analysis of point data sets. Int J Geogr Inf Syst 1:335–358

Openshaw S, Charlton M, Craft A (1988) Searching for leukaemia clusters using a geographical analysis machine. Papers Reg Sci Assoc 64:95–106

Openshaw S, Cross A, Charlton M (1990) Building a prototype geographical correlates exploration machine. Int J Geogr Inf Syst 4:297–311

Openshaw S, Brunsdon C, Charlton M (1991) A spatial analysis toolkit for GIS. EGIS 91, Proceedings Second European Conference on Geographical Information Systems, Brussels, Belgium, pp. 788–796

Peuquet DJ (1984) A conceptual framework and comparison of spatial data models. Cartographica 21:66–113

Peuquet D (1988) Representations of geographic space: towards a conceptual synthesis. Ann Assoc Am Geogr 78:375–394

Pielou EC (1977) Mathematical ecology. Wiley, New York

Rayner JN (1971) An introduction to spectral analysis. Pion, London

Ripley B (1981) Spatial statistics. Wiley, New York

Stringer P, Haslett J (1991) The spatial distribution of ill-health and material deprivation: an exploratory analysis using interactive graphics. Northern Ireland Regional Research Laboratory, Dublin

Tobler W (1979a) Cellular geography. In: Gale S, Olsson G (eds) Philosophy in geography. Reidel, Dordrecht, pp 379–386

Tobler W (1979b) Smooth pycnophylactic interpolation for geographic regions. J Am Stat Assoc 74:519–536

Tobler W (1989) Frame independent spatial analysis. In: Goodchild MG, Gopal S (eds) The accuracy of spatial databases. Taylor and Francis, London, pp 115–122

Tukey JW (1977) Exploratory data analysis. Addison-Wesley, Reading, MA

Walker PA, Moore DM (1988) SIMPLE: an inductive modelling and mapping tool for spatially-oriented data. Int J Geogr Inf Syst 2:347–354

Warren RE (1990) Predictive modelling of archaeological site location: a case study in the Midwest. In: Allen KMS, Green SW, Zubrow EBW (eds) Interpreting space: GIS and archaeology. Taylor and Francis, London, pp 201–215

Williams I, Limp WF, Briuer FL (1990) Using geographic information systems and exploratory data analysis for archeological site classification and analysis. In: Allen KMS, Green SW, Zubrow EBW (eds) Interpreting space: GIS and archaeology. Taylor and Francis, London, pp 239–273

# Using geographic information systems in urban planning and policy-making

**Michael Batty**

National Center for Geographic Information and Analysis, State University of New York at Buffalo, 301F Wilkeson Quad, Buffalo, NY 14261-0001, USA

**Abstract.** In this chapter, we discuss the various ways in which the emergent technology of geographic information systems (GIS) can best be utilized within the urban planning process. One of the major difficulties of developing effective integration between GIS and the functions of the planning process involves the comparative crudity of the present array of functions within GIS when compared with those functions which are required in urban planning. We first present typical GIS functions and discuss their comparative crudity, and then we illustrate the sorts of functions which are necessary to the urban planning process in the form of models and spatial analytic techniques. This serves to identify the functions which GISs require if they are to be used most effectively within planning. The rational planning process used to organize these functions is outlined and the ways in which the various phases and functions are influenced and influence the use of GIS noted. Different ways in which computer models might be embedded in GIS and into planning processes are defined and the conditions under which these different strategies can be employed are noted. We conclude the chapter with some speculations concerning the future of GIS and computer technology more generally within the planning process.

## 1. Introduction: GIS and the visual paradigm of urban planning

There is little doubt that the late twentieth century will be remembered as a time when visual paradigms spawned by the development of the computer became significant once again in the development and application of science. For some applications areas, the development of computer graphics is leading to a revival of previous approaches which were strongly based on visual media for analysis and communication and this is nowhere more apparent than in current applications of computers in urban planning. Since the 1950s, the development of computation in urban planning although regarded as an essential medium for dealing with spatially extensive data, has been fraught with problems of both a technical and conceptual nature. But the recent development of geographic information systems (GIS) and their potential for providing a general computational framework for such planning promises to resolve many of these longstanding difficulties in that the older traditions of visual thinking and communication in urban planning are at last being addressed through computational approaches which rely upon the same logic. In this, GIS lies at the vanguard of such change.

To date, the development of computers has been largely dominated by miniaturization although recently there has been rapid progress in the convergence of computers and telecommunications. Nevertheless it is miniaturization which has led to the revolution in graphics of which the development of GIS is now so important to urban planning. Like most features of the computer revolution, current developments reflect very long gestation times. It was at least thirty years before computers became full interactive and forty before computer graphics really took off. Miniaturization is still proceeding apace and could well cross further thresholds which will trigger new applications as yet unenvisaged. Geographic information systems seem to be following a similar path. There has been a concern with data banks, information systems and computer mapping since the 1960s, but only in the last decade has the development of GIS really taken off. Much of this can be traced to the development of machines which exploit their memory for graphics applications but the provision of more and more digital data, the pervasive effect of the microcomputer in developing greater computer literacy as well as the development of many more map-based applications in practice are all features which have contributed to this rapid growth.

This is perhaps surprising given the rather mixed experiences agencies have had with spatial modeling and GIS since the 1960s. The desire to develop data-intensive models for public policy and for spatial planning originated in the late 1950s when the notion that social and city systems might be treated quantitatively was first embraced. But the early experience with building large models on small computers with little data was problematic and this first wave of interest in using computers for such applications lost its momentum (Brewer, 1973; Greenberger et al., 1976; Kraemer et al., 1989). Nevertheless in the intervening years, several of the problems which plagued these early applications have disappeared. For example, very good data for planning purposes now exists online. The US Census Bureau's TIGER files and the USGS DLG files provide fairly comprehensive geographical coverage of boundaries and features down to the block scale. Other types of data covering transportation and employment also exist in machine readable form and although much of this data may not be in quite the right form for applications, it provides a basic starting point for GIS and spatial modeling.

Despite the fact that applications of GIS are growing dramatically and that new software is still coming onto the market, the actual usage of such systems is more problematic. To explore this, we need to have yet another working definition of a GIS in this book, and to this end, we shall consider such a system one in which data has some spatial or geographical referent, this data being ordered according to this referent, and displayed by software in such a form that some spatial analysis is possible. This distinguishes GIS from computer mapping software and other forms of geoprocessing. However, despite the fact that a GIS must have some functionality relating to analysis, many applications to date have tended to emphasize their use for elaborate computer mapping, storage and retrieval of spatial data but not for the sorts of analytical purposes for which they are ultimately intended (French and Wiggins, 1990).

In fact, the sorts of functions which are required in urban planning and spatial policy analysis are usually absent from the archetypal GIS whose functionality embodying spatial analysis and modeling is weak (Newton et al., 1988). This relates to the origins of the current software systems. Many GISs have emerged from areas of concern involving subsystems of the urban and regional system, from utilities companies, from land use zoning, from natural resource management and so on. These types of application have invariably developed with physical rather than socio-economic concerns in mind. Consequently such software contains methods for overlaying data, for generating maps or coverages by synthesizing other spatial data, for identifying areas of the geographical space which meet certain criteria and for tracing links in networks. These types of function although of some use in urban planning, tend to be less important than those functions based on models, methods, design protocols and such like which provide the lifeblood of such planning. Hence the promise of such systems is often not borne out when they come to be applied in planning contexts. The emphasis on using them for mapping then becomes explicable.

Because the rate of development of GIS is so great at the present time, it is difficult to provide a definitive summary of the state-to-the-art concerning the use of GIS in planning. Some of the criticisms which have been leveled against the current generation of GISs may no longer apply in the immediate future as such systems come in incorporate or develop links to demo-economic modeling, spatial interaction, urban growth models and so on. At present, massive diversification of GIS is taking place as new functions are identified and incorporated. In this chapter, we will concern ourselves with ways in which GIS can be linked or embedded in the scientific and political environments in which urban planning and policy analysis exist, rather than providing a catalog of useful functions which such systems should offer. In any case on the way to our goal, we will recite the typical functions which urban planning requires and note how these relate to the present generation of GISs.

We will begin by describing the typical technical functions of GIS and how these are related to those required in planning. The rational planning process provides a useful vehicle on which to mount these functions and we will begin our analysis of how GIS might best be incorporated into planning using this model. We will sketch the pertinent issues as to the ways in which GIS might incorporate planning through extensions of its functionality in terms of relevant planning models and how planning might embed elements of GIS into its process through models which in turn contain those GIS functions of relevance. We will not conclude with any particular design for the use of GIS in planning largely because the purpose of this chapter is to highlight the issues and begin the debate. Moreover we will not say much about organizational and political questions relating to the successful use of GIS in planning, and in this sense our debate will be limited. Indeed, there is a popular model of the development of computation which suggests that organizational questions – questions of 'orgware' as they have been termed – are more important than hardware and software, and this is certainly borne out in studies of the use of information systems in development planning (Batty, 1990; Elam, 1990, Nijkamp and Rietveld, 1984). Here we will be simply content to emphasize the technical issues.


## 2. The functions of GIS and the science of planning

It is important to be clear about the scope and definition of GIS before we embark on a more detailed discussion of the various functions contained in such systems and the need for extending such functionality towards urban planning. Our concern here is with that software which encompasses the storage, retrieval and display of spatial-geographical data with the added requirement that such systems display such data in two-dimensional map form and that such systems enable analysis of such maps to take place. Some definitions emphasize the particular requirement that a GIS must generate new maps or coverages from the basic spatial data which is input, although this definition is a little narrow for our purposes (Scholten and Stillwell, 1990). In this context, we consider GIS to cover all those systems from the most detailed site-specific scale based on the cadastre to the thematic scale at the regional-national level to the global level where map generalization and projection are central. This definition thus includes land information systems based on property rights and urban information systems in which spatial and aspatial data exist side-by-side but it excludes software which is purely for mapping which we refer to more generally as constituting geoprocessing software. In short then, we consider that two features of GIS which are central to the types of software of concern here are an ability to display data in spatial form as maps or like variants and an ability to perform some rudimentary analysis of such data. In fact, as we will argue, GISs are particularly appropriate to this mapping function having been largely derived from such concerns, and particularly deficient with respect to spatial analysis and modeling which is the focus of our discussion here.

The software of GIS is usually based upon representing spatial data either in grid form from a raster or in terms of its more detailed topology enabling such data to be aggregated to various sizes of zone or polygon. Increasingly raster systems provide vector conversion and vice versa but the ability to merge vector and raster maps is an all important property of such systems and much of their geoprocessing capabilities are given over to such operations. Accordingly, such systems must have elaborate peripheral as well as central hardware to enable images and maps to be input and output, and this raises the cost of acquisition considerably over other types of hardware-software system. In terms of the organization of their software, spatial data is usually stored as relational two-dimensional tables from which various cross-classifications can be derived, and which are highly suited to the manipulation of spatial data.

The typical functions of a GIS involve the following operations on spatial data: topological operations which transform and generalize two-dimensional spatial data, thus producing maps at various levels of aggregation, overlay analysis techniques which combine maps as layers and enable various visual, statistical and logical operations on the resulting coverages, buffering and related spatial subdivision methods which identify areas conforming to various criteria, elementary statistical operations involved in describing, smoothing, and developing methods of error and bias in data. Another related domain to which these functions we have just described might be applied, concerns the application of similar operations with respect to one-dimensional, network data, while there are also developments in which three-dimensional data can be handled in terms of its representation and display.

However, there are many functions at the level of description which have not been incorporated as yet into contemporary GISs and we will list these below. There are ingenious developments in linking GIS to other functions which exist in terms of their own specialist software and various methods of importing and exporting data between GISs and other software are being actively developed. Whether or not this is the best strategy is unlikely to depend entirely upon analytical and scientific considerations but on the type of problem which is being informed through the medium of a GIS. Nevertheless, in terms of conventional spatial analysis, many useful functions do not yet exist within an archetypal GIS. For example, Openshaw (1990) demonstrates that the following basic functions are absent from most systems and should probably be added through the extension of such systems. These are: basic spatial statistics such as correlation indices, various moments of spatial probability distributions, procedures to enable exploratory spatial analysis, methods for tracking error, procedures for optimal spatial aggregation, methods of interpolation and smoothing necessary to forecast missing data and massage spatial data into consistent forms, econometric and regression analysis procedures, pattern recognition devices, data simplifiers – the list could be expanded much further but the point is that most GISs have not been evolved with explicit spatial analysis in mind and thus, for the most part, lack the functions which are essential to good analysis and best practice.

There are also some key problems of generalization which dominate contemporary GISs. The problem of scale is central to geographical analysis and cartography. One obvious function of a GIS is to enable procedures for aggregating and disaggregating spatial data with the added functionality that such changes in spatial detail are well understood in terms of the degree of invariance or scale dependence assumed in the data. In fact, the spatial aggregation problem is not well-defined or understood at the present time and thus GISs can act as a focus around which to explore such issues. When maps are aggregated, disaggregated or generalized to different scales, the major problem is one of changing information which must somehow be communicated effectively, especially if such systems have an active practical use. There are both statistical and visual-presentational-communication types of problems here but there is also the problem of designing systems whose geometric features and attributes change as scale is changed and data aggregated. This is a central research problem in spatial analysis and cartography and one which limits the development of truly general GISs.

In short, GISs which are built for thematic mapping, for cadastral mapping, and for network representation are still likely to be significantly different in their treatment of geometry and topology to require quite different basic functions, hence different software. There are also very few GISs which can be adapted to handling temporal or space-time data. Such data pose their own problems which although in some sense analogous to spatial series, are different enough to require their own development. So far most GISs have been designed and marketed as providing software and functionality for immediate planning over very short time horizons or for comparative-static analysis where the time frame is indefinite. There are few systems which have been adapted to monitoring urban change and policy impacts while there are few, if any systems which have survived longer than the decennial population census period. Many systems have been developed around the population census but there are few documented examples of systems which have been used to update one census to the next.

Existing GISs thus contain only the most rudimentary of functions relevant to the sort of analytical, simulation, prediction and design tasks which dominate strategic planning or the monitoring and query needs for systems of ongoing data collection and planning control. With respect to this differential between planning functions based on strategic planning at one end of the spectrum and control at the other, the applications of GIS are diverse. Scale from national-regional to urban to local can make a difference in terms of the types of data which are relevant and the way in which data pertaining to all these scales are aggregated. Differences between thematic and cadastral have been noted but there are also differences between systems which deal predominantly with physical data which is often represented as a grid or raster and systems in which data pertains to differently sized and positioned areas which inevitably involve treatment in terms of vectors. Systems which are partial − designed for a single sector and a single problem − have different demands from those which are more comprehensively focused. Systems which are to be used for and/or contain mainly data about the public sector can be substantially different from those designed to be useful to the private sector. There are obvious differences between information systems which deal predominantly with spatial data, those that deal with aspatial and those that deal with nonspatial. We have already noted systems which discriminate the phenomena in terms of scale or in terms of points and lines − utilities networks for example, in comparison to those dealing with points, lines and areas. As yet we do not have a well-developed classification of GISs in practice, and it is clear that single systems are being used in a fairly blunt manner for many different applications. The comparative crudity of the adaptation involved is well-remarked by the fact that many proprietary GISs such as ARC-INFO, SPANS, SYSTEM 9, etc. are marketed for a wide variety of 'stylized' problems, rather than being developed with a particular context in mind. Moreover the crudity of present systems is such that it is fairly difficult to adapt most of them to diverse problems.

There are a number of general conclusions which can be drawn from this discussion. It is not too bold or sweeping a statement to argue that one single GIS is never likely to be applicable to all problems in the context of physical planning. There is a tendency in the industry to develop the all-purpose GIS although close analysis of the issues indicated above suggests that different problem contexts are different enough to require their own purpose-built systems. In some senses, this is what is happening in the development of computer-based infrastructure in physical planning in that the most appropriate modules for the task in hand are assembled by the planners as users, rather than by using a general purpose package which contains all the required functions.

In another sense too, the development of GISs by the various vendors is beginning to take on a modula-unit-based focus. Special purpose GIS, for example TRANSCAD, are being developed for transport problems where the emphasis on networks rather than zones is much stronger. Vendors are developing systems which can be modularized and purchased as different composites or bundles while systems are being carefully fashioned to enable easy input and output of geometric and attribute data to and from other systems. Finally, the long term goals of the vendor are important in influencing the types of systems which are developed and sold. For example, there are those vendors such as IBM, Prime etc. who see GIS as an opportunity to push certain hardware platforms. On the other hand, there are vendors like ESRI who depend entirely on the hardware platforms available and who develop many versions of their software for different machines. These too are issues which affect the functionality of a GIS. And there are firms such as Intergraph who develop a blend of hardware-software in which the user is often caught both ways.

We will now begin to map out the functionality which GISs contain to inform different tasks. In essence at the urban planning level, we are dealing with models and data whose basic unit would either be the street network or a set of relevant attributes associated with some geometric area such as the block group or census tract. Various types of aggregations of zones would be required while different sorts of network are needed for different transport modes and facilities. There are also demands in this kind of analysis for data and models which deal with physical features and map areas. To summarize and list some of these, a typical strategic urban planning process would require: demographic, economic, land use, transportation data, data concerning urban economic structure, all cast in the form of employment, population by age-sex, and interactions in terms of people, goods, money and information flows (Echenique, 1983). Physical data pertaining to the quality of the landscape, topography, soils, geology, land use type, housing structures and so on, would also form part of this kind of system. At present in fact, there are few GISs which could embrace this range of data, store, retrieve, and map it consistently let alone develop this data for the various models and forecasting purposes which are required. There is extensive functionality here which in principle could well-reside inside a GIS but in practice the state-of-the-art is such that there have been few extensions to these domains and vendors remain content to develop more general functionality suitable for a wider range of problems at a more elemental level.

The real issue is how to approach the development of much more sensitive and useful GISs for urban planning in particular, other tasks in general (Couclelis, 1989; Harris, 1988). It is not clear at all whether the strategy should be to extend current GIS packages further, noting the fact that many such packages are somewhat of a hotch-potch already or whether to proceed in a completely modular fashion, building a seed-bed for the development and use of appropriate functions. Another alternative would be to build a hierarchy of GISs using various forms of model as buttresses and filters between their use in the general planning process and their application in the real world. In Fig. 1, we sketch the basis of this argument in the following three possibilities: a) the idea that there is one all-encompassing GIS, parts of which relate to different models; b) the no-

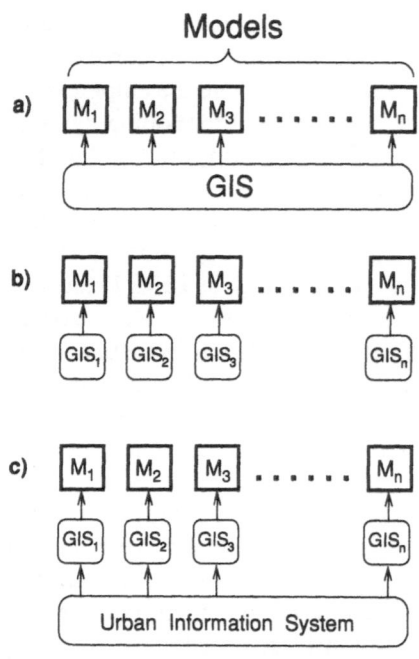

**Fig. 1.** Relationships between models and information systems

tion that for each model, there might be a tailormade GIS; and c) the idea that there might be a hierarchy of ISs leading to GISs which provide the seed-bed in which this landscape of models flourishes. One last point is worth making: in the following discussion, we will assume that the link between a GIS and its application to the planning process is through the medium of some sort of model or formalized function, an issue which we develop and extend when we outline the ways in which planning processes and information systems might be embedded within one another.

## 3. The technology of the planning process

There are many very different varieties of planning process ranging from those which are implicit in the decision-making behavior of the key actors involved in making decisions, to those highly formalized and explicit processes based on strictly ordered problem solving protocols. Here where we are dealing with urban planning, the processes of relevance are largely explicit in that they depend upon some legislative mandate for the preparation of plans or the execution of controls which are directly or indirectly based upon some adopted plan. In this context then, we will assume a two-fold process, the first a process of highly-formalized strategic planning based on an explicit process of rational decision which is conducted on a cycle of months or years in contrast to the complementary process of negotiating and bargaining which is a feature of the control mechanisms used to keep the strategic plan on course. This latter process is operational on a day-to-day or week-by-week cycle. GIS and variants on these types of information system are regarded as being central to such processes (Harris, 1988).

The processes in question can also be operated across several different hierarchical levels or scales although here we implicitly adopt an exemplar based on planning at the urban-metropolitan level. Moreover, we must also be conscious of the fact that any rational process of planning is infinitely recursive in that the planning process can be used to design a planning process at the next level and so on until the process involves the substantive systems and problems of interest. Sometimes these processes which depend on the same processes of decision at a lower level are called meta-planning processes (Chadwick, 1971). We will also assume here that the highest level process of interest is based upon technically informed protocols which involve resolution of the substantive problems of interest related to urban and metropolitan spatial structure.

The rational model of decision emerged formally in the 1950s and 1960s in a variety of decision contexts – urban planning, economic allocation, psychological problem-solving, even in scientific inquiry itself (Chadwick, 1971). It is based on the simple notion that the planning process embodies an analytic phase in which the problem is explored, followed by a synthetic phase in which the solution is devised or generated. Problems are informed by data and survey relating to the issue in question and progress towards their solution is cast in terms of general goals and more detailed objectives. The analysis involved, which in turn is based on the information collected, is oriented towards generating a requisite understanding of the problem, thus enabling some prediction of the immediate future and nature of the problem to be assessed. In some contexts, the problem might be seen to resolve itself and therefore no action is required but usually planning problems are so interconnected that it is necessary to develop some solution involving explicit intervention in the urban system.

The synthetic phase is based on the design of a solution or resolution of the problem. This involves some structured intuition which might be supported by various formal design and modeling techniques. Usually a range of alternative solutions are generated and then evaluated simultaneously or sequentially against the earlier stated objectives. On the basis of this evaluation, a decision is made to implement one of these alternative plans or to reiterate the process in the search for better plans. In fact, throughout the process there is reiteration between various stages which enables the problem to be better defined and solutions to be better refined. This reduces uncertainty about appropriate outcomes but it also acknowledges that the process is one of to-ing and fro-ing between problem and solution and in this sense, cannot be considered as mechanical or deterministic in any way (Boyce et al., 1970).

The problem-analysis-science phase of the process which is followed by the solution-synthesis-design phase might be reiterated many times but at some point, the exigencies of the process determine that some form of plan will emerge. Then begins a process of implementation which might involve a similar planning process but of a different kind, and in turn as plans are further implemented and used to control and stimulate urban development, this technical process may be reworked many times in different contexts. Although the meta-process is likely to repeat itself on a regular cycle which is usually mandated in some legislative context, if the process is conceived in terms of many similar processes at different levels, the plan-making and decision-making activities are carried out continuously. In this way, the set of processes can be considered as part of the problem-solving infrastructure which exists in any planning organization.

The motivation to operate the process however is driven by various techniques, models and methods which support the problem-solving sequence in diverse ways. Alongside the phases and stages of the process, there exist a variety of decision-support systems which inform our ability to define and redefine problems and goals and objectives, to understand and predict the future nature of the problems in question, and to ensure the generation of imaginative planning solutions (Manheim, 1986). In turn, these decision support systems are driven by various information systems which enable the understanding, analysis, prediction and prescription of the questions of interest. In Fig. 2, we show how this process interacts with these decision support systems which in turn interact with systems which portray and capture relevant information about the problem. In more traditional terms, decision support systems are in essence models of the urban system in various guises but by referring to them as decision support, this identifies them as contingent upon the planning process itself and not necessarily models which exist in their own right outside the wider confines of the planning process (Densham and Rushton, 1988).

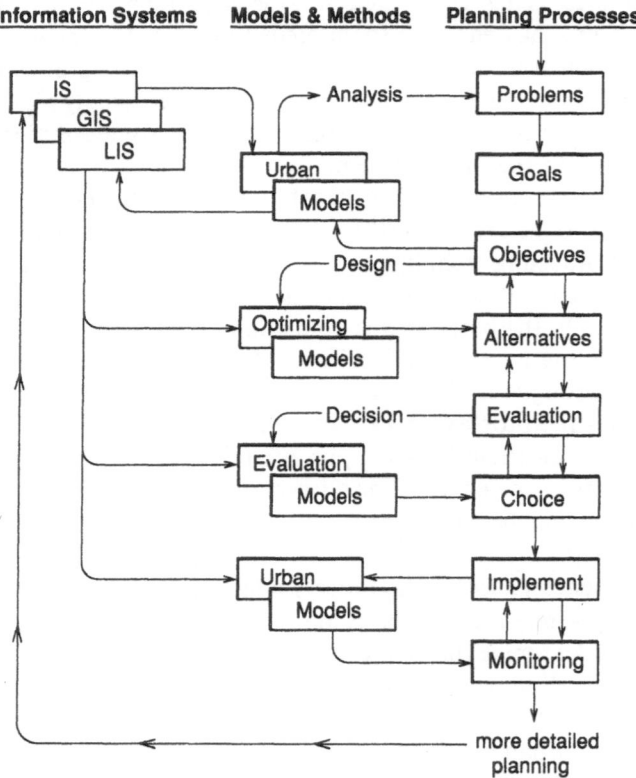

**Fig. 2.** Planning support systems incorporating models and information systems

This combination of planning process, decision support and information systems in Fig. 2 are similar to what Harris (1989) has defined as a 'planning support system'. The key feature of this system of processes is the way they relate to one another and the way in which appropriate urban and planning theories are embedded within. We have made an assumption here which we indicated in the previous section and that is that in the development and use of information systems, models in the form of decision support systems are used to blend and filter the information so that it becomes informative and useful to the planning process itself. In other words, we are defining decision support systems as the interface between information systems and planning processes. The implication is that whenever an information system is developed and used in any planning activity, there is some analytical function of the information system which enables the information to be useful to the problems in question. We will see that various uses of information systems in planning always require some formal or at least informal analytic functionality ranging from simple descriptive statistical analysis which is part of the information system itself to full-fledged modeling and simulation which might stand apart from the information system but is consistently linked to it.

In Fig. 2, we indicate that there may be several types of information system linked in an unspecified way but informing and supporting different analytical perspectives on the problem context and the process used in its solution. With such a diverse set of processes and problem elements, it is most unlikely that all the information necessary for formal use in decision support can or should be embodied in a single information system. In fact, there may be the need for only one GIS while several non-geographic information systems may be necessary in representing facets of the problem. Given this type of structure, it would seem unlikely that the process could rely on an amorphous information system serving all aspects of the problem-solving. In the same way, there is unlikely to be a single comprehensive model which might support the process. During the last decade, there has in fact been a trend towards the break-up of comprehensive models, in the development of many partial models which can be used as building blocks to produce an appropriate planning support system. In later sections we will explore the idea that an appropriate planning support environment is highly modularized with respect to the information systems and models used, and that an appropriate system for any particular problem will inevitably involve the assemblage of individualized components.

It is worth reflecting a little on the sorts of data and information system which are useful to the strategic planning process. If we assume that data relevant to this type of planning cover the demographic, economic, transportation, educational, housing, and health subsystems and that these subsystems give rise to activity data as well as physical data on land use, man-made condition, energy, economy, and physiography, then there are various ways of developing information systems for this data. In fact, it is likely that there will be more than one geometry relating to the map form of this data — based on networks, and on areas — while the idea that there is a common geographical unit of representation for physical and activity data is not realistic. If this range of data is required, then there are several ways in which this can be put together. In fact, there is also the problem of missing data, inconsistencies and errors in data which need to be evaluated before any general GIS can be constructed and it would thus appear that a fully integrated system is neither necessary nor desirable and in any event is probably impossible to construct.

In terms of the ways in which we develop planning support systems, we have indicated so far that it is unlikely that any attempt to design a single GIS to support such a task is desirable. However this is flying in the face of what is happening in practice. For example, vendors are attempting to make their GIS systems much more flexible, embracing a variety of tasks and there is little doubt that an increasingly wide range of end users are being sought for the industry standard products such as ARC-INFO, SPANS, and so on. For example, the use of ARC-INFO for health planning, social services planning, site selection across a variety of scales, population forecasting, and so on indicates just how wide the actual as well as potential usage of these general purpose systems might be. However, the systems themselves are evolving and becoming more modular as the vendors learn more about their potential client demands. In the medium term, it would seem that GISs will become more diverse in that new systems will be developed for new purposes while existing systems will be modularized in the quest to develop tailor-made bundles of modules relevant to a wide variety of client needs.

## 4. Embedding GIS into planning and planning into GIS

The view that we have been espousing throughout this chapter is that GIS is part of the emergent infrastructure of the post-industrial society and that it constitutes the landscape upon which post-industrial planning will evolve. As such, it is unlikely that this landscape will be homogeneous in any sense. It is more likely to be somewhat anarchic with many types of data and information, methods and model competing with one another for use in active decision-making. Moreover, the ways in which information systems, models and planning processes come to be linked will depend upon the particular circumstances of the issue in hand. In short, the way theory and model are built into the planning process or into the information system is likely to vary quite widely. In this section, we will examine these three activities – information, models and planning processes and consider the implications of different embeddings and couplings of any one of these into any other. We will also refer to these three areas in the shorthand of I (information systems – GIS), M (models) and P (planning) in our assemblage of functions and the way these are linked together. In the figures which follow, we will use the following geometric forms for each activity: triangles (I), rectangles (M) and circles (P).

First we will consider the relationship between models and planning. In the way the rational decision model has evolved during the last 30 years, the planning or problem-solving process is considered quite separate from the models and methods which support it as is portrayed in Fig. 2. If the range of model types is examined, then models may be orientated towards description, prediction or prescription. This latter category represents models which enable problems to be solved, goals and objectives to be optimized, and thus they represent in themselves planning or problem-solving processes. In conventional urban planning however, an implicit assumption is always made that its problems and processes are so ill-defined that no single model, optimizing or otherwise, can be developed to capture the essential nature of the planning task. The process is seen as separate from any of the models which inform it, even though some of these models might be optimizing, but pertaining to only a small fraction of the decisions which constitute a realistic plan. In Fig. 2, the fact that these two streams are separate implies this distinction.

In examining the coupling of models to the planning process, there are three possible types. First the one we have been implying is based on a loose coupling in which models inform planning and planning informs the development of these models. The second possibility which in fact is the one most likely to be adopted in practice is one in which models are subsumed or embedded into the planning process itself. This reduces the role of models to that which is often seen in practice. The third possibility is at the other extreme; the planning function is subsumed within the model system itself. In fact, in many systems theories, models are seen as describing 'natural' change within any system with planning regarded as the control mechanism steering the system to some goal but organized very definitely as a subset of the overall model itself. Sometimes, where planning is dominated by professionals who have not been trained in policy-making but in systems modeling, this is often the role relating planning to models which is assumed. The three embeddings are shown in Fig. 3 (a), (b), and (c).

We can develop the same ideas of embedding for the relationship between GIS and models, and GIS and planning. However, in an earlier section, we argued that the way the planning process relates to GIS is through the medium of some model. In short, for GIS to be useful in a policy context, then some analytic function is necessary to produce information in the right guise; this we assume is a 'model' in its most catholic sense. Accordingly we will not examine the direct relation between GIS and planning for this is through the intermediary of a model. Thus we need only examine the relation between GIS and model and then concatenate this with this relationship between model and planning process. In the same way as previously, GIS can be loosely coupled to models through links which simply allow the importing and exporting of key data and results between the two. In fact, this is the usual way of linking the two in most applications to date.

At one extreme, we can also consider that models are entirely embedded within the GIS and this is the strategy favored by the designers of such systems. In this case, relevant models are regarded as simply additional functions to a GIS. In some cases, such a development may be possible although as we have argued earlier, this is unlikely to be easy to achieve. Any GIS which is extensive enough to embrace a suite of relevant urban planning models and methods is unlikely to be an integrated package and must thus be a set of relatively independent modules which can be coupled together in diverse ways. At the other extreme, models may entirely incorporate the GIS. Again this is unlikely in that most GISs contain many more functions that are relevant to any single model or set of models. There is little point in embedding statistical functions into a GIS which in turn is used in a model to map, present and generalize data at different map levels say. How-

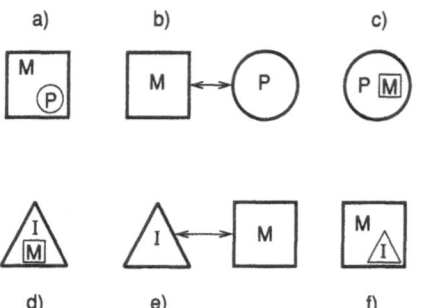

**Fig. 3.** Embedding urban models into geographic information systems and planning

ever, it is likely that there are many models which need a limited amount of GIS functionality for display and aggregation purposes and thus it may be necessary to uncouple elements of a standard GIS for these purposes. These three alternatives are presented in Fig. 3 (d), (e), and (f).

There are several applications which attempt to embed models into GISs such as SPANS, ARC-INFO, TRANSCAD and so on. In contrast, the number of applications in which GIS functions are embedded into models is less easy to identify. Many such applications contain these functions but are rarely considered in the terminology of GIS. For example, much of the graphics concerned with land use-transportation models and the various statistical descriptions which are computed as by-products are simply regarded as necessary methods for displaying and communicating model data and results. Transportation models, especially those implemented on microcomputers, present useful examples of where such functions have been implemented within models (see Young et al., 1989). We must also note at this point, before we try to explore the continuum between GIS, models and planning, that there are many freestanding packages concerned with elements of GISs; these would not, however, be regarded as constituting GISs because their purpose is somewhat narrower than GIS per se. Mapping packages and other geoprocessing software as well as applied statistical packages are examples of such software.

We can now begin to put the continuum of the GIS-model-planning process together. In Fig. 4, we have cross-classified the 'model-planning' relation against the 'model-GIS' linkage indicating the sorts of connections which result. The main diagonal elements of this table range from the two extremes where planning is embedded in models which in turn are embedded in GIS to the opposite form where GIS are embedded into models which are embedded into planning. The complete loose-coupling of GIS-model-planning occupies the central cell of this table. The three upper off-diagonal elements of this table involve linking GIS to the planning process but with models embedded first into GIS or into planning and thus violating our assumption that the link from GIS to planning must be through some form of model. In fact although models may be embedded in some other function, there may still be direct links from a model which mediate between GIS and planning, and thus we have not completely excluded these possibilities from our consideration.

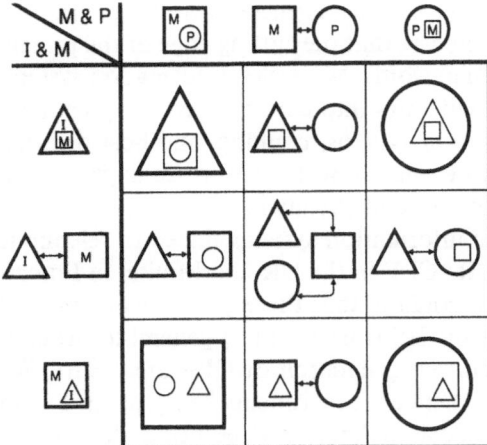

**Fig. 4.** Models and planning processes related to models and information systems

The three lower off-diagonal elements involve the two possibilities where models are part of GIS and models are part of planning with weak coupling between planning and GIS in both cases. The last case involves GIS and planning being both subsumed within modeling and in some respects, this mirrors a view of the world which is often that adopted by the model builder. If we further relax our restriction that a model should act as a mediator or interface between GIS and planning, then there are many other possible combinations we can develop. Moreover, if we make this analysis more detailed emphasizing types of model, GIS, and planning process, then the number of possible forms becomes very great. All these can be seen as ways in which the link between information and planning through models can be designed. The ultimate choice will depend on many issues and thus it is not possible to specify ideal types. Our preference here is for processes of planning which embrace information and models in contrast to GISs into which models and planning are integrated but this is clearly the urban planner's perspective and is no more valid than any other in a priori terms. These ideas only provide guidelines for linking GIS to planning. However, to make progress on such issues, more case studies are needed and more experimentation with models and GISs is urgently required.

## 5. Conclusions: whither GIS in planning?

It is tempting to think of the present dissemination of GIS as being the equivalent for picture processing as wordprocessing was a decade or so ago. This however is a little misleading for GIS is considerably more specialist than wordprocessing and also more professionally rooted in substantive skills. Picture processing in its general sense extends much further than GIS in that CAD, image processing, computer animation, computer graphic design and the whole range of visual computation define this type of computing. Yet in some senses, the dramatic development of GIS and its positive embrace by most agencies involved in graphics and mapping of some sort, do indeed indicate that the cutting edge of the present revolution is now in pictures rather than in words. Moreover, information systems containing graphics are now finding their ways into countless everyday contexts such as the automobile, public information systems, banking, and host of other ordinary service functions. In the next decade, we will begin to see digital visual media enter the home through television sets and already in several western countries, there are virtually costless public information systems linked to telephone and television such as CEEFAX and ORACLE in the UK and Minitel in France. Once broadband transmission becomes standard, then the population at large will have access to countless different types of GIS from the most general to the considerably more specific of which the current generation of GISs are a part. For example, at present in New York State, there are experiments being developed to implement online query systems for social welfare services using the current generation of PC-ARC-INFO.

What is also clear is that as workstations reduce in price and increase in power to meet the expanding power of microcomputers coming from the other direction, hardware will become less of a problem and many agencies will then have access to incredibly cheap software enabling all those who wish to use GIS to do so. Already PC-ARC-INFO is available throughout the British University system for a fraction of its list price and such economies of scale are being repeated in many applications across a wide range of software products. Increasing GIS use is also dependent upon better and more data too. More and more data is becoming available electronically online. Recent population censuses in 1990 in the US and 1991 in Canada and the UK will be available mainly in digital form while other data sources are increasingly available online. Satellite data and developments in the quality of remote sensing and interpretation are also providing new data sources for mapping and detailed thematic work with population, urban growth, land use control and so on. Data on individual behavior with respect to household expenditures, income, education, health and suchlike are now available from the online credit and banking infrastructure and despite problems of privacy and confidentiality which are ever present, such data is becoming available as value-added services. The list of new digital data sources is still growing and many different types of information system are necessary simply to store and access the data. GIS will be central to such concerns.

The emergent infrastructure will be composed of hardware, software, data and orgware and for a time yet, this will be ever changing as new developments in hardware and new demands for applications continue to have an impact on software and orgware. It is difficult to predict what will happen with GIS, whether existing systems will grow and extend or break into more specialist parts, or whether the existing set of vendors will radically change in response to new markets. These types of scenario will have a profound impact on practice however. Past experience in software and hardware suggests that within a decade, some of the vendors such as Intergraph, ESRI, Tydac, Prime and so on are likely to have disappeared, their products absorbed into others as new types of applications and ideas in software emerge and as some vendors find it impossible to adapt. In a decade, it will be interesting to see who will lead the hardware industry and what the hardware will be on which GIS is mainly based. It is likely that from the present major group of suppliers – IBM, DEC, Apple, SUN and so on – some will have gone while others who are now unknown or do not exist will have emerged. The rate of change is still so fast that software systems will have changed out of all recognition and it is even possible that GIS systems will have been supplanted by other spatial-geographical systems with a more analytic flavor. The examples of such change from the last decade are clear. A decade ago the industry standard operating system for micros was CP/M. What, I hear you say, is CP/M?

Yet perhaps the most exciting prospect in computer-aided planning relates to the emergence of a rich environment where dramatically innovative applications can be fashioned out of existing modules of software and new modules can be fashioned out of splitting and aggregating existing ones. Already there are suggestions that software objects called 'knowbots' can be designed like computer viruses to crawl through networks picking up relevant software and delivering it to users who can then fashion it into useful applications (Cerf, 1991). The notion of the 'computer as the network' and the idea that software modules can be regarded as primitives will be central to applications in the next decade. GISs will need to adapt to these developments if they are to survive and expand. For example, existing systems are in the process of adapting to the new generation of windowing systems such as X and close on its heels will come object-oriented programming. In such a competitive context, it is often easier for new firms to enter the market then for existing firms to make the transition.

The other more substantive development which concerns the form of scientific planning relates to the use of models and other quantitative techniques. The initial developments in the 1960s floundered on the problems posed by computation, data availability, oganizational inertia and changing priorities in practice. The first demise of this style of planning has now begun to reverse itself and in the 1990s, various types of model will again constitute the cutting edge of urban planning (Boyce, 1988). In a sense, the emergence of an information culture and infrastructure is a process which is likely to have many false starts and thus it should not be surprising that such applications wax and wane. It is still likely that during the 1990s, many applications in GIS will remain experimental while software will be abandoned and some of it not replaced with new. Nevertheless, the computer revolution will still force the pace in new applications areas in fields such as urban planning.

At the same time, a new generation of transportation studies and projects is likely to begin and this will further stimulate computer applications and the development of GIS. Thus the need for guidelines in the development of information systems is as important as ever, and if there is a clear conclusion to this chapter, it is that detailed case studies of the development of GIS in planning are urgently required. Demonstration projects, ideal forms of computer environment and laboratory, experimental designs in a theoretical and practical context – all these are required if GIS is to become ever more relevant to planning applications. At the present time, it is necessary to experiment with different types of GIS and although practice is rarely able to adopt such modes of experimentation explicitly, these are bound to develop as different systems emerge and as GISs are extended to different types of applications.

*Acknowledgements.* This chapter represents work that is related to research being carried out at the National Center for Geographic Information and Analysis, supported by a grant from the National Science Foundation (SES-88-10917). The Government has certain rights in this material. Support by NSF is gratefully acknowledged. Any opinions, findings and conclusions or recommendations expressed in this material are those of the author and do not necessarily reflect the views of NSF. A longer version of this paper was first presented at the Twentieth Anniversary Symposium of the United Nations Centre for Regional Development (UNCRD) in Nagoya, Japan, in November, 1991.

# References

Batty M (1990) Information Systems for Planning in Developing Countries, Volume 2, in Information Systems and Technology for Urban and Regional Planning in Developing Countries: A Review of UNCRD's Research Project, United Nations Centre for Regional Development, Nagoya, Japan, pp 103–236

Boyce DE (1988) Renaissance of Large-Scale Models. Papers Reg Sci Assoc 65:1–10

Boyce DE, Day ND, McDonald C (1970) Metropolitan Plan-Making, Regional Science Research Institute, University of Pennsylvania, Philadelphia, PA

Brewer GD (1973) Politicians, Bureaucrats and the Consultant: A Critique of Urban Problem Solving, Basic Books, New York

Cerf VG (1991) Networks, Scientific American 265:72–81

Chadwick GF (1971) A Systems View of Planning: Towards a Theory of the Urban and Regional Planning Process, Pergamon Press, Oxford, UK

Couclelis H (1989) Geographically Informed Planning: Requirements for Planning-Relevant GIS, A paper presented at the Thirty Sixth North American Meetings of the Regional Science Association, UCSB, Santa Barbara, CA

Densham P, Rushton G (1988) Decision Support Systems for Locational Planning. In: Golledge RG, Timmermans H (eds) Behavioural Modelling in Geography and Planning, Croom Helm, London, pp 56–90

Echenique M (1983) The Use of Planning Models in Developing Countries: Some Case Studies. In: Chatterji L, Nijkamp P (eds) Urban and Regional Policy Analysis in Developing Countries, Gower Publishing Company, Aldershot, Hampshire, UK, pp 115–158

Elam JJ (1990) Information Technology in Urban and Regional Planning, Volume 1, in Information Systems and Technology for Urban and Regional Planning in Developing Countries: A Review of UNCRD's Research Project, United Nations Centre for Regional Development, Nagoya, Japan, 1–102

French SP, Wiggins LL (1990) California Planning Agency Experience with Automated Mapping and Geographic Information Systems. Environ Plann B 17:441–450

Greenberger M, Crenson MA, Crissey BL (1976) Models in the Policy Process: Public Decision Making in the Computer Era, Russell Sage Foundation, New York

Harris B (1988) Planning Technology: A Bridge Between Theory and Practice, A paper presented to the 1988 Annual Meeting of the Association of Collegiate Schools of Planning, held at SUNY, Buffalo, September 1988

Harris B (1989) Beyond Geographic Information Systems: Computers and the Planning Professional. J Am Plann Assoc 55:85–90

Kraemer KL et al. (1989) Managing Information Systems: Change and Control in Organizational Computing, Columbia University Press, New York

Manheim ML (1986) Creativity-Support Systems for Planning, Design and Decision Support, Microcomputers in Civil Engineering 1:14–3

Newton PW, Taylor MAP, Sharpe R (eds) (1988) Desktop Planning: Advanced Microcomputer Applications for Physical and Social Infrastructure Planning, Hargreen Publishing Company, Melbourne

Nijkamp P, Rietveld P (eds) (1984) Information Systems for Integrated Regional Planning, North-Holland Publishing Company, Amsterdam

Openshaw S (1990) Spatial Analysis and Geographical Information Systems: A Review of Progress and Possibilities. In: Scholten HJ, Stillwell JCH (eds) Geographical Information Systems for Urban and Regional Planning, Kluwer Academic Publishers, Dordrecht and Boston, MA, pp 153–163

Scholten HJ, Stillwell JCH (1990) Geographical Information Systems: The Emerging Requirements. In: Scholten HJ, Stillwell JCH (eds) Geographical Information Systems for Urban and Regional Planning, Kluwer Academic Publishers, Dordrecht and Boston, MA, pp 3–14

Young W, Taylor MAP, Gipps PG (1989) Microcomputers in Traffic Engineering, John Wiley and Sons, New York

**Part B**
**Technical issues**

# Developments in areal interpolation methods and GIS

Robin Flowerdew[1] and Mick Green[2]

[1] North West Regional Research Laboratory
[2] Centre for Applied Statistics, Lancaster University, Lancaster LA1 4YB, UK

**Abstract.** This paper is a review and extension of the authors' research project on areal interpolation. It is concerned with problems arising when a region is divided into different sets of zones for different purposes, and data available for one set of zones (source zones) are needed for a different set (target zones). Standard approaches are based on the assumption that source zone data are evenly distributed within each zone, but our approach allows additional information about the target zones to be taken into account so that more accurate target zone estimates can be derived. The method used is based on the EM algorithm. Most of the work reported so far (e.g. Flowerdew and Green 1989) has been concerned with count data whose distribution can be modelled using a Poisson assumption. Such data are frequently encountered in censuses and surveys. Other types of data are more appropriately regarded as having continuous distributions. This paper is primarily concerned with areal interpolation of normally distributed data. A method is developed suitable for such data and is applied to house price data for Preston, Lancashire, starting with mean house prices in 1990 for local government wards and estimating mean house prices for postcode sectors.

## 1. Introduction

A key theme in geographical information systems is the integration of different data sets, and one of the key problems in data integration is the diversity of areal units in use for different purposes. Frequently it is desirable to compare two or more sets of regional data which are available for different zonal systems. A common example concerns administrative zones, for which data are collected by national censuses, government departments and by the administrative units themselves, and postal zones, which are the easiest way to aggregate data for most commercial purposes, such as records of sales or client contacts. In order to compare data for administrative and postal zones, or any other pair of incompatible zonal systems, methods must be found for estimating what one set of data would be like if it were available for the other zonal system. This problem is referred to as the areal interpolation problem (Goodchild and Lam 1980).

With colleagues at the North West Regional Research Laboratory, we have been engaged in a research project to develop improved methods for areal interpolation, in particular by using other available information to help improve our estimates. This paper develops and applies a method suitable for normally distributed data, and should be regarded as a companion to earlier work on areal interpolation for count data (Flowerdew and Green 1989, 1991; Flowerdew et al. 1991).

## 2. A review of the problem

Although some, perhaps most, researchers who have encountered the areal interpolation problem have given up, regarding it as an insuperable obstacle, there have been a number of attempts to overcome it. One approach, best exemplified by the work of Tobler (1979), has been to regard data for discrete zones as manifestations of an underlying continuous density surface. For data on population, for example, Tobler postulates the existence of a continuous population density surface over the entire study area, regarding the populations of specific zones as the integrals of that surface over the region defined by the zonal boundaries. This surface must have what Tobler calls the pycnophylactic property, in other words it must yield the actually observed data for the original set of zones. If a surface can be specified given data for one set of zones, it is possible to calculate the population for any other set of zones, however defined, by integrating the surface over the new zones. This is an effective procedure where it is sensible to regard the data being modelled as varying smoothly over space. For many types of data of social and economic interest, however, there are abrupt discontinuities in spatial distribution, making this approach and its variants less effective.

The main alternative approach is based on the assumption that data are likely to be uniformly distributed within the zones for which they are available. If we refer to the zonal system for which we have data as source zones, and the system for which we want to estimate values as target zones, then data for target zones can be estimated as a weighted average (or weighted sum) of data for the source zones with which they intersect. Weighting is in proportion to the area of the zones of intersection. This method will be referred to here as 'areal weighting'; a more precise account is given below.

In their account of the areal weighting method, Goodchild and Lam (1980) distinguish between extensive and intensive variables. If a zone is divided into a set of subzones, then a variable is described as extensive if its value for a zone is the sum of its values for the subzones; it is intensive if the value for the zone is a weighted average of the values for the subzones. Some variables, like relative relief, are neither extensive nor intensive. For an extensive variable, therefore,

$$Y_t = \sum_s Y_{st} \, ,$$

where $t$ is a target zone which is divided into a set of zones of intersection $st$ by the boundaries of the source zones. For an intensive variable,

$$Y_t = \sum_s \frac{Y_{st} A_{st}}{A_t} \, ,$$

where $A_{st}$ is the area of intersection zone $st$.

The areal weighting method also requires a method of relating source zone values to intersection zone values. In the case of intensive variables, it would normally be assumed simply that:

$$Y_{st} = Y_s$$

but in the case of an extensive variable it is assumed that the share of the total value accruing to each intersection zone is proportional to the area of the intersection zone:

$$Y_{st} = \frac{Y_s A_{st}}{A_s} \, .$$

The areal weighting method is thus based on the assumption that the values of the variable of interest are evenly distributed within the source zone. It seems to be the best method when there is no information available to suggest the distribution might be uneven. In practice, however, there is often plenty of information available which would lead us to expect this distribution to be far from even. Population distribution, for example, is likely to be strongly influenced by slope and elevation, and more directly by the distribution of land use types. A method for areal interpolation is needed which goes beyond simple areal weighting to allow researchers to use their geographical knowledge for improving their estimates of likely target zone values.

In earlier papers (Flowerdew and Green 1989, 1991; Flowerdew et al. 1991) we have developed an approach to 'intelligent' areal interpolation in which other information is taken into account to give better estimates than are possible with areal interpolation. This information may be available for the target zones, or for a third set of zones which may be called control zones. Essentially the method works through establishing a regression relationship between the variable of interest and one or more ancillary variables (the 'other information' above). Once this relationship has been established, it can be used, along with area, to estimate values for the variable of interest for the target zones.

The regression relationship mentioned above must be estimated in a manner appropriate for the probability distribution the variable of interest is assumed to have. We have therefore been developing a series of methods appropriate for different distributions of this variable. Most of the early work was on extensive variables, especially count data for which the Poisson distribution is applicable (Flowerdew and Green 1989, 1991). More recently, work has been extended to deal with binomial and other distributions (Flowerdew et al. 1991; Green 1990). The approach taken can be regarded as a special case of the EM algorithm (Dempster et al. 1977). In this paper, we develop and apply an algorithm suitable for continuous variables. We assume that these variables are means of a set of observations in each source zone. In the example, these are the means of prices for houses on the market in each zone.

## 3. Algorithm for continuous variables

Consider a study region divided into source zones indexed by $s$. For source zone $s$, we have $n_s$ observations on a continuous variable with mean $y_s$. We wish to interpolate these means onto a set of target zones indexed by $t$.

Consider the intersection of source zone $s$ and target zone $t$. Let this have area $A_{st}$ and assume that $n_{st}$ of the observations fall in this intersection zone. In practice $n_{st}$ will seldom be known but will, itself, have to be interpolated from $n_s$. In what follows we may assume that $n_{st}$ has been obtained by areal weighting:

$$n_{st} = \frac{A_{st} n_s}{A_s} \ .$$

More sophisticated interpolation of the $n_{st}$ would increase the efficiency of the interpolation of the continuous variable $Y$ but this increase is likely to be small as quite large changes in $n_{st}$ tend to produce only small changes in the interpolated values of $Y$. We will consider $n_{st}$ as known.

Let $y_{st}$ be the mean of the $n_{st}$ values in the intersection zone $st$, and further assume that

$$y_{st} \sim N(\mu_{st}, \sigma^2/n_{st}) \ .$$

Now

$$y_s = \sum_t n_{st} y_{st}/n_s$$

and

$$\begin{bmatrix} y_{st} \\ y_s \end{bmatrix} \sim N\left( \begin{bmatrix} \mu_{st} \\ \mu_s \end{bmatrix}, \begin{bmatrix} \sigma^2/n_{st} & \sigma^2/n_s \\ \sigma^2/n_s & \sigma^2/n_s \end{bmatrix} \right) \ .$$

Clearly if the $y_{st}$ were known we would obtain $y_t$, the mean for target zone $t$ as:

$$y_t = \sum_s n_{st} y_{st}/n_t \ ,$$

where

$$n_t = \sum_s n_{st} \ .$$

The simplest method would take $y_{st} = y_s$ to give the areal weighting solution. Here we wish to allow the possibility of using ancillary information on the target zones to improve on the areal weighting method. Adopting the EM algorithm approach we would then have the following scheme:

*E-step:*

$$\hat{y}_{st} = E(y_{st}|y_s) = \mu_{st} + (y_s - \mu_s) \ ,$$

where

$$\mu_s = \sum_t n_{st}\mu_{st}/n_s \ .$$

Thus the pycnophylactic property is satisfied by adjusting the $\mu_{st}$ by adding a constant such that they have a weighted mean equal to the observed mean $y_s$.

*M-step:*
Treat $\hat{y}_{st}$ as a sample of independent observations with distribution:

$$\hat{y}_{st} \sim N(\mu_{st}, \sigma^2/n_{st})$$

and fit a model for the $\mu_{st}$

$$\mu(s,t) = X\beta$$

by weighted least squares.
    These steps are repeated until convergence and the final step is to obtain the interpolated values $y_t$ as the weighted means of the $\hat{y}_{st}$ from the *E*-step, i.e.

$$y_t = \sum_s n_{st}\hat{y}_{st}/n_t \ .$$

In practice it is often found that it can take many iterations for this algorithm to converge. Thus although this approach is relatively simple it may not be computationally efficient. However, since we are dealing with linear models we can use a non-iterative scheme as follows. Since ancillary information is defined for target zones only we have

$$\mu_{st} = \mu_t \ , \quad \text{all } s$$

and

$$\mu(t) = X^{(t)}\beta \ ,$$

where $\mu^{(t)}$ is the vector of means for target zones and $X^{(t)}$ the corresponding design matrix of ancillary information.
    Since

$$\mu_s = \sum_t p_{st}\mu_{st} \ ,$$

where

$$p_{st} = n_{st}/n_s$$

then

$$\mu^{(s)} = PX^{(t)}\beta \, ,$$

where

$$P = [p_{st}] \, .$$

Thus we can estimate $\beta$ directly using data $y_s$ and design matrix $X^{(s)} = PX^{(t)}$ by weighted least squares with weights $n_s$. The final step is to perform the $E$-step on the fitted values $\hat{\mu}_{st} = \hat{\mu}_t = $ computed from $X^{(t)}\hat{\beta}$ and form their weighted means to produce the interpolated values $y_t$.

## 4. Application

House price data were collected for the borough of Preston in Lancashire during January–March 1990 by sampling property advertisements in local newspapers. The sample used in this exercise included 759 properties, each of which was assigned to one of Preston's 19 wards on the basis of address. Sharoe Green ward contained 170 of these properties, Cadley contained 70, Greyfriars 69, and the others smaller numbers, with under 20 in Preston Rural East (3), Preston Rural

**Table 1.** Mean house prices in Preston wards (source zones), January–March 1990

| Ward | Population | Area [ha] | Mean house price [£] | Number of cases |
|------|-----------|-----------|----------------------|-----------------|
| Preston Rural East | 5827 | 5779 | 140667 | 3 |
| Preston Rural West | 9218 | 4597 | 145825 | 6 |
| Greyfriars | 6526 | 270 | 91206 | 69 |
| Sharoe Green | 9402 | 800 | 81632 | 170 |
| Brookfield | 7262 | 176 | 48176 | 21 |
| Ribbleton | 6993 | 542 | 54810 | 20 |
| Deepdale | 6466 | 124 | 47823 | 30 |
| Central | 4320 | 168 | 40700 | 6 |
| Avenham | 6015 | 149 | 48569 | 29 |
| Fishwick | 7381 | 322 | 42031 | 39 |
| Park | 7426 | 134 | 34762 | 25 |
| Moorbrook | 5269 | 92 | 38200 | 25 |
| Cadley | 6681 | 193 | 83986 | 70 |
| Tulketh | 6538 | 101 | 45262 | 58 |
| Ingol | 4564 | 158 | 60294 | 54 |
| Larches | 6801 | 177 | 52973 | 26 |
| St Matthew's | 6072 | 112 | 37671 | 26 |
| Ashton | 6105 | 276 | 63275 | 54 |
| St John's | 5551 | 102 | 35991 | 28 |

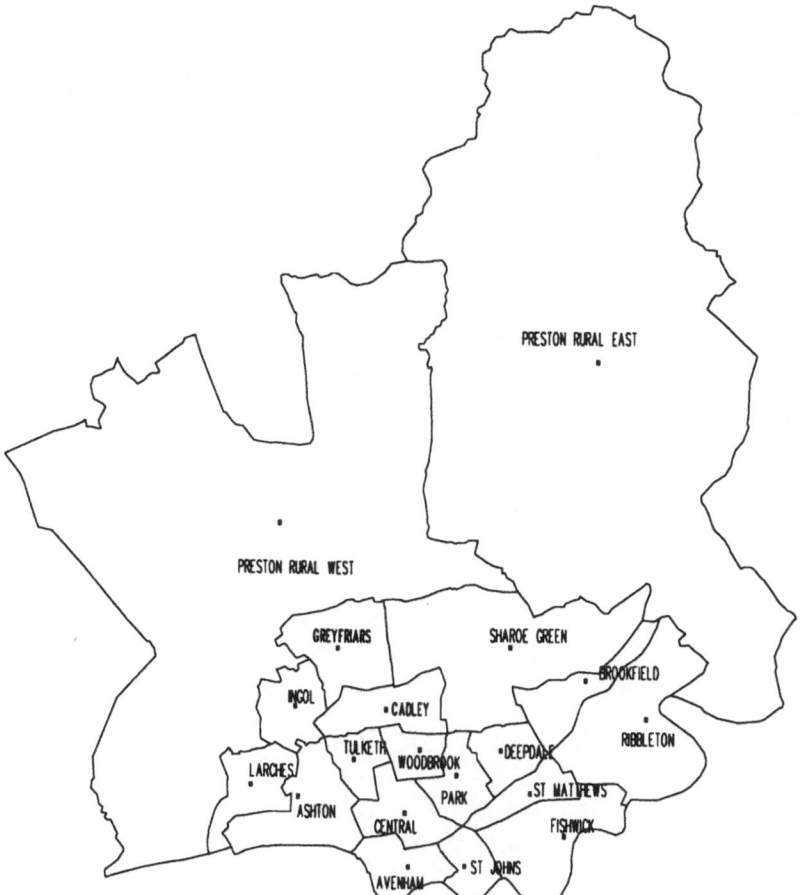

**Fig. 1.** Preston wards

West (6) and Central (6). The mean house prices for the wards ranged from £ 145,825 in Preston Rural West to £ 34,762 in Park. Table 1 shows the area, population and mean house price for the wards, and Fig. 1 shows their locations.

Wards were taken as source zones, with postcode sectors being the target zones. A set of 20 postcode sectors covers the borough of Preston, 6 of them also including land ouside the borough (Fig. 2). In some places, such as the River Ribble and some major roads, ward boundaries and postcode sector boundaries are identical, but in general they are completely independent. Ward and postcode sector coverages for Preston were overlaid using ARC/INFO to determine the zones of intersection and their area. There were many very small zones, some of which may have been due to differences in how the same line was digitised in each of the two coverages. The smallest ones were disregarded up to an area of 9 ha. This left 73 zones of intersection (Fig. 3).

Ancillary information about postcode sectors was taken from the June 1986 version of the Central Postcode Directory (CPD), the most recent version available to the British academic community through the ESRC Data Archive (Man-

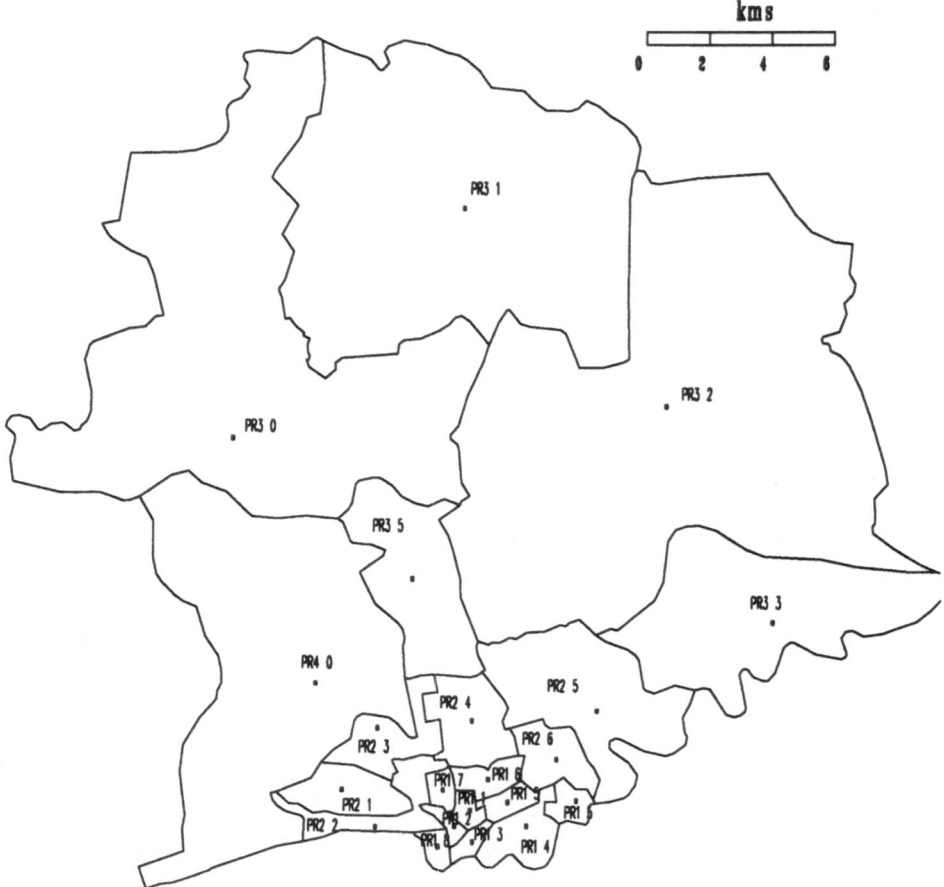

**Fig. 2.** Preston postcode sectors

chester Computing Centre 1990). This source lists all the unit postcodes, and gives a limited amount of information about them: the dates when they were introduced and terminated, and whether they are 'large user' or 'small user' postcodes. Most postcodes in the Preston area have been in existence since the establishment of the CPD in 1981 and are still in use. However, new postcodes are needed in areas of new residential development. Postcodes may go out of use in areas undergoing decline, or in areas where the Post Office has reorganised the system. Large user postcodes usually refer to single establishments which generate a large quantity of mail, such as commercial and major public institutions, while groups of private houses are small users.

The CPD can therefore generate a number of potential ancillary variables, including the number of individual postcodes in a postcode sector, the proportion of postcodes belonging to large users, the proportion of new postcodes and the proportion of obsolete postcodes. Each of these might be expected to reflect aspects of the geography of Preston. Although none of them are directly linked to house prices, they may be correlated with factors which do affect house prices.

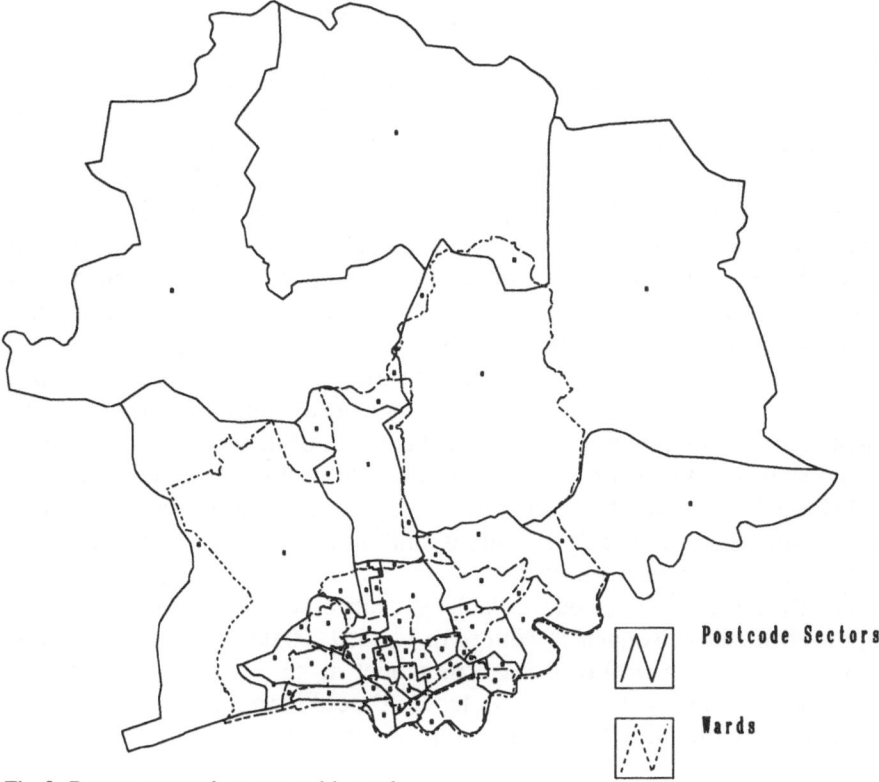

**Fig. 3.** Preston postcode sectors with wards

The most useful one may be the proportion of large user postcodes, which are likely to be most common in central or commercially developed areas. To some extent, these areas are likely to have less valuable housing stock, so a negative relationship to house prices is anticipated. For the purposes of the paper, this 'large user' variable was used as an ancillary variable in two forms. A binary variable was created separating those postcode sectors with more than 10% of postcodes being large users from those with under 10% large users; and a continuous variable was created representing the proportion of postcodes belonging to large users. The latter ranges from 0.03 in postcode sectors PR2 6 and PR3 0 to 0.60 in postcode sector PR1 2.

Two models were fitted, according to whether the ancillary variable was treated as binary or continuous. In all cases, the first stage was to estimate how many sample points were located in each intersection zone. This was done, as suggested above, by an areal weighting method. The relevant output from the models comprises the set of estimated house prices for the target zones, the relationship between house price and the ancillary variable, and the goodness of fit of the estimation procedure. The EM and direct estimation methods give identical answers, but the EM method takes longer to reach the solutions – in the binary case, 15 iterations were needed before target zone prices converged to the nearest pound.

## 5. Results

Table 2 shows the interpolated values for the 20 postcode sectors, together with the ancillary data. It can be seen that there are substantial differences between the areal weighting method and the estimates made using ancillary information. There are also major differences between the estimates reached according to the form of the ancillary variable. We do not have information about the true values for the target zones, so it is not possible to evaluate the success of the methods directly. However, the estimated values using the ancillary variable in its continuous form seem unrealistically low for sectors PR 1 1, PR 1 2 and PR 1 3: there are few houses anywhere in England with prices as low as £ 20,000.

The goodness of fit statistics produced in fitting the models give further information for evaluating them. The house price data are assumed to be normally distributed; a model can therefore be evaluated in terms of the error sum of squares. The total sum of squares for house price values defined over the source zones is 335,600,000,000 (the numbers are very large because prices are expressed in pounds, rather than thousands of pounds, and because they are weighted by the number of houses in each ward).

When the large user postcode variable is incorporated in the model in binary form, the error sum of squares is reduced to 143,600,000,000. This yields a coefficient of determination ($R^2$) of 0.572. The model suggests that postcode sectors with high proportions of large user postcodes should have mean house prices of £ 44,056 and sectors with low proportions should have mean house prices of

**Table 2.** Interpolated house prices (£) for Preston postcode sectors (target zones)

| Postcode sector | Area [ha] | % large users | Areal weighting | Binary variable | Continuous variable |
|---|---|---|---|---|---|
| PR1 1 | 82 | 44 | 37326 | 35513 | 14578 |
| PR1 2 | 62 | 60 | 41896 | 44592 | 21816 |
| PR1 3 | 68 | 57 | 43554 | 42489 | 20425 |
| PR1 4 | 349 | 23 | 40284 | 38682 | 41322 |
| PR1 5 | 258 | 17 | 43164 | 37784 | 43611 |
| PR1 6 | 175 | 16 | 41216 | 38777 | 44093 |
| PR1 7 | 87 | 13 | 40167 | 40762 | 51096 |
| PR1 8 | 99 | 41 | 48569 | 48569 | 55580 |
| PR2 1 | 413 | 4 | 93164 | 73882 | 69150 |
| PR2 2 | 469 | 27 | 73465 | 44758 | 47295 |
| PR2 3 | 453 | 5 | 91700 | 76745 | 76857 |
| PR2 4 | 567 | 9 | 84618 | 96625 | 87439 |
| PR2 5 | 1725 | 22 | 110793 | 52784 | 66205 |
| PR2 6 | 364 | 3 | 54406 | 75946 | 67160 |
| PR3 0 | 183 | 3 | 145013 | 147321 | 149497 |
| PR3 1 | 223 | 9 | 140667 | 149021 | 140570 |
| PR3 2 | 4437 | 6 | 140781 | 148923 | 144435 |
| PR3 3 | 183 | 11 | 140667 | 106685 | 138045 |
| PR3 5 | 1162 | 4 | 145640 | 147260 | 148283 |
| PR4 0 | 2912 | 6 | 143509 | 97002 | 96692 |

£ 86,392. The estimated mean values in Table 2 are produced by adjusting these figures to meet the pycnophylactic constraint – in other words, ensuring that the observed source zone means are preserved.

It might perhaps have been expected that the crudeness of measurement of the large unit postcode variable would reduce the goodness of fit of the model, and hence that using the proportion of large user postcodes as a continuous variable would improve the model. In fact, however, this model had an error sum of squares of 229,300,000,000, with an $R^2$ value of only 0.317. The estimating equation involved a constant term of 84,560 and a coefficient of $-126,260$; in other words, estimated mean house price declined by £ 126,260 for a unit increase in the proportion of large user postcodes – more realistically, it declined by £ 1,263 for a unit increase in the percentage of large user postcodes. As noted above, this resulted in unrealistically low values for those postcode sectors with a high proportion of large user postcodes.

Examination of plots indicates that the relationship of house price to the ancillary variable levels off for higher values and use of a linear relationship can severely underestimate house price for zones with a high proportion of large user postcodes. This is confirmed by incorporating a quadratic term in the relationship which improves the fit ($R^2 = 0.365$) and lessens the underestimation problem. Using a logarithmic transformation gives a further small improvement. This highlights the general point that when using continuous ancillary variables careful choice of the form of the relationship may be necessary.

## 6. Conclusion

The poorer fit obtained when the ancillary variable was used in continuous rather than binary form is surprising, and suggests the need to experiment with other functional forms. It may also be worthwhile experimenting with other ancillary variables, either singly or in combination. Some other variables may be obtainable through the use of the CPD and digitised postcode sector boundaries, such as the density of small user postcodes, which might be expected to relate to house prices. In many practical applications, variables more directly related to house prices may be available, including of course things like client addresses, provided they are postcoded.

Nevertheless, the improvement in fit of the models discussed in this paper does suggest that taking advantage of ancillary data available for target zones can do much to improve areal interpolation. Without knowledge of mean house prices for postcode sectors, we are unable to quantify the improvement.

As stated in the introduction, this work is part of a more general project designed to develop better methods of areal interpolation. Unfortunately, different types of data require slightly different methods, but we hope that we have given some indication of how ancillary data can be used to improve areal interpolation for normally distributed data. Work is in progress to develop comparable methods where those described here are not applicable.

*Acknowledgements.* This research was supported by the British Economic and Social Research Council (grant R 000231373). We are indebted to the North West Regional Research Laboratory for use of their GIS facilities, to John Denmead for use of advanced INGRES facilities he has developed, and to Evangelos Kehris and Isobel Naumann for research assistance. We also thank Susan Lucas for use of her data, collected as part of a postgraduate research project at the Department of Geography, Lancaster University.

## References

Dempster AP, Laird NM, Rubin DB (1977) Maximum likelihood from incomplete data via the EM algorithm. J R Stat Soc B 39:1−38

Flowerdew R, Green M (1989) Statistical methods for inference between incompatible zonal systems. In: Goodchild M, Gopal S (eds) Accuracy of spatial databases. Taylor & Francis, London, pp 239−247

Flowerdew R, Green M (1991) Data integration: statistical methods for transferring data between zonal systems. In: Masser I, Blakemore M (eds) Handling geographical information. Longman, London, pp 38−54

Flowerdew R, Green M, Kehris E (1991) Using areal interpolation methods in geographical information systems. Papers Reg Sci 70:303−315

Goodchild M, Lam NS-N (1980) Areal interpolation: a variant of the traditional spatial problem. Geo-Process 1:297−312

Green M (1990) Statistical models for areal interpolation. In: Harts J, Ottens HFL, Scholten HJ (eds) EGIS '90 Proceedings, vol 1. EGIS Foundation, Utrecht, pp 392−399

Manchester Computing Centre (1990) Post Office Central Postcode Directory (POSTZON file). CMS 628, Manchester Computing Centre

Tobler WR (1979) Smooth pycnophylactic interpolation for geographical regions. J Am Statistical Assoc 74:519−530

# Using information based rules for sliver polygon removal in GISs

**Krysia Rybaczuk**

Department of Geography, Trinity College, University of Dublin, Dublin 2, Ireland

**Abstract.** Of all the roles that GISs perform, their ability to successfully integrate cartographic information from a wide range of sources and scales to create a unified data base is perhaps one of the most important. Ensuring that information held is of a substantive quality however, is very much up to the user.

When data is obtained from several sources, a variety of topographical and categorical differences may be observable between the representations. Processes responsible for such disparities include survey errors, data capture techniques, machine processing to ensure that the data 'fits' the current data model being utilised and the temporal infidelity associated with cartographic material.

When each representation is viewed in isolation, such 'discrepancies', are often insignificant, however, when the data sets are integrated using polygon overlay techniques, such mismatches become flagged as errors, known as sliver polygons. Generally these correspond to a particular geometrical shape, and so the accepted policy among the majority of the large software vendors is to use this geometrical value as a basis for 'removal'. Problems arise however when the geometry of the 'sliver' polygons correspond to polygons of differences between data sets that are worthy of retention, notably change, such as land use variation.

If the information or attributes of the sliver polygon are taken into account, rather than simply their geometry, the degree of uncertainty associated with their removal will be reduced and the quality of the overlaid material maximised. This contribution looks at the problems of data integration and sliver polygon removal and possible alternatives based on user defined rules.

## 1. Introduction

As Geographic Information Systems (GISs) become more widely used by a wider section of the community, and data for input into such systems become assimilated from a greater variety of sources, the issue of information quality becomes a vital one if the worth of any GIS analysis is to be accepted. The accuracy of spatial data depends on a variety of factors such as positional accuracy, completeness and currency. These factors are then further compounded by the processes of assimilating cartographic material into a digital data base, and they become very apparent when the process of GIS overlay is performed. Thus the presence of 'error' within digital cartographic material generated by GISs is a factor of both weaknesses in the information itself and weaknesses of current approaches to dealing the data and their associated uncertainty.

## 2. Weaknesses of the data for assimilation into GISs

Data for digital cartographic integration is derived from many sources, both in terms of its type and its origin and as such is prone to variation. Data may be in the form of:

● Maps/cartographic/remotely sensed material,
● Statistics,
● Non-statistical 'soft' data.

A GIS however is only capable of utilising a relatively restricted subset of information about the real world consisting of *only* geometry and hard statistics, thus forcing 'analysis' to operate only on those items and offering the user a relatively limited set of tools.

In addition to the differences in data type, there are also differences in data sources. Thus each individual piece of information, be it even of the same type, is likely to have undergone different processing routines and will have followed different quality control procedures. Even in the same agency where procedures and quality regulations are consistently maintained, human errors may influence the result, as may the varying quality of the surveyed material.

Cartographic data for GIS entry are derived from either primary or secondary sources (Table 1), both of which are predisposed to a variety of error components in their acquisition.

**Table 1.** Sources of cartographic information

---

● Primary data source
    ● Survey material
        positioning
        baseline definition
        primary, secondary, tertiary triangulation
        local surveys (plane surveying & thematic surveys)
    ● Remote sensing
        orbital
        sub-orbital
● Secondary data source
    ● Map
        geometry
        features/attributes
        context

---

● Primary data sources – Surveyed material

Surveying techniques rely on the procedures of measurement, registration, recording of data, processing of data, analysis of data and finally its presentation. Each of these stages will have an error component attached to it. The extent of such errors will be determined by the observational skill of the surveyor and interpreter, but also by the precision of the instruments and techniques used to acquire and analyse the data.

● Primary data sources – Remotely sensed data

The term remotely sensed imagery incorporates a wide range of mappable information, ranging from visible air photographs to satellite imagery responsive to various components of the electromagnetic spectrum. Air photographs have been used in various degrees of sophistication for most of this century, for military reconnaissance work[1] and then increasingly for both topographic and thematic mapping. In the last twenty years, more sophisticated satellite retrieval techniques have evolved, and these too have followed a similar route of military development followed by potential geographic and environmental use.

Problems equated with remotely sensed imagery, be it aerial photography or satellite imagery, depend on the skill and technology with which the image is both obtained and interpreted. In the case of suborbital platforms, flight operated scanners are prone to image distortion as a result of aircraft instability such as pitch roll and yaw, in addition SLAR data suffers from the intrusion of topographic effects. Problems of the photogrammetrist and aerial photograph interpreter have been extensively documented elsewhere[2], and in the most part rely on the quality of the image, the skill of the photogrammetrist and the precision of the equipment being used.

When satellite remote sensed data is utilised as either a topographic or thematic cartographic information source, the quality of the data will initially be determined by the conditions in which the image was obtained. This includes the weather conditions prevailing at the time, the angle of the satellite and the resolution of the satellite receiver. Furthermore however, quality will also be affected by the post processing techniques undertaken to produce 'satisfactory' data for information systems.

These techniques attempt to extrapolate the maximum amount of detail from the image, by positional re-adjustment, thematic interpretation and thematic enhancement. As with ground surveyed material, classification involves a degree of generalisation, as no natural thematic feature will abruptly stop at a defined boundary of a specified width and scale. Here however, rather than using visually observed criteria, the user is reliant upon an ability to distinguish remotely sensed spectral responses using a variety of statistical techniques.

---

[1]    A heavily highlighted feature of the Gulf War in 1991 was the reconnaissance capability of the allied forces.
[2]    E.g. Burnside (1979)

● Secondary data sources

For most users unable to afford the resources required for primary surveys, most digital cartographic information will be obtained from maps, and will therefore be prone to the inaccuracies involved in the map making process, such as drafting errors, scaling, generalisation and symbology. Thus the representation of criteria on the map will be both real (in terms of positional locations) and fuzzy (in terms of attribute defined locations, such as the boundary between two soil types). GIS databases are capable of acquiring both these sets of information and storing them and their relationships for both geometry and features (Table 2). A third level of information that a map contains is its contextual information, such as what the map was designed for and the message it is trying to convey. Current GIS database storage techniques do not allow the assimilation of such information, and therefore the map as such can never be fully incorporated into the database system.

Geometric data input into GISs can be subdivided into three main groups. Firstly there are those which deal with the representation of topographic features in the real world, such as the location of a triangulation point, the centreline of a road, or the shape of a football stadium. Similar to these cartographic elements are artificial features which have their boundaries rigidly defined. The most common usage of such data is for thematic operations, in which postcode sectors, administrative units, or most national borders may be delimited and used as the basis for attribute depiction. There may be a level of generalisation in the drawing of the areas themselves, but their location in space, although artificial, is an accepted one. The final group of cartographic material that may be used as input into a GIS, are data which deal with subdividing spatial information into categories. This is often termed categorial or fuzzy data (Chrisman 1982). This type of data tends to be primarily areal in nature and usually deals with landscape descriptors, although it has a more general application in contour and isopleth mapping. Such data may represent features such as different vegetation zones, climatic zones, soil types, land use, purchasing power or spheres of influence. As these attributes are difficult to partition or demarcate in terms of a linear delimitation, their borders always remain fuzzy. The reality is that often the boundary between such areas does not exist as a clearly defined linear feature, but rather as an area of intermixing in which one of the neighbouring features might predominate, but not necessarily. Thus when the boundary between two zones needs to be defined from multi-source data input, problems arise. As most GISs are incapable of holding *fuzzy* data, some decision has to be reached as to which of the sources should be defined as having the accurate and accepted boundary.

**Table 2.** Levels of information held in a map and GIS database

|  | Real | Fuzzy | Behavioural/cognitive aspects |
|---|---|---|---|
| Geometry | map AND GIS | map AND GIS | |
| Features | map AND GIS | map AND GIS | |
| Context | | | map ONLY |

This will then form the 'baseline' onto which other material can be attached. If the wrong decision is made at this point, then all further analysis using the data, may be error-prone.

## 3. Weaknesses of techniques for information assimilation into GISs

Source material errors are further compounded by the processes governing their transfer into the digital environment. Digitising or scanning techniques produce their own problems of imprecision, inadvertent generalisation and feature displacement. Transforming data from 'spaghetti' into the data model used by the GIS may also cause feature displacement and further generalisation. For example post digitising routines often cause the location of a linear feature in space to shift in order to permit perfect geometric node matching. A further consequence of these routines is to create tiny triangles in the linear coverage as small lines collapse. Unless all these errors are removed during post-digitising verification, they will have uncertain and potentially serious impacts during subsequent analysis.

Thus any integration of digital cartographic data from several sources will inevitably produce situations in which the same feature in space (such as natural barriers or administrative boundaries) will be represented by lines in different physical location. The true location on the ground will undoubtedly lie somewhere in between the cartographically portrayed ones. Such errors or inconsistencies are inevitable because of the number of sources being used, and the widely differing range of processes that might have been used to create each one.

## 4. Data overlay and integration within GISs

Integration within a GIS implies more than simply overlaying cartographic lines. For rather than dealing with an increasingly complex single geography, integration looks at multiple layers of geographical representation, and tries to combine the data from them into a single, unified and easily queried information base. Data suitable for input into such a system, will usually be derived from various sources and in some cases the data may actually consist of different 'geographies'. For example, different enumeration units may have been utilised for the acquisition of socio-economic information,[3] or enumeration areas may give a large area the characteristics of a smaller zone within it, by making assumptions of equal density distribution throughout the area.[4] The general tendency however, is for the same geographical phenomena to be represented inconsistently in different sources as a consequence of the inherent error processes present in data collection and representation.

The most common form of digital cartographic integration occurs as a geometric overlay. Overlay operations within Geographic Information Systems are based on the principles of set algebra. Sets of information can be manipulated

---

[3]    Flowerdew and Green (1989) discuss this problem for the integration of data collection units.
[4]    Openshaw (1984) outlines this in his discussion of the modifiable areal unit problem.

to produce new data in which certain conditions hold true. Most GISs offer the user the ability to use methods such as UNION, INTERSECTION and IDENTI-TY. However, as most GISs concentrate on an accumulation of information, few allow the user to retain areas that do not hold true for certain conditions as part of the overlay process. Such operations may be performed by more complex querying. Other facilities provided often include buffering and 'cookie cutting', whilst some systems such as SPANS allow for equations representing models of spatial information to be input into the overlay processor.

However the process of overlay creates problems of imperfect linear feature matching. When lines that are overlaid do not match perfectly, two possibilities present themselves. Firstly, the mismatch may be a result of some change in the data set, secondly, the mismatch may be a result of two map outlines that have been subjected to all the subjective processes described in the preceding two sections, and the mismatches may simply be a symptom of the error processes involved. In many respects therefore, unless the data is perfectly matched by hand prior to the use of the overlay routines, the overlay process is inherently error prone.

The small errors that exist between differing representations of the same feature are termed sliver polygons. As they actually exist within a coverage, they cannot be ignored in any analysis that might then be undertaken as the space they occupy is real world space, which now has a dubious value attached to it.

## 5. Weaknesses of current systems in dealing with uncertainty

The initial weakness lies in the data structures that are used within most current GIS packages; notably the relational data base structure, in which the world is described in a sub-optimal manner relying on just geometry, attributes and geo-codes. In this version of the world, no fuzzy boundaries between categories are allowed to exist. Only absolute boundaries are tolerated. This creates problems for data which might represent areas, but may have been interpolated from points, or for categories which share a 'mixed' zone with their neighbouring category.

Thus in such systems, zones of uncertainty, such as sliver polygons require removal. Effectively removing sliver polygons entails accurate identification. The characteristics inherent to sliver polygons can be discriminated on the basis of their size, shape and feature class. In general, sliver polygons are very small in area, and can often be identified in the database by their relatively insignificant size as compared with other areal features. In addition to their small area, sliver polygons often have an extensive perimeter, as they are areas which have been created by marginal mis-alignment of lines. It the two characteristics are combined, then the average sliver polygon will have the properties of a small area combined with a large perimeter. There is therefore, the possibility of designating polygons on the basis of their area to perimeter ratio. A further characteristic that can be utilised is that of class or attribute. Particular combinations of classes can be isolated as being untenable in a particular coverage and therefore worthy of removal.

The difficulty in identification arises when polygons that need to be retained display similar traits to the sliver polygons. For example roads, drainage ditches and other linear features often exhibit large perimeters in comparison to area, whilst small units such as farmsteads may fall below the allowable areal tolerance. The situation becomes further complicated when change is being investigated, as discriminating sliver polygons from change can often be a demanding process, given that both features will often display similar dimensions and shape (Fig. 1). Change rarely arises in large uniform polygons, rather it manifests itself in small incremental zones very similar to the way slivers arise.

Various routines exist in each of the major systems to overcome the problems posed by sliver polygons and on the whole these rely on elimination rather than subjective reassignment. In ARC/INFO, the advised procedure for the removal of slivers relies on the removal of areas on the basis of features in the attribute table falling below a user-defined criterion:

> *eliminate* reduces the number of polygons or lines in a coverage by merging selected features with one of their neighbours...
>
> ARC/INFO PC overlay users Guide, section 4.

Advised methods for selection rely on the ratio between the area and the perimeter of sliver polygons, although the suggestion is that the area divided by the perimeter should be less than $10^5$.

The problem with such methods is twofold. Firstly, where change occurs in urban land use, or areas of intensive resource usage, it often does so in small areal units which may easily be grouped in with sliver polygons by a standard GIS 'sliver removal' algorithm. The cumulative effects of the process can also be

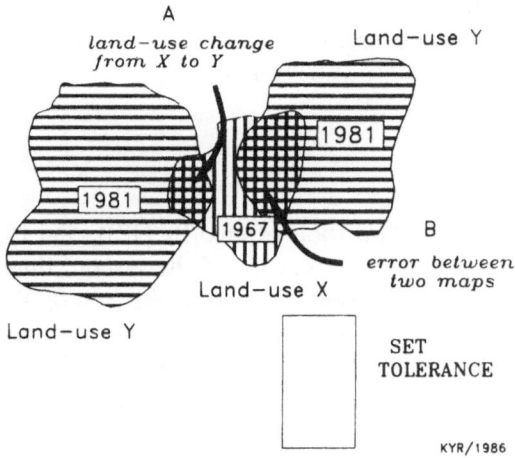

**Fig. 1.** Tolerancing affecting. Land use change ($A$), error ($B$)

---

[5] This is a strange ratio to pursue, as it would result in the elimination of a large number of areas which do not display the properties of sliver polygons; many robust areas display that type of ratio. A superior ratio would be to suggest that the perimeter divided by the area should be greater than 10.

quite drastic, as the more coverages that are overlaid, the greater the opportunity for a gradual depreciation in the quality of the data and in the reliability of any subsequent results. Secondly, the reassigment of sliver polygons once they have been eliminated via some geometrical technique is often undesirable. The sliver polygon is rarely viewed as an areal entity in its own right.

Instead, it is the boundaries defining the polygon that are seen as the problem, and the solutions to the elimination of such slivers rely on the removal of one, or some, of the bounding lines. In removing one of the bounding lines of the polygon, the area that was once held within becomes exposed and the small sliver will merge its geography with one of its adjacent neighbours. The method for reassigning such areas within GISs usually relies on a merge with one of the surrounding areas based on a chosen geometrical property, for example in ARC/INFO...

"The longest arc and label points of each selected polygon are eliminated."
ARC/INFO PC overlay users Guide, section 4.

Therefore the fate of such polygons is in many respects random, as they are not assigned to the most suitable neighbouring area, but on the magnitude of one of its boundaries. In a geometrical sense, with no other option possible, this type of routine is preferable to a random assignment as it at least relies on a modicum of intelligence. The basis for this is that the sliver polygon will have most in common with the region with which it shares the longest boundary. Unfortunately this method does not take account of the fact that one boundary may be only fractionally longer than the other, or that the longest boundary may not be the most suitable.

The most significant problem associated with the nearest largest neighbour approach is that areas become mis-assigned. These are usually borderline areas, which are often the ones of the most interest to geographers. It is the designation of the boundaries that usually causes the most conflict, especially in categorical or 'fuzzy' data, such as soil maps or vegetation distribution.


## 6. Alternative approaches

An alternative approach is to ignore the standard tools offered by GISs, and to utilise the information held within the database component of the system, thus developing methodologies which do not use the options such as ARC/INFO's *eliminate*. Such methods require a reorientation of the way slivers are viewed. Rather than seeing them as areas requiring identification and 'elimination', this approach deals with the fuzziness of data by recognising that sliver polygons should be merged with the most suitable of their neighbours.

Several options exist when using this type of approach. Firstly, a matrix of allowable movement can be created, thus change is only allowed in certain directions, which are defined by a set of predefined rules. Secondly, spurious polygons failing a geometric test can be merged with their neighbours on the basis of probability matches. Again these are defined by the user for the data at hand. Thirdly, if changes through time are being modelled, a best fit history approach can be implemented. This looks at data in terms of time, rather than in terms of the current and past classification only. Sliver polygons are merged with bounding areas that have 'acceptable' histories.

If one of the bounding arcs that make up an areal feature is 'dropped', the feature will merge with whichever of its neighbours the line borders. Thus, if a suitable neighbour can be found for the suspected sliver polygon to merge with, the line separating them simply needs to be removed. This ensures that the error polygon will merge with a neighbour of the user's choosing rather than one with which it merely shares the longest boundary.

## 7. Using predefined rules based on attributes for sliver polygon removal

In a combined coverage representing change between two time periods, making the elimination of an error polygon successful requires three factors. Firstly, that the polygons of error are removed. Secondly, that small original polygons remain, and thirdly, that polygons of change remain. To ensure that the correct polygons are retained and their rivals removed the characteristics of each of the groups need to be identified and searched for in the data set. Error polygons can be identified by their size, shape, new identification class and their old identification class, whilst small original polygons can be identified by their attribute, in that they will have the same class value in both the original and overlay coverage. In this way they are mutually exclusive from the error polygons.

The use of attributes will depend upon the degree of mutual exclusively that exists between the features for retention and those for removal. For example if two coverages represent a change over time, the initial rule could be that the second coverage is more accurate than the first. This would hold true for the entire coverage, with the exception of certain pre-determined cases governed by a series of rules, that limit the changes polygons can make through time. If for example, it was desired to freeze certain features from the original data set, these could be considered to be outside the analysis. For example, the contiguity and compatibility of the polygon attributes can be used to formulate a series of hypothetical rules which gives precedence to roads, footpaths, grazing and housing as defined in the original coverage, unless they had been altered in certain allowable directions. Otherwise, the overlay coverage can be regarded as the more accurate of the two. These rules are shown in Table 3.

This method allows the user to exercise subjective control over the resulting data set, as in some cases the original data set will prove to be the more accurate, whilst in others the overlay data set will be the more reliable. Without such a mechanism, it is always the new, overlaying data set that is deemed to be the more accurate of the two, with any discrepancies between the two data sets being attributed to update. In employing such a methodology, the user is in effect creating a matrix of allowable change. In simple cases this might be a simple yes/no matrix, but in more complex cases, probabilities could be assigned to the various change options. In the above example a matrix of allowable movement would

**Table 3.** Rules which could be applied to a classified coverage to preserve certain features

---

DATABASE RULES
Rule 1: Anything that used to be a road and which is not now, let it become a road again.
Rule 2: Anything that used to be a footpath, and which is not now, let it become a footpath again.
Rule 3: Anything that used to be grazing, which is not housing, road or grazing now, let it be grazing again.
Rule 4: Anything that used to be a house which is not a road, let it be a house again.
Rule 5: Accept the other changes as viable.

---

**Table 4.** A matrix of allowable movement

| From            to | House | M'ket G'dn | Wheat | Barley | Road | Grazing | Foot-path |
|--------------------|-------|------------|-------|--------|------|---------|-----------|
| House              | Y     | N          | N     | N      | Y    | N       | N         |
| Market Garden      | Y     | Y          | Y     | Y      | Y    | Y       | N         |
| Wheat              | Y     | Y          | Y     | Y      | Y    | Y       | N         |
| Barley             | Y     | Y          | Y     | Y      | Y    | Y       | N         |
| Road               | N     | N          | N     | N      | Y    | N       | N         |
| Grazing            | Y     | N          | N     | N      | Y    | Y       | N         |
| Footpath           | N     | N          | N     | N      | Y    | N       | Y         |

look similar to that shown in Table 4. Thus some features would become fixed once they had been implanted into the system using the original coverage.

## 8. Using selective neighbourhood matching for sliver polygon removal

In adopting a methodology of selective neighbourhood coalescence, the user acknowledges that small suspected error polygons will have to be eliminated using a variety of techniques, but the emphasis, is on *where* those areas go, rather than *whether* they should go. If the user can direct the fate of areas to be removed, then their elimination does not induce as a great a measure of concern and uncertainty as in the previous cases, given that they will be merged with suitable areas. Admittedly, some degree of detail will be lost, but at least the user has the satisfaction of knowing that it will not be redirected into false, or random avenues. Thus this approach deals with error minimisation on 'fuzzy' boundaries, rather than error removal. Damage is minimised by relocating areas of uncertainty to probable polygon partners and neighbours.

If one considers a land use cover consisting of several classified categories, a probability matrix could be defined which would determine the probabilities of a particular land use being merged with all other land uses in that analysis. Probabilities may be given in terms of percentages which are independent of each other; or alternatively, probabilities for all land uses may be given in terms of proportions, which when summed total one (or percentages which when summed equal 100%). If a polygon fails to pass an areal tolerance test (such as an area/

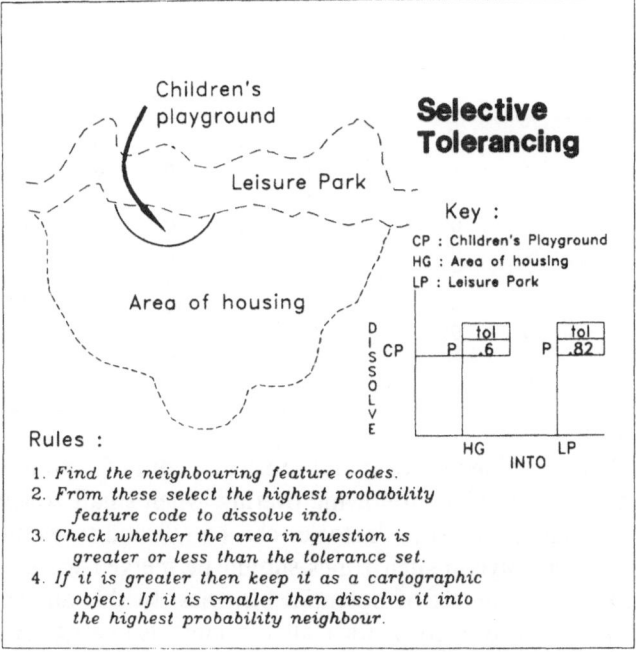

**Fig. 2.** A methodology for selective neighbourhood matching for sliver polygons

perimeter ratio), then the matrix will suggest which of its neighbours the polygon should merge with given the available options (Fig. 2).

This system might be usefully employed in an analysis of land use change, in which the aim was to define zones of change rather than update existing material. As each zone would not be defined as 'maize' or 'grassland', but as 'maize to maize' or 'meadow to moor', slivers would not automatically share the same parent class as one of its neighbours. The problem the user would face in a large coverage is the large number of potential changes that might exist. With a coverage of only 4 classes, for example, the potential number of changes through time is 16, which would then have to be matched with each other bringing the total number of neighbour match assessments required to $16^2$. For a classification system with 10 classes the matrix would require $10^4$. Not only would this be time consuming to build, but it would also take time to check and access. The amount of effort such a system might require may well counteract any benefits to be gained from its application.

## 9. Using polygon 'histories' for sliver polygon removal

An alternative approach, which would reduce the size of the matrix required for reassignment, involves making use of the histories attached to each polygon. If the land use cover is being assessed in terms of its change through time, then each polygon will have a series of land uses attached to it. Instead of viewing these as

**Table 5.** Matrix of probable histories

| From \ To | | 1 | 2 | 3 | 4 | 5 | 6 | 7 | 8 | 9 | 10 |
|---|---|---|---|---|---|---|---|---|---|---|---|
| Beets | 1 | 1 | 1 | 0 | 0 | 0 | 0 | 1 | 0 | 0 | 1 |
| Wheat | 2 | 1 | 1 | 1 | 1 | 1 | 0 | 1 | 0 | 0 | 1 |
| Bare | 3 | 0 | 1 | 1 | 0 | 1 | 0 | 1 | 0 | 1 | 1 |
| Rapeseed | 4 | 0 | 0 | 1 | 0 | 1 | 0 | 1 | 0 | 0 | 1 |
| Stubble | 5 | 0 | 1 | 1 | 0 | 1 | 0 | 1 | 0 | 0 | 1 |
| Road | 6 | 0 | 0 | 0 | 0 | 0 | 1 | 0 | 0 | 0 | 0 |
| Building | 7 | 0 | 0 | 0 | 0 | 0 | 0 | 1 | 0 | 0 | 0 |
| Beans | 8 | 0 | 0 | 0 | 0 | 0 | 0 | 1 | 1 | 0 | 1 |
| Potatoes | 9 | 1 | 1 | 1 | 1 | 1 | 0 | 1 | 0 | 0 | 1 |
| Footpath | 10 | 0 | 0 | 0 | 0 | 0 | 1 | 1 | 0 | 0 | 0 |

separate entities, they could be combined to form an historical record of land use within that defined polygonal space. Thus, following overlay, when small zones of change are suspected of being slivers, their histories can be referred to as an indicator of which neighbouring polygon such zones should be merged with.

This would involve setting up a matrix of potential histories. In this all the land use categories from $t_0$ can be plotted against all the land use categories from $t_1$. The matrix can then be filled in with either a 0 or a 1. A zero denotes that given the choice between the land use classification at $t_0$ and at $t_1$, the most probable present class for the polygon will be that existing at $t_0$. In the same way a 1 indicates that the most suitable present class for the polygon will be that existing at $t_1$. Thus for agricultural change from winter to summer the matrix shown in Table 5 might be appropriate.

A tolerance value can be set to define slivers, which can then be tested against the matrix to see which of their two histories are the most acceptable. If for example, the previous state is deemed to be the most acceptable, then the neighbouring polygons of the sliver are searched and the first one found to be sharing a common previous state is flagged as the polygon most suitable for merging with the sliver. The merge is carried out by replacing the 'history' of the sliver polygon with that of the polygon it is to be merged with, and the line between the two polygons is dropped.

This can be illustrated using a small hypothetical test coverage consisting of both small zones and long thin zones defining land use (Fig. 3). If this is overlaid with another coverage in which zone boundaries have changed (Fig. 4), a complicated polygon coverage ensues (Fig. 5). Using a traditional geometric eliminate algorithm removes a substantial number of true sliver polygons (Fig. 6), such as polygon number 24 and 28 (Fig. 5), but it has also removed areas of change that are worthy of retention such as polygon numbers 9, 10 and 12 (Fig. 5), furthermore original features such as polygon number 13 (Fig. 3) have also disappeared.

Fig. 7 illustrates the same overlay as Fig. 5 but shows the assigned 'history' of each polygon. Thus a value of 309 would indicate an initial attribute value of 3 and a subsequent attribute value of 9. A value of 606 implies no change. Applying the best fit history approach in this case produces a more acceptable result (Fig. 8). Here those polygons removed include all the slivers removed by the

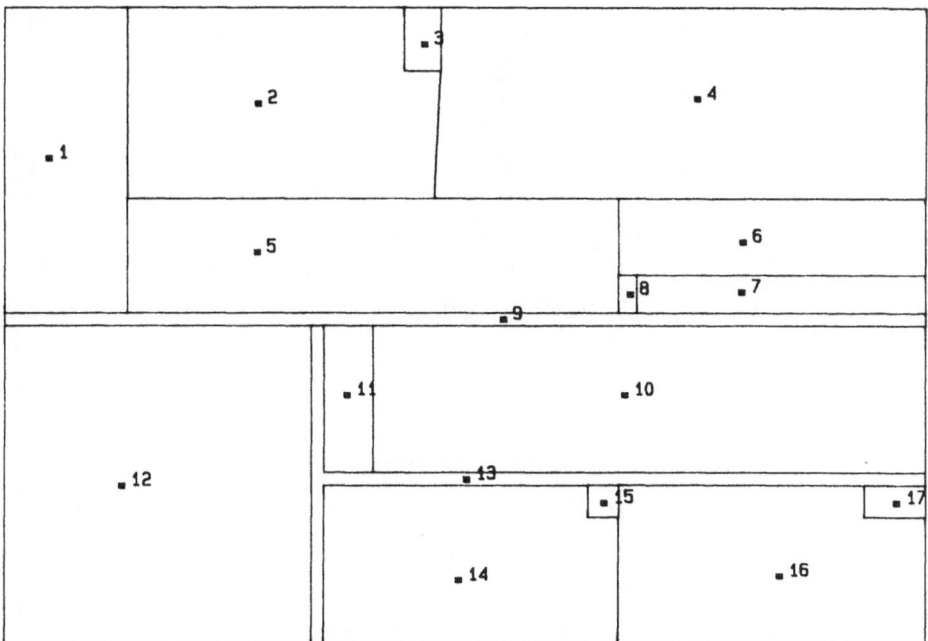

**Fig. 3.** A test coverage consisting of small zones and long thin zones

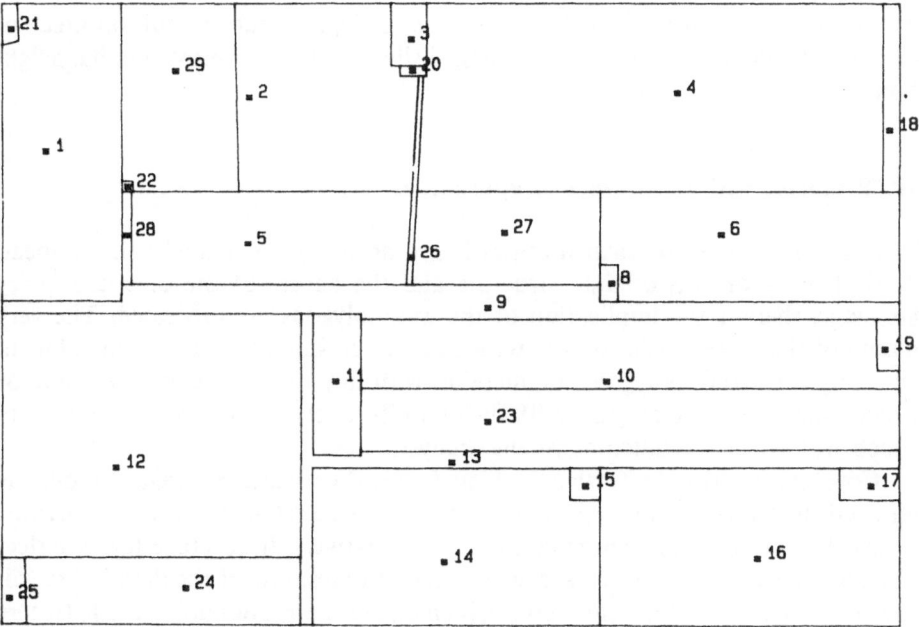

**Fig. 4.** Coverage of change

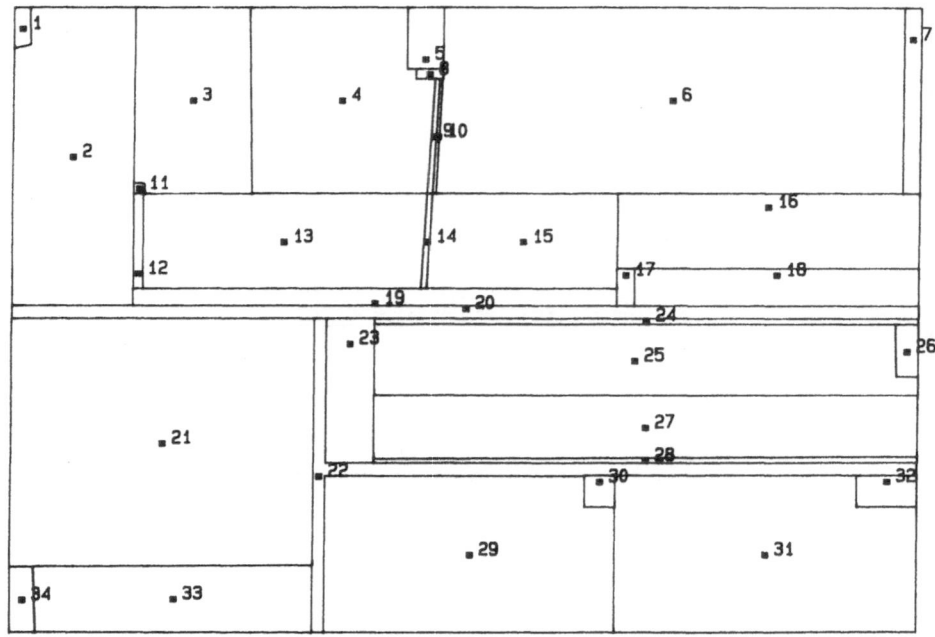

**Fig. 5.** Overlay of the original coverage and the coverage of change

geometric algorithm used in the previous case (Fig. 7), but in addition areas of change remain and assignment has merged slivers with their most suitable neighbours.

## 10. Problems with these types of approaches

The implementation of such routines is not as simple as it might first appear. Aside from problems of data structure and the nature of the database query language, there is the implication of the way polygons are processed. The very nature of the system relies on the sequential processing of information. Thus in choosing a starting polygon, for merging with any of its neighbours (some of which may require merging as well), the user effectively enforces a spatial priority which will affect the outcome of the analysis.

Coverage priority is a further problem facing the user. If coverages for overlay reflect different spatial units then one of them may be set as the accurate baseline, to which other coverage layers converge. Alternatively, if coverage layers reflect different versions of the same space at different times, one of the data layers will have to be set as the base time onto which changes are overlaid. Thus $t_0$ (where $t$ represents time) might be set as the original time, whilst further time snaps $t_{(n-1)}$ might be regarded as less accurate in terms of bases geometry. Solving such problems requires the ability for spatiotemporal modelling. Yet most existing GIS technology can only identify geometric change. It cannot easily identify *when*

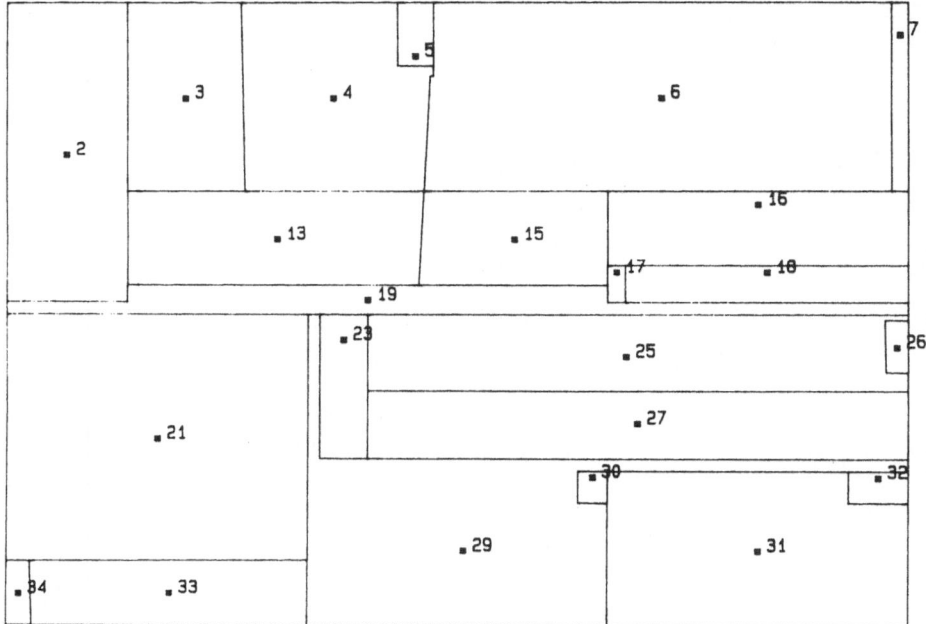

**Fig. 6.** ELIMINATION on the basis of AREA/PERIMETER less than 0.1 units

change occurred merely that it has. A further drawback of this type of technique is that it requires prior knowledge about the nature and behaviour of the attributes.

In terms of processing, a large coverage would require a large matrix and a long processing time. Another potential problem is defining the matrix. How sure can the user be about particular changes through time? To what extent can user preconceptions preclude changes in previously unexpected directions? In terms of land use change such deterministic delimitations are probably less likely to run into problems than they would in the field of resource management. Furthermore, given that the alternatives are geometrically biased, most cases of operator knowledge intervention should at least produce a considered result, rather than a randomly determined one.

## 11. Conclusions

The current methods and concepts behind the elimination of spurios sliver polygons have their foundation in regarding such areas as geometric abnormalities. If such anomalies can be swept away, then somehow, the coverage will become accurate. Rather than furnishing the user with the expectation that such intrusions into cartographic purity are to be expected, sliver polygons are often seen as irritating by-products of a powerful analysis procedure, and the attitude towards them in proprietary GISs is quick removal, usually involving geometric solutions.

100

K. Rybaczuk

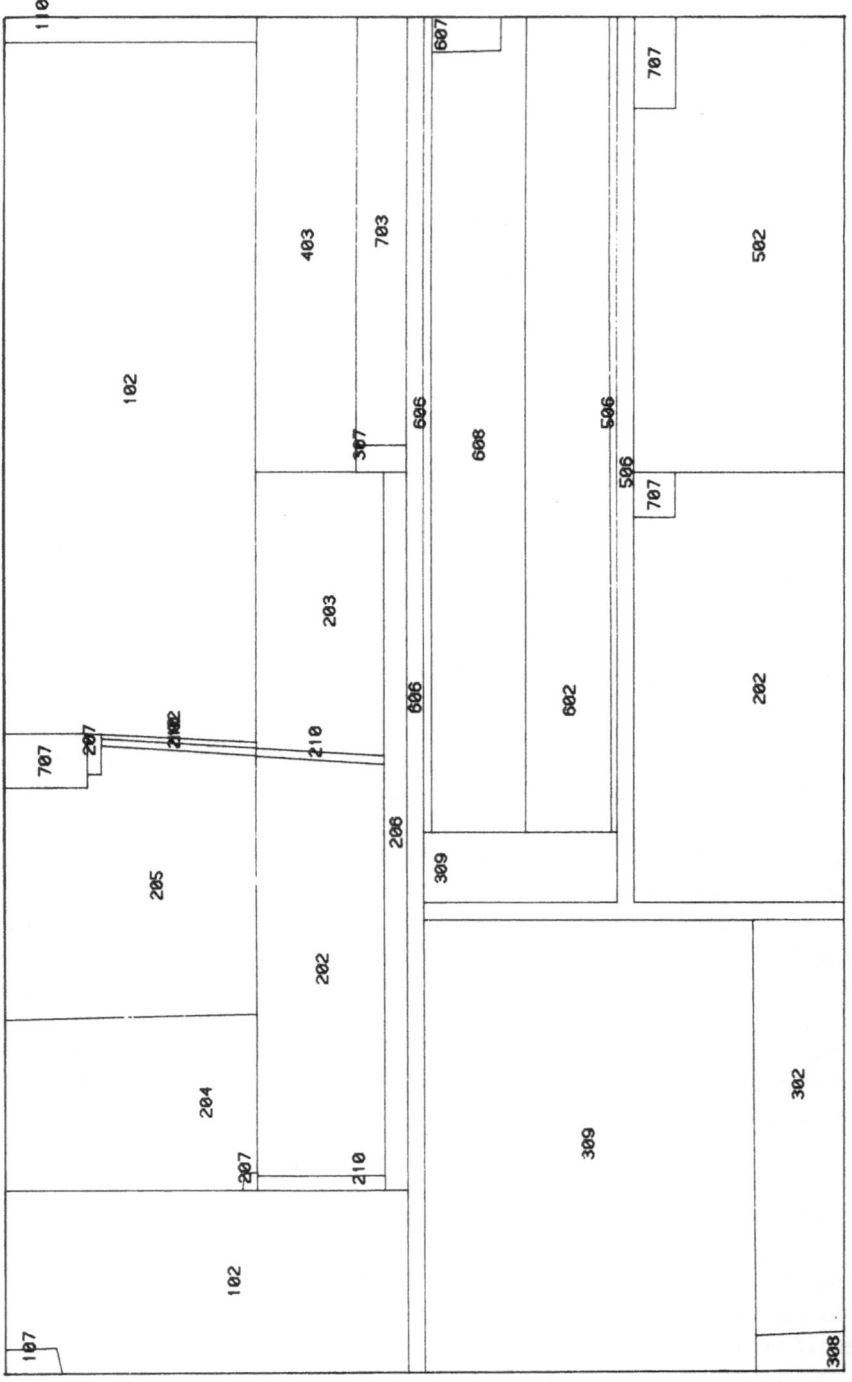

**Fig. 7.** The overlaid coverages showing changed histories as defined by the matrix of probable histories

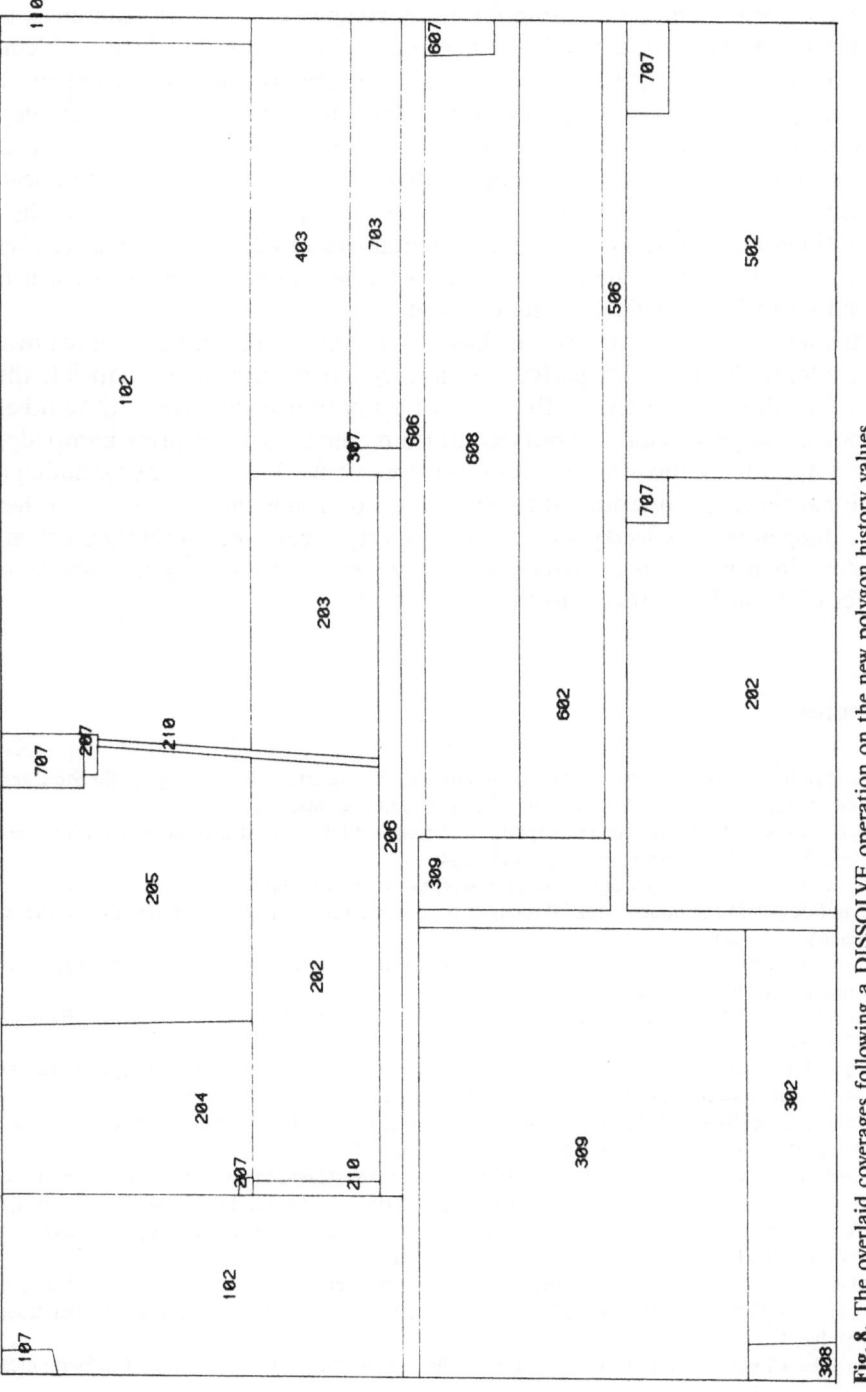

**Fig. 8.** The overlaid coverages following a DISSOLVE operation on the new polygon history values

As the presence of error within GISs is inescapable, if the user continues to view the removal of inaccuracies as one of error elimination, in which problems of mismatches are conclusively solved, then the user will always be left in an uncertain position in which a proportion of the coverage is erroneous. An alternative philosophy is to accept that a degree of uncertainty will persist within coverages that have been created from less than perfect data sets and that in dealing with this, emphasis should be placed on the reassignment process, rather than on the elimination process itself. Other alternatives to allay overlay mismatches can be derived using methods which stress selective neighbourhood coalescence, as opposed to forced neighbourhood matching.

Attributes can therefore form the basis of a more intelligent polygon removal methodology, albeit a still imperfect one, as they give the user some control in the form of an ability to determine the result of polygon reassignment. Unlike other methods of polygon removal however, it also requires a greater prior knowledge of the data, forcing the user to cover every eventuality. Furthermore, assuming a superior accuracy for one coverage over another, or of one time state over another can be dangerous, as coverages often have varying accuracies within their extent. Therefore in using such a methodology, the onus for accuracy assessment is transferred from the software to the user.

## References

Beard K (1989) Use error: the neglected error component. Auto-Carto 9, Proceedings of the 9th International Symposium on Computer Assisted Cartography, pp 808–817

Burrough PA (1987) Multiple sources of spatial variation and how to deal with them. In: Chrisman, NR (ed) Proceedings, Auto-Carto 8, pp 145–154

Burnside CD (1979) Mapping from aerial photographs. Wiley and Sons, New York

Chrisman NR (1982) Methods of spatial analysis based on error in categorical maps. Ph. D. thesis, University of Bristol

Chrisman NR (1984) The role of quality information in the long-term functioning of a geographic information system. Cartographica 21:79–89

Chrisman NR (1987) The accuracy of map overlays: a reassessment. Landscape and Urban Planning 14:427–439

Chrisman NR (1989a) Error in categorical maps: testing versus simulation. Auto-Carto 9, Proceedings of the 9th International Symposium on Computer Assisted Cartography, pp 521–529

Chrisman NR (1989b) Modelling error in overlaid categorical maps. In: Goodchild MF, Gopal S (eds) Accuracy of spatial databases. Taylor and Francis, New York

Easterfield ME, Newell RG, Theriault DG (1990) Version management in GIS: applications and techniques. In: Harts J, Ottens HFL, Scholten HJ (eds) EGIS '90, Proceedings of the First European Conference on Geographical Information Systems, EGIS Foundation, Utrecht, pp 288–297

ESRI (1987) ARC/INFO Overlay manual. ESRI, California

Flowerdew R, Green M (1989) Data integration: statistical methods for transferring data between zonal systems. Paper presented at the 29th European Congress of the Regional Science Association, Cambridge, UK

Langran GE, Chrisman N (1988) A framework for spatiotemporal information. Cartographica 25:1–14

Openshaw S (1984) The modifiable areal unit problem. Catmog 38

# Which spatial statistics techniques should be converted to GIS functions?

## Daniel A. Griffith

Department of Geography, Syracuse University, H. B. Crouse Building, Syracuse, NY 13244-1160, USA

**Abstract.** A fundamental difference between spatial and aspatial data is that observations for geo-referenced data are correlated strictly due to their relative locational positions. This self-correlation activates complications in the statistical analysis of geo-referenced data that lie dormant in the statistical analysis of traditional data comprised of independent observations. Seeking a remedy for this problem has helped motivate the development of an array of procedures labelled spatial statistics. This contribution seeks to identify existing spatial statistical tools that would be routinely of value to the GIS user community – a major analyzer of geo-referenced data – and that currently are theoretically and conceptually ready to be made available through GIS packages (e.g., ARC/INFO, IDRISI). Technical and computational implementation issues are addressed. Current expert opinion, gleaned from two recent international conferences, is summarized. Gateways are identified for spatial statistical tools to be introduced into a GIS. The principal conclusion is that there is evidence from a growing number of initiatives, as well as testimony from spatial scientists and geographic analysis practitioners, supporting the need to convert spatial statistics techniques into GIS functions.

## 1. Introduction

Presently an overwhelming amount of data that are analyzed on a computer are geo-referenced or spatial data (see the Chorley Report, 1987; National Research Council, 1990a; IBM, 1991); such data result from entities being aggregated (usually) into locationally referenced groupings, yielding attributes that have some geometric coordinate system (e.g., Cartesian) attached to them. For example, remotely sensed data involve points on the earth's surface being aggregated into pixels, spectral measures of light reflectance made on these pixels (attributes), and the configuration of these pixels recorded (relative positioning). Government census data serve as a second example; households are aggregated into census tracts, for instance, household characteristics are measured for each tract, and tract centroids often are established as coordinates. In recent years this near preponderance of spatial data analyses has lead to a scholastic awareness, in the scientific community at large, of complications attributable to the geographic dimension of data, especially in terms of the validity of traditional statistical analyses. Accordingly, increasing attention has begun to focus on geographic information systems (GISs) as well as the general field of spatial statistics, spatial econometrics, and geostatistics. For example, the announced goals in the solicitation of proprosals for the National Science Foundation Engineering Research Center, entitled the National Center for Geographic Information and Analysis (1987), included the objective of promoting advances in spatial statistics within the context of GISs. The Board on Mathematical Sciences of the National Research Council (1990b) has targeted spatial statistics as one of twenty-seven topics of national concern in mathematics (its rank is 17). And, as a cooperative effort, the Department of Mathematics at Lancaster University and the North West Regional Research Laboratory in Great Britain have initiated a project to integrate statistical and GIS software, the U.K. Economic and Social Research Council (ESRC) funded an 'experts' workshop on this same theme in 1991[1], and the International Geographical Union (IGU) hosted an educational GIS workshop on this theme in 1991[2].

A GIS is comprised of hardware and software that, together with spatial data, are organized and interfaced efficiently and effectively; it embodies three components, namely (1) a specially organized geo-referenced database, (2) a data processing system, and (3) concepts and techniques of spatial analysis. The function of a GIS is to store, manipulate, and facilitate the analysis of geo-referenced data by combining tabular attribute data with computerized maps. It encompasses a unique combination of elements, including high-resolution graphic displays, large-capacity electronic storage devices, efficient and rapid data storage, retrieval and management capabilities, high-volume communication channels, specialized algorithms for spatial analysis, and specialized computer languages for making spatial queries. Spatial statistics is concerned with the statistical analysis of geo-referenced data, and differs from classical statistics in that the observations analyzed are not independent (violation of this single assumption is the crux of the problem). Spatial statistics interfaces with GIS through spatial analysis. But to date little spatial statistics can be found in GISs. Rather, GIS spatial analysis functions have been limited to selected geographic operations research techniques (e.g., minimum path algorithms), and simple operations such as map overlays (based upon set theory) and proximity determination.

The primary goal of this chapter is to identify existing spatial statistical tools that would be routinely of value to the GIS user community, and that currently are theoretically and conceptually ready to be made available through GIS packages. Addressing of this goal is accompanied by a discussion of technical and computational problems.

---

[1]    The ESRC workshop was entitled "Spatial Analysis for GIS," and hosted by the Department of Geography, University of Sheffield, during March 18–20, 1991
[2]    The IGU workshop was entitled "Merging GIS and Spatial Statistics," and hosted by Masaryk University, Brno, Czechoslovakia, on April 22, 1991

## 2. Prominent differences between spatial and aspatial data

A fundamental difference between spatial and aspatial data is that observations for geo-referenced data are correlated strictly due to their relative locational positions (this is referred to as spatial autocorrelation), resulting in spill-over of information from one location to another. Actually this spill-over results in redundant information being present in data values, with this redundancy increasing as the degree of locational dependence increases. This duplication of information activates complications in the statistical analysis of geo-referenced data that lie dormant in the statistical analysis of traditional data comprised of independent observations; in other words, invoking an assumption of independent observations suppresses potential data complexities. These complications are similar to those found in time series analysis; but, they are exacerbated by the multi-directional, two-dimensional nature of spatial dependence (time series entails dependencies that are uni-directional along a single dimension). The net result is that classical statistics applied to geo-referenced data fail to capture locational information, raising questions of estimator sufficiency. Most parameter estimators affiliated with variation are biased; some simple linear combinations of the sample observations (e.g., the sample mean) remain unbiased though. In addition, questions of efficiency and consistency of estimators are raised, too.

Although the literature is replete with documentations pointing out serious inferential consequences attributable to a disregarding of spatial dependence, overlooking or ignoring this latent dependence more often than not is what is done (see Anselin and Griffith, 1988; Anselin and Hudak, 1991). Three approaches can be taken to the statistical analysis of geo-referenced data that go beyond ignoring spatial dependencies, and hence violating at least one assumption of classical statistical models. First, a researcher can concentrate solely on the locational information component of geo-referenced data (e.g., point pattern analysis). Second, a researcher can exploit the redundant information contained in geo-referenced data for interpolation purposes (e.g., geostatistics). Third, a researcher can seek to orthogonalize the information content of observations, in order to avoid "double counting" of information (e.g., spatial autoregression analysis), and then proceed with conventional analyses. This third approach is analogous to removing multicollinearity from a set of variables by subjecting the set to, say, principal components analysis, and then working with the constructed component scores.

For the most part, the scientific community is not interested solely in locational information, and hence has found little value in divorcing locational information from attribute information. Furthermore, the goal of many data analyses is other than interpolation; but, packages such as GEO-EAS, GEOPACK, BLUEPACK-3D, GEOSTAT, and GS+[3] have been developed to aid in research

---

[3]　GEO-EAS (Geostatical Environmental Assessment Software) is a software package that does kriging, and is available from the USEPA Environmental Monitoring Systems Laboratory in Las Vegas, Nevada. GEOPACK (Geostatics for Waste Management) is a software package that does kriging, and is available from the USEPA Research Laboratory in Ada, Oklahoma. BLUEPACK 3-D (estimating and gridding of natural resources) is a software package that does kriging, and is available

endeavors involving kriging (a more comprehensive list is provided in Legendre, 1991). Frequently scientists are interested in linking their data analyses to standard statistical techniques. The spatial autoregressive approach enables this avenue to be pursued, especially since many classical statistical techniques can be translated into a regression model format. Therefore, the emphasis of this contribution will be on spatial autocorrelation and autoregression analysis.

The second salient difference between spatial and aspatial data, beyond the presence of information redundancy, has to do with the numerical intensity required by its analysis. This numerical intensity arises from three sources. First of all, geo-referenced data sets sometimes are comprised of massive numbers of observations. A single remotely sensed image, for example, might include 10000000 pixels. Second, orthogonalizing information content in order to handle redundant information requires the use of spatially lagged variables; conceptually the value of an attribute in areal unit i is cast as a weighted average of attribute values taken on by other areal units. Computation of spatially lagged variable values requires the determination of which areal units are dependent. This phenomenon is represented by an $n$-by-$n$ spatial weights matrix, say $C$; for a very local spatial dependency structure, the entries of matrix $C$ commonly are defined such that $c_{ij} = 1$ if areal units $i$ and $j$ are juxtaposed, and $c_{ij} = 0$ otherwise. Next this matrix usually is converted to its stochastic version, say $W$, in order to incorporate the notion of a weighted average (i.e., $w_{ij} = c_{ij} / \sum_{j=1}^{n} c_{ij}$); value $y_i$ is cast as a function of $\sum_{j=1}^{n} w_{ij} y_j$. Hence, for some $n$-by-1 attribute data vector $Y$, the corresponding spatially lagged variable is $WY$.

As is the case with principal components or factor analysis, orthogonalizing the information content of geo-referenced data involves the computation of eigenvalues. Here, rather than a $p$-by-$p$ matrix (for $p$ variables included in a principal components analysis, say), eigenvalues are needed for the $n$-by-$n$ matrix $C$ or $W$, depending upon the spatial autoregressive model being employed. For even moderate size spatial data sets (e.g., the 363 census tracts of Houston), then, this computational task becomes very intensive. Griffith (1990c) reports that simulation experiments involving up to 5625 areal units had roughly 93% of CPU time devoted to the calculation of these eigenvalues. Griffith (e.g., 1990a,b, 1992) has begun to uncover latent systematic patterns in this set of eigenvalues that suggests an approximation technique for which their calculation no longer would be required.

The second source of numerical intensity stems from spatial autoregressive models being nonlinear in nature. The aforementioned eigenvalues are needed to compute a normalizing factor that ensures that the normal distribution probability density function integrates to unity, regardless of whether data are located in an autocorrelated space or its orthogonalized, unautocorrelated counterpart. In-

---

[3]  (continued)   from Geovariances International, Fontainebleau, France. GEOSTAT is a software package that does kriging, and is available from Systems GEOSTAT International Inc., Montreal, Canada. GS+ (professional geostatistics for the agronomic and biological sciences) is a software package that does kriging, and is available from Gamma Design Software, Plainwell, Michigan

clusion of this normalizing factor converts the analysis into a nonlinear one. Suppose that this normalizing factor is denoted by $J(\varrho)^2$, where $\varrho$ is the spatial autocorrelation parameter. Then for a simultaneous spatial autoregressive model the spatial autoregressive equation becomes (see Griffith, 1988b)

$$Y/J(\varrho) = X\beta/J(\varrho) + \varrho(WY - WX\beta)/J(\varrho) + \xi/J(\varrho) \ ,$$

where vector $\xi$ conforms to all of the conventional statistical assumptions. A pure spatial autoregressive model is given by the equation

$$Y/J(\varrho) = \mu 1/J(\varrho) + \varrho(WY - \mu 1)/J(\varrho) + \xi/J(\varrho) \ ,$$

where $1$ is an $n$-by-1 vector of ones. A spatial autoregressive response model is given by the equation

$$Y/J(\varrho) = X\beta/J(\varrho) + \varrho WY/J(\varrho) + \xi/J(\varrho) \ .$$

Parameter estimation for these spatial autoregressive equations must be done using nonlinear optimization techniques (see Upton and Fingleton, 1985). Therefore, because estimation involves iterative computation, it becomes numerically intensive.

This nonlinearity also results in the presence of messy analytical derivatives. Variable $Y$ appears on both sides of this equation. Parameter $\varrho$ appears in both the numerator and the denominator of various terms in this equation. Thus, the algebraic expression of partial derivatives with respect to $\varrho$ can be very messy. Therefore, the nonlinear optimization involved with parameter estimation often is best done utilizing numerical rather than analytical derivatives, which of course is numerically intensive. Of note is that the aforementioned approximation of $J(\varrho)$ helps simplify the affiliated analytical derivatives.

Consequently, the two basic differences between spatial and aspatial data are that geo-referenced data tend to contain considerable redundant information (which cannot be ignored), and spatial statistical techniques necessary for geo-referenced data analysis are numerically intensive. Recognition of these two features is critical to the selection of spatial statistical techniques for inclusion in GISs.

## 3. Emerging expert opinion

Although some spatial statistics macros exist for ARC/INFO, and IDRISI has some built-in spatial autocorrelation functions, spatial analysis in SPANS, Geo/SQL, and topoLogic presently fail to include any spatial statistics options. Expert opinion presently being expressed concerns strategies for overcoming this omission by making available to the GIS user community techniques of spatial analysis that already have been developed; attention is being paid to implementation, rather than the development of new methods of spatial data analysis. Formally acknowledging the emergence of this opinion at this point in time subsequently should prove useful to scholars who study the history of spatial statistics, GIS, and the such. The current situation has been characterized by participants of an ESRC workshop as requiring statistical analysis associated with GIS to be done by exporting geo-referenced data to standard statistical software packages (e.g., SAS, SPSS-X, MINITAB, GLIM). SAS has standard mapping capabilities together with several spatial statistics macros being available for it, SPSS-X has only standard mapping capabilities, and SYSTAT has some unique spatial functions (e.g., Voronoit-esselation, projections). The major drawback of this implementation strategy concerns the inability to preserve spatial structure in such a way that output from standard statistical software packages can be imported back into a GIS. Four different approaches, whose implementation has been scattered geographically, have been taken to rectify this situation. Some specialist software has been developed (e.g., SAAP, SpacStat[4]); some functions have been embedded in selected GIS software (the Moran Coefficient − a product moment type index of spatial autocorrelation − function in IDRISI); some macros have been developed that are accessible by but not integrated into GIS software (see Kehris, nd; Bivand, 1990); and some macros have been developed that tease spatial statistics out of standard statistical packages (see Griffith, 1988b, 1991; Bivand, 1989). The major drawback to this strategy is that in the limit one would need to replicate, for instance, SAS inside of ARC/INFO, or vice versa. Not only would this duplication of software and effort be questionable and costly, but given that both packages already require multi-volume user manuals, a formidable size encyclopedia of user manuals also would be necessary. A second important drawback is that none of the standard statistical software packages presently has a spatial statistics module. Therefore, a prudent course of action to pursue seems to be to formulate spatial statistics algorithms for incorporation into GISs, and simultaneously devise user-friendly interfaces between GIS databases and standard statistical software packages, for easy access to traditional statistical techniques.

The basic message of a recent IGU workshop addressing this same theme was more pedagogical in nature. Its co-convenors posed the following two questions, around which revolves issues of implementation of spatial statistics in GISs:

- What can spatial statistics do for GIS?
- What can GIS do for spatial statistics?

These two questions allude to scientific versus management treatment of data; the scientist tends to be concerned with quality, reliability, consistency, and uncertainty features, whereas the manager tends to be concerned with availability, accessibility, timeliness, and cost.

This first question may be answered, in part, by exploring the statistical estimation issues of bias, consistency, efficiency, and sufficiency, as well as statis-

---

[4]   SAAP (Spatial Autocorrelation Analysis Program) is software that calculates the Moran Coefficient and the Geary Ratio, and is available from Exeter Software. SpacStat is software utilizing GAUSS that does spatial econometrics, and is available from Dr. Luc Anselin, Geography Department, University of California/Santa Barbara

tical properties of geographic sampling in terms of both constraints and the nature of phenomena being sampled. As mentioned earlier, some estimators remain unbiased in the presence of spatial dependence, while others (especially those associated with variances) do not. Griffith (1988a) has demonstrated that classical statistics tend not to be sufficient because they overlook latent spatial dependence when applied to geo-referenced data (they do not summarize all of the locational information that is present). He also has demonstrated that the traditional (OLS) estimators tend to be less efficient, with this efficiency decreasing as the level of spatial dependence increases, but increasing as sample size increases[5]. With regard to geographic sampling, Overton and Stehman (1990) have produced a useful document reporting findings about systematic random sampling being superior to unconstrained random sampling when geo-referenced data are being studied. Certainly GISs will be able to facilitate the implementation of such a sampling design. Accordingly, much more needs to be learned about sampling distribution properties of this particular sample design, in order for spatial statistics to be able to help GISs include appropriate spatial analysis techniques. Moreover, then, one important opinion about spatial dependence highlighted here is its ability to obscure statistical features of geo-referenced data that tend to be far more conspicuous for other types of data.

The second question may be answered by distinguishing between attribute and geometric descriptors. Attribute descriptors may be viewed as being in a feature space that leads to tabulated classification results, whereas geometric descriptors may be viewed as being in a mapping space that leads to tabulated geo-referenced results. A combination of these two descriptor types yields graphical results. GIS can provide *tools*, especially graphical ones, to extract information on positional relationships among database entities. GIS software, in general, is capable of and efficient for various procedures, e.g., selection, distance measurement, which can be useful for characterizing spatial/topological relationships that can be viewed as a basis for further quantitative analysis. Thus, for example, GISs offer capabilities for more easily implementing the sort of systematic random sampling design previously mentioned. Or, for instance, GISs offer a vehicle for avoiding the tedium of manually constructing geographic connectivity matrices like those included in the previously stated spatial autoregressive equations; IDRISI already offers capabilities to this end, as do macros for ARC/INFO (see Kehris, nd).

Jointly the answers to these two questions reflect upon the theme of merging GIS and spatial statistics. A GIS needs spatial statistics diagnostics in order to at least warn a user of potential data analysis problems, and an advanced spatial statistical toolbox for data processing. Spatial statistics needs a GIS to make much of its numerically intensive requirements transparent to the user.

In conclusion, one key concern is software integration; the emergence of interfaces most likely will be demand driven (from the user community), will need to be user-friendly, and will require developers to have access to different software source codes. In terms of stand-alone capabilities, GISs need an internal

---

[5]    A more comprehensive study of this efficiency issue presently is being completed as a collaborative effort between Griffith, and Cordy and Overton of the Statistics Department, Oregon State University

algorithm that generates spatial weights matrices directly from spatial databases (such an $n$-by-$n$ matrix would be computer memory intensive); once constructed, this matrix should be able to be exported to standard statistical software packages, or accessed by GIS functions that utilize internal spatial statistics algorithms. These algorithms should include (1) a standard ordinary least squares (OLS) multiple regression procedure (without all of the frills that usually accompany such a procedure nowadays), (2) a test for spatial autocorrelation in regression residuals (e.g., the Moran Coefficient), and (3) a nonlinear regression procedure designed specifically for estimating spatial autoregression parameters. They should constitute the core of a spatial statistics toolbox for GISs. In addition, user-friendly interfaces between standard statistical software packages and GISs are needed.

## 3.1 Two immediate gateways into GIS for spatial statistics

The preceding discussion acknowledges that at least some GISs already include the option of calculating the Moran Coefficient (MC). But this statistic is useful mostly in an inferential context, meaning that it needs to be accompanied by calculation of its standard error. This calculation is numerically intensive for regression residuals, and would require the inclusion of a goodly number of additional numerical capabilities within a GIS; on the other hand, a simplified approximation of it for inclusion in a GIS is presented in this section. Any model that can be rewritten with a multiple linear regression equation can be cast as a spatial autoregressive model; and it should be if MC detects spatial autocorrelation in its errors. But the most bothersome numerically intensive feature of such a model has to do with handling the accompanying normalizing factor, denoted earlier in this contribution by $J(\varrho)$. An approximation of it for inclusion in a GIS is outlined in this section. The computational feasibility of the general linear model (GLM) should allow more sophisticated multivariate techniques to be added to the battery of tools already available through regression modelling (e.g., the simple variable mean, N-way ANOVA, two groups discriminant function analysis, and the correlation coefficient). Here the error vectors are equivalent to those from individual regressions. But, correlations of these errors together with convolutions of these correlations with varying levels of spatial autocorrelation merit exploration. The cross-MC, which would facilitate multivariate geo-referenced data exploration, also is ready for inclusion in a GIS, and is the third spatial statistic presented in this section.

First, the MC standard error for regression residuals computationally can be approximated asymptotically quite easily (for $k = 1$, where $k$ is the number of regressor variables). Rather than

$$\text{VAR(MC)} = \{2n^2/[(1^t C\, 1)^2(n-k)(n-k+2)]\}\left[1^t C1 + tr\{[(X^t X)^{-1}X_t CX]^2\}\right.$$
$$\left. - 2tr[(X^t X)^{-1}X^t C^2 X] - \{tr[(X^t X)^{-1}X^t CX]\}^2/(n-k)\right] ,$$

where $tr$ is the matrix trace operator, under an assumption of normality, the limit as $n \to \infty$ for this expression reduces it to $2/1^t C1$, which is much simpler to compute. The cost of using this asymptotic value does not seem to be very high, a contention suggested by the following empirical comparison based upon Puerto Rico agricultural data ($n = 73$):

*mean, $k = 1$:* $0.0737^2$ (exact) versus $0.0762^2$

*one-way ANOVA, $k = 5$:* $0.0679^2$ (exact) versus $0.0762^2$

*two-way ANOVA, $k = 10$:* $0.0685^2$ (exact) versus $0.0762^2$;
　　　　　　　　　　　　　　$0.0668^2$ (exact) versus $0.0762^2$.

This asymptotic value is ready to be used by the GIS user community. Theoretically this approximate value needs to be evaluated over a wide range of possible geographic partitionings, perhaps through simulation experimentation, in order to assess its general reasonableness.

Second, the normalizing factor of the transformation from an autocorrelated observational space to an unautocorrelated observational space must be simplified in order for spatial statistics to be properly introduced into a GIS. Approximations of this term for either regular or irregular lattices allow considerable reduction in computational requirements for estimating spatial autoregressive models, and enable standard statistical software packages to deal more effectively and sometimes efficiently with spatial statistical models. These approximations are ready for users of GIS, but with a cautionary note attached, and are of the form

$$\alpha_{1,n} LN(\delta_{1,n}) + \alpha_{2,n} LN(\delta_{2,n}) - \alpha_{1,n} LN(\delta_{1,n} + \varrho) - \alpha_{2,n} LN(\delta_{2,n} - \varrho) ,$$

where $LN$ is the natural logarithm, and $\alpha_{j,n}$ and $\delta_{j,n}$ ($j = 1, 2$) depend upon the type of spatial autoregression model in question, and can be read from stored tables. One theoretical need here concerns evaluation of these approximations over a wide range of different geographic distributions and surface partitionings. Their impact on properties of estimators (bias, sufficiency, efficiency, consistency) needs to be explored. Computationally the regular, square lattice findings need to be extended to other surface partitioning schemes, especially irregular lattice ones (presumably the most common type of surface partitioning encountered in a GIS database).

Third, the cross-MC may be defined, for two variables $X$ and $Y$, as

$$(n/\mathbf{1}' C\mathbf{1})(Y - \bar{y}\mathbf{1})' C(X - \bar{x}\mathbf{1})/[(Y - \bar{y}\mathbf{1})'(Y - \bar{y}\mathbf{1})(X - \bar{x}\mathbf{1})'(X - \bar{x}\mathbf{1})]^{1/2} .$$

In the absence of spatial autocorrelation, the expected value of this term equals $-\varrho_{XY}/(n-1)$, which reduces to the familiar $-1/(n-1)$ term when $Y \equiv X$. This index should help GIS users understand multivariate geo-referenced data sets. On the one hand, when spatial autocorrelation levels are low across a set of variables, then the structure revealed by principal components analysis of the correlation matrix $R$ should be the same as that for the cross-MC matrix; this was the finding for Buffalo crime data. On the other hand, this result indicates that orthogonalization of data will set cross-MCs to zero *only if* spatial autocorrelation is absent in the variables; this is the finding for a MANOVA of Puerto Rico agricultural data. Theoretically this index value needs to be evaluated for various non-zero spatial autocorrelation schemes.

In conclusion, so that numerical intensity of calculations is held to a minimum, (1) the standard OLS regression procedure included in a GIS should include an MC function to test for spatial autocorrelation in residuals, coupled with the aforementioned asymptotic variance for hypothesis testing purposes, and (2) the spatial autoregression algorithm should include the normalizing factor approximation. Furthermore, a standard product moment correlation function should be included in a GIS toolbox, together with a cross-MC function. This last pair of functions should be linked to the map overlay operation.

## 4. Implications

Consequently, there is both evidence and prevailing expert opinion indicating that spatial statistics techniques need to be converted to GIS functions (also see Anselin and Getis, 1991). The evidence may be gleaned from a growing number of initiatives to modify, adapt, or the such, existing GIS software for spatial statistical purposes (e.g., Kehris, nd); evidence also is suggested by the increasing number of specialized software packages appearing that solely do spatial statistics in one form or another (e.g., SpacStat, GEO-EAS, SAAP). These developments furnish new directions for interfacing GIS and spatial modelling through spatial statistics. Those candidates most suitable for initial inclusion in GIS software encompass indices of spatial autocorrelation, and selected spatial autoregression models, which are major components of the aforementioned dedicated spatial analysis software. The spatial statistics toolbox, through which these functions would be accessed, should include an elementary multiple linear regression function, a spatial weights generator, and an MC function to test for spatial autocorrelation in regression residuals; the asymptotic variance of MC should be substituted for the exact variance, in order to minimize numerically intensive computations. The spatial autoregressive function should be based upon an algorithm that approximates the normalizing factor, in order to minimize the numerical intensity of computations. In order to support exploratory multivariate analyses, a standard product moment correlation function, and a cross-MC function should be included in this toolbox. Finally, the development of this spatial statistics toolbox should be paralleled by the development of transparent, user-friendly interfaces between GISs and standard statistical software packages.

On the one hand, new spatial analysis applications should naturally emerge once spatial statistical procedures become readily available in GISs. This expectation is based on an analogy with an explosion in the number of time series applications that followed the dissemination of user-friendly Box-Jenkins type of software, especially within standard statistical software packages. It also is exemplified by the Dependent areal Units Sequential Technique, developed by Arbia, for selecting the optimal location of samples for surveying purposes. On the other hand, a number of technically sophisticated researchers already are supplying new applications, especially in terms of interfaces between GISs and standard statistical software packages. Besides the SAS and MINITAB codes mentioned earlier, several fascinating and successful projects have been undertaken incorporating FORTRAN subroutines into GLIM that allow it to access directly (ARC/)INFO databases[6]. Unfortunately, at present these interfacings between GISs and standard statistical software packages require great ingenuity on the part of the research.

In the more distant future, this spatial statistics toolbox should be expanded to include, in order of importance, (1) geostatistics, especially kriging, (2) contemporary point pattern analysis techniques, (3) spatial statistics regression diagnostics, and (4) optimal geographic sampling procedures. Because GISs best support scientific visualization activities, the resulting enhancements should be consistent with the type of exploratory data analysis described and promoted by Haining (1990).

## References

Anselin L (1989) Spatial regression analysis on the PC: spatial econometrics using GAUSS, unpublished manuscript, Geography Department, UC Santa Barbara

Anselin L, Getis A (1991) Spatial statistical analysis and geographic information systems, paper presented at the 31st European Congress of the Regional Science Association, Lisbon, Portugal, August 27–30

Anselin L, Griffith D (1988) Do Spatial Effects Really Matter in Regression Analysis? Papers Reg Sci Assoc 65:11–34

Anselin L, Hudak S (1991) Spatial econometrics in practice: a review of software options, paper presented at the 38th North American meetings of RSAI, New Orleans, November 7–10

Bivand R (1989) User interfaces for geographical software tools: a case study in regression modelling, paper presented at the Sixth European Colloquium on Theoretical and Quantitative Geography, Chantilly, France, September 6–9

Bivand R (1990) Spatial statistics: front-end inference support for GIS, paper presented at the Third Scandinavian Research Conference on Geographical Information Systems, Helsingor, Denmark, November 14–16

Chorley R [chairman] (1987) Handling Geographic Information. Her Majesty's Stationary Office, Report to the Select Committee on GIS, London

Griffith D (1988a) Advanced Spatial Statistics. Kluwer, Dordrecht

Griffith D (1988b) Estimating spatial autoregressive model parameters with commercial statistical packages. Geogr Anal 20:176–186

Graffith D (1989) Spatial regression analysis on the PC: spatial statistics using MINITAB, Discussion Paper No. 1. Institute of Mathematical Geography, Ann Arbor, MI

Griffith D (1990a) A numerical simplification for estimating parameters of spatial autoregressive models. In: Griffith D (ed) Spatial Statistics: Past, Present, and Future. Institute of Mathematical Geography, Ann Arbor, MI, pp 185–195

Griffith D (1990b) Simplifying the normalizing factor in spatial autoregression, invited paper presented at the International Geographical Union Commission on Mathematical Models Workshop on Spatial Modelling, Boston University, November 8, 1990

Griffith D (1990c) Supercomputing and spatial statistics: a reconnaissance. Profess Geogr 42:481–492

---

[6]    See R. Flowerdew and M. Green (1989), "Statistical methods for areal interpolation: the EM algorithm for count data," *Research Report 3*, and (1990) "Interference between incompatible zonal systems using the EM algorithm," *Research Report 6*, North West Regional Research Laboratory, Lancaster University

Griffith D (1991) Spatial regression analysis on the PC: Spatial statistics using SAS, unpublished manuscript, Department of Geography, Syracuse University

Griffith D (1992) Simplifying the normalizing factor in spatial autoregressions for irregular lattices, Papers Reg Sci 71:71–86

Haining R (1990) The use of added variable plots in regression modelling with spatial data. Profess Geogr 42:336–344

IBM (1991) Exploring new worlds with GIS, Directions 5 (#2, Summer/Fall), 12–19

Kehris E (nd) Spatial autocorrelation statistics in ARC/INFO, Research Report No. 16, ESRC North West Regional Research Laboratory, Lancaster University

Legendre P (1991) Spatial autocorrelation: trouble or new paradigm? Paper presented at the Statistics Canada Methodology Symposium '91, Spatial Issues in Statistics, Ottawa, Canada, November 12–14

National Research Council (Mapping Science Committee; Commission on Physical Sciences, Mathematics, and Resources) (1990a) Spatial Data Needs: The Future of the National Mapping Program. National Academy Press, Washington, DC

National Research Council (Board on Mathematical Sciences) (1990b) Renewing US Mathematics: A Plan for the 1990's. National Academy Press, Washington, DC

National Science Foundation (1987) Solicitation: National Center for Geographic Information and Analysis. Directorate for Biological, Behavioral, and Social Sciences. Washington, DC

Overton S, Stehman S (1990) Statistical properties of designs for sampling continuous functions in two dimensions using a triangular grid, Technical Report 143. Department of Statistics, Oregon State University

Upton G, Fingleton B (1985) Spatial Data Analysis by Example, Vol 1. Wiley, New York

# The accuracy of digital representations of 2D and 3D geographical objects: a study by simulation

**Chris Brunsdon** and **Steve Carver**

North East Regional Research Laboratory, CURDS, Newcastle University, Newcastle Upon Tyne NE1 7RU, England

**Abstract.** Simulated vector coastlines of known and varied complexity are rasterized at various levels by the quadtree method. The rastering error for each combination of coastline and raster size is calculated by a simple boolean overlay method. The relationship between line complexity, raster size and rasterizing error is investigated and a method of selecting the most appropriate raster size based on the complexity of the source data and the required level of accuracy is forwarded.

## 1. Introduction

Representing complex coastlines by means of a dense raster database can be inefficient and costly in terms of processing and storage. Similarly, representing terrain surfaces in a large sample DEM can also be very inefficient, especially in areas of mixed relief. The use of vector based systems and TINs solves this problem to a certain extent, but can lead to a loss of accuracy in some areas depending on the sampling strategy and the tolerances used. Defining these is usually left to the judgement of the user. Important factors influencing sampling strategy are: line or terrain complexity; storage overheads; and the accuracy required.

The accuracy of spatial databases and error propagation in GIS operations is currently the subject of much attention, if only in the academic research sector. One area where accuracy as an issue is particularly important is in the choice between vector and raster based systems. Putting aside problems relating to the accuracy of the source data, digitising error and attribute uncertainty, then the primary concern to the user is how to trade-off positional accuracy against computational efficiency.

Vector based systems offer greater positional accuracy for the storage of map-based data, but many of the algorithms used for analysing this kind of data are extremely time-consuming. In addition to the positional data itself, information about the topological relationships between the geographical entities must also be stored and processed. In the raster case, however, topology is inherent in the data structure. As a result of this, and the employment of quadtree representation (Samet, 1984), equivalent algorithms for raster data require much less computational time. Unfortunately there is a price to pay for this since raster representation of data is generally much less accurate, due to its higher disk storage requirements. This dilemma holds true not only in the realm of 2D GIS, but also in that of 3D or DTMs (digital terrain models) although much depends on the users choice of tolerances and sampling strategy.

Little attention seems to have been given to how tolerances chosen by the user can affect the overall accuracy of the results gained from GIS operations. Few GISs, if any, provide the user with anything more than arbitrarily determined defaults for important tolerances governing the results of basic operations such as 'cleaning' input data, buffering and converting vector data to raster form. In particular, this contribution concentrates on those tolerances governing the representation of lines and surfaces as quadtrees and DTMs, respectively.

The following sections of the chapter cover methodology and analysis of the results before making some conclusions on how empirical investigation of the relationships between accuracy and sampling strategy can provide users with more objective and reliable means of specifying important tolerances. This is a significant topic in spatial analysis for two reasons. Firstly, any technique which processes spatial data in some way is only as reliable as the data it uses as input. Thus issues of data reliability and robustness of methods given map data containing errors are of importance whenever practical applications take place. Secondly, the occurrence and diffusion of error in geographical data may itself be viewed as a spatial process, and perhaps one of which there is little understanding, thus demanding a considerable amount of future investigation.

## 2. Methodology

It has long been suggested by numerous authors working in the field of cartography and GIS that the complexity of a geographical feature greatly affects the accuracy at which it can be represented in either map (analogue) or digital form (e.g. Baugh and Boreham, 1977; Chrisman, 1982; and Goodchild and Mark, 1987). Put in simple terms, the more complex the feature then the greater the difficulty of re-creating an accurate copy. In the context of digital mapping, this relationship is complicated by the need to minimise storage overheads. This in turn creates the need for some means of resolving the problem of how to reduce storage space to a minimum whilst retaining the required level of accuracy.

Here, the relationship between feature complexity and the accuracy of its digital representation using different sampling strategies is investigated. Both 2D and 3D features, in the form of coastlines and landscapes, are artificially generated using fractal mathematics. A number of authors have eluded to the fractal nature of geographical features and their use in generating and quantifying landscape form (e.g. Goodchild and Mark, 1982; Shelberg et al., 1982; and Muller, 1986).

More recent opinion would suggest that fractals are less useful than previously thought. The main objection lies in the assumption that cartographic line features can be modelled by a statistically self similar process. This would suggest that the level of line complexity does not appear to change regardless of the scale at which the line is observed. However, in recent years there has been some controversy as to whether this is the case (see for example Buttenfield, 1989; Carpenter, 1980; and Jasinski, 1990). It seems more likely that variation in shape is in fact scale-dependent, so that magnification of lines could reveal different patterns of complexity.

Despite this, it has been generally established that although many natural features are not self-similar at *all* scales, they may be over some reasonably large ranges of scale. The simulation methods given here in an initial investigation do generate curves which are statistically self-similar, but scale dependency could be added to the techniques at some future point. This could be done, relatively easily in the case of the coastline simulation, as will be discussed in the next section.

## 2.1 Generating artificial coastlines and landscapes

In carrying out a study of this kind, it is useful to be able to simulate geographical feature data of controlled complexity. In this way a sample set of data having a broad and well-balanced set of shapes can be created. A further advantage of this methodology is that the complexity of each object is known exactly. If field data is used, there is the added problem of estimating shape description indices from sample data which is subject to digitising error. However, before simulation can take place, it is necessary to specify a model as a basis. A reasonable starting point is the concept of the fractal landscape as mentioned above.

Underlying the fractal model is the concept of self-stimilarity. Put simply, this implies that 'zooming in' to any portion of an object may reveal a replica of the whole object. This is assumed to happen to an arbitrary level, so each object contains an infinite number of nested replicas of itself. In geography this is never known to happen with exact replication, but a statistical form of this phenomena, where each replica may not be perfect but drawn from the same probability distribution, has been considered (Mandelbrot, 1967). This has proved reasonably successful in emulating real phenomena.

A model of the cartographic line based on this principle has been used in the current study. Consider the line $AB$ in Fig. 1 a. If this line is bisected, and a small perpendicular displacement $a$ is applied to the bisection point $C$, giving point $D$, then a more complex curve $ADB$ is obtained. If the displacement were in the opposite direction, to point $E$, then the curve $AEB$ would be obtained. Two facts need now be noted. Firstly, the choice between $ADB$ and $AEB$ could be made using a random number generator, giving a stochastic process. Secondly, this process can be applied recursively to the two line segments $AD$ and $DB$ (or $AE$ and $EB$) and to their resultant sub-segments *ad infinitum*. In this way the random process replicates itself onto subdivisions of the line with decreasing scale. Thus,

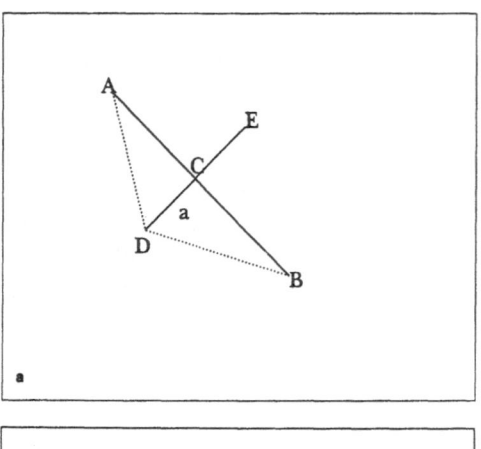

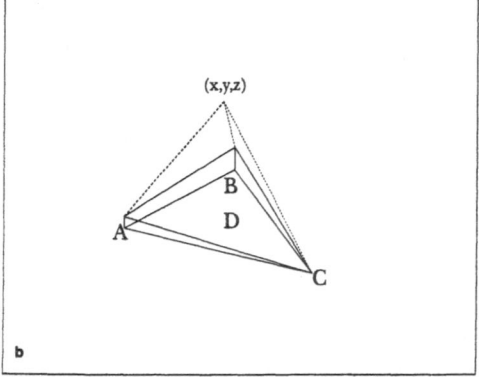

**Fig. 1a,b.** Method of generating fractal curves and surfaces. **a** 2D; **b** 3D

the product of this procedure is a fractal line curve. The fractal dimension (as defined by Mandelbrot, 1977) can be shown to be a function of $a$, the amount of displacement $CD$ considered as a proportion of $AB$, which is kept fixed.

Using a method such as that given above yields some simulations of coastlines which are at least visually satisfying. An example of lines of various dimensions is given in Fig. 2. Clearly, in practice the curves generated are not true fractals (the authors' computing equipment is not yet capable of completing an infinite process!), but are the result of a finite number of iterations of the random displacement technique. This implies that after a certain amount of 'zooming in', the curves consist entirely of straight line segments. In practice this has little effect, since this is also the case when digitising true map line segments.

The method could be extended to give coastlines which are not self-similar as follows. The ratio of the lengths $AB : CD$ is kept constant in the self-similar case, but could be specified as a function of the initial length $AB$. This would allow the relative amount of perturbation to vary with scale. In this context, the concept of fractal dimension may no longer be well defined, but lines of different complexity could still be generated under control, and the linkages between other line complexity indices (such as the average angle between line segments in Section 3) could be investigated.

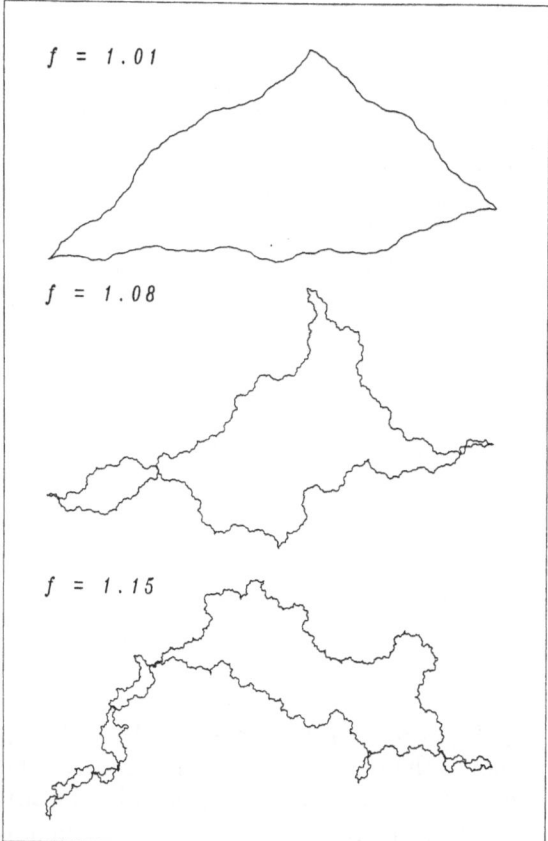

*f* = 1.01

*f* = 1.08

*f* = 1.15

**Fig. 2.** Simulated coastlines of known fractal dimension (*f*)

The chosen method can be extended into three dimensions, to simulate terrain, without much further effort. In this case, the self similarity model is applied to a triplet of (x, y, z) coordinates whose projection onto the xy plane forms a triangle *ABC*. Let *D* be the coordinates of the centroid of *ABC*. Then generate a z-value for *D* from a distribution whose mean is the mean of the z-values for *A, B* and *C*. This gives one iteration of the process. The same can now be recursively applied to the triangles *ADB, CDA* and *BDC*, and to the triangles formed with their respective centroids until the required level of accuracy is obtained. This is illustrated in Fig. 1b.

To obtain a DEM from this process, a Triangulated Irregular Network (TIN) is built around the vertices, and the altitude of each point on the DEM sampling grid calculated by interpolation. Finally, it is found by the authors that the edges between triangles tend to give unnaturally sharp changes in terrain slope. To correct for this, a spatial smoothing operator (Schalkoff, 1990) is applied to the grid height matrix calculated for the DEM. An example of the final terrain simulated by this method is given in Fig. 3.

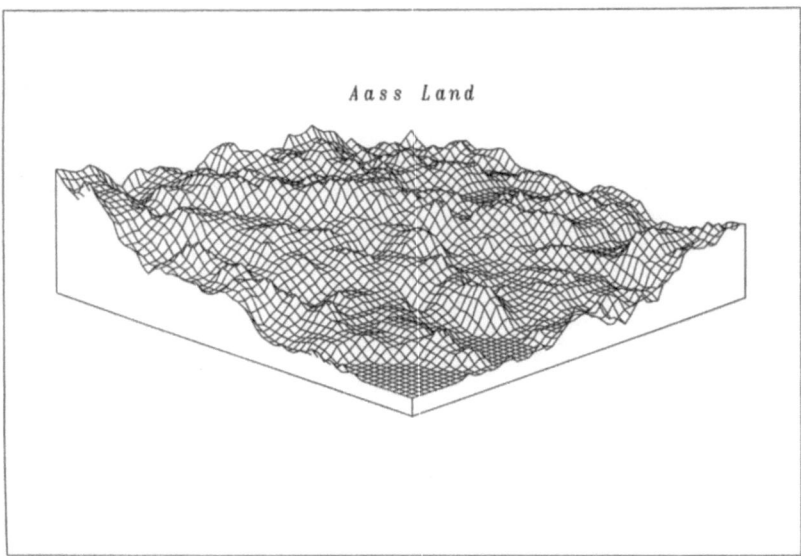

**Fig. 3.** Simulated terrain surface

## 2.2 Measuring errors in 2D

The error involved in converting a polygon of known fractal complexity into a quadtree representation is defined here as the total area of overlap between the original polygon vector and its quadtree facsimile. This is illustrated in Fig. 4d. Calculating the error involved in the conversion process is carried out here totally within the ARC/INFO GIS. A total of 21 different coastline polygons of known fractal dimension (f) between 1.0 and 1.2 are artificially generated in the manner described above and input into ARC/INFO. Once in the GIS as topologically 'clean' vector polygons, each is 'quaded' at successively higher quad levels. Although Arc/Info has no quadtree handling capabilities, 'quadding' is effectively achieved by converting the vector polygons to grid format using the POLY-GRID command with grid cells of exponentially decreasing size, the size of quad level 1 being determined by the maximum dimension of the vector polygon. The resulting series of single variable files (SVFs) are then converted back to vector polygons using the GRIDPOLY command. The polygon coverages thus generated in effect resemble the outline of the original polygon had it been converted to quadtree format at increasingly higher quad levels in, say for example, Tydac Technology's SPANS GIS. This is illustrated in Fig. 4b. It is then possible to overlay the original vector polygon and its quadtree versions (as shown in Fig. 4c and 4d) and so calculate the area of overlap and output these figures for analysis. The Arc Macro Language (AML) is used to automate this process.

## 2.3 Measuring errors in 3D

In measuring the amount of error involved in DTM sampling strategies, work here has to date concentrated on the VIP (very important point) algorithm used to

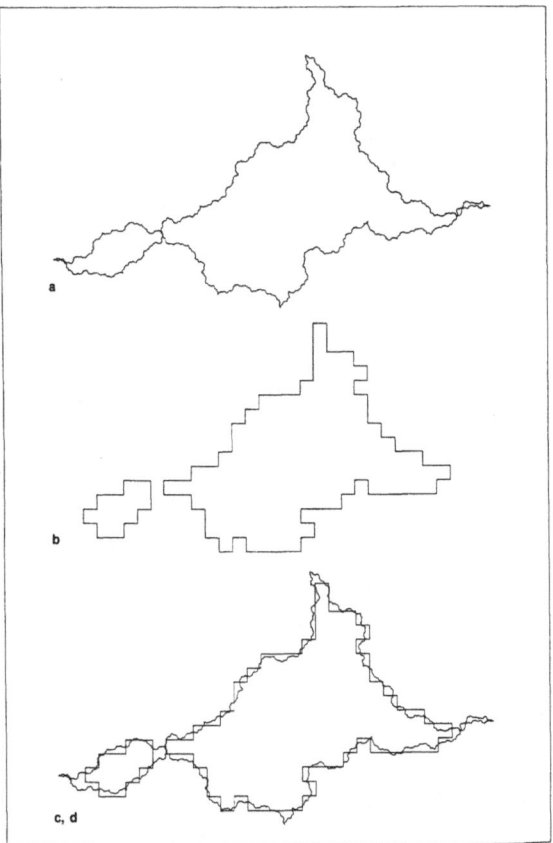

**Fig. 4a–d.** Quadding vector polygons and calculating resulting error. **a** Original vector polygon ($f = 1.08$); **b** Quadtree version at quad level 6; **c** Overlay of vector and quadtree polygons to calculate area of mismatch; **d** Inset: showing mismatch error

determine the number of significant points needed to describe a given surface. The implementation of the VIP algorithm in ARC/INFO allows the user to specify what proportion (as a percentage) of the data points in the original DTM data structure they wish to include in creating a new, more generalised version (see Fig. 5b). The VIP command in ARC/INFO works on the TIN (triangular irregular network) lattice files (essentially DEM's) to create a fresh set of points selected for their significance in reflecting the general morphology of the surface (as determined by their 'VIP' value). The resulting point coverage is used to build a new TIN and TIN DEM lattice. The lattice file derived from the original fractal landscape DEM shown in Fig. 3 is processed here using the VIP command (i.e. 'VIPed') 10 times using exponentially lower percentage selection levels (i.e. 50%, 25%, 12.5%, ... etc.). Using the ARC/INFO LATTICEOPERATE command it is possible to perform mathematical operations on one or two lattice files. In this manner, the original lattice file and its VIPed derivatives are subtracted and values in the resulting lattice file squared to give error surfaces from which an overall rms error can be calculated for each percentage selection level used in the VIP process. Example error surfaces are shown in Fig. 5c and Fig. 6.

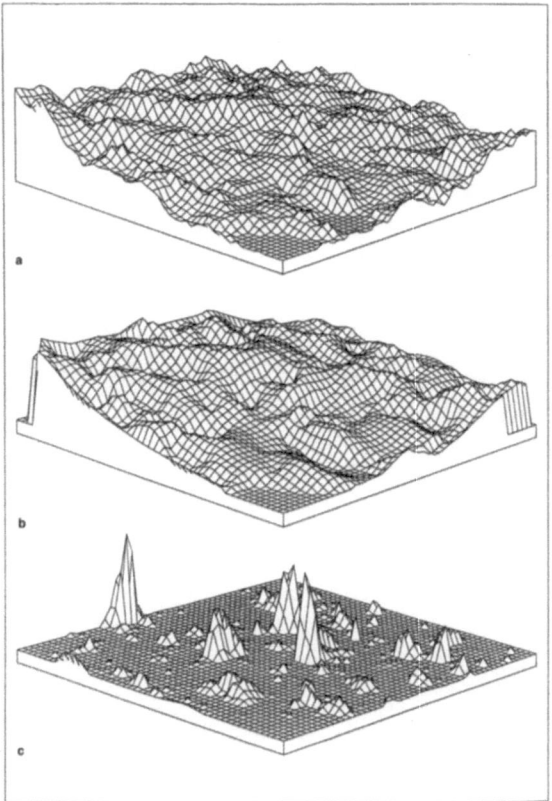

**Fig. 5a – c.** Terrain surface generalised using VIP algorithm showing resulting error surface. **a** Original terrain model; **b** Generalised version (VIPed at 1.56%); **c** Error surface

## 3. Analysis of results

### 3.1 2D results

Using the method described in 2.1, coastlines of fractal dimension ranging from 1.00 (straight lines) to 1.20 are generated, in intervals of 0.01. Each coastline needs to be a closed polygon, so that 'quadding' can take place. For each coastline, quadding from level 1 to level 13 is attempted. The error (in terms of area of mismatch) is noted in each case. These are listed in Table 1. In some of the more complex examples, ARC/INFO fails to fit a quad level of 13, so values associated with these are marked as missing. The results are also illustrated in graph form in Fig. 7. At very low quad levels, the error values for each coastline show no distinct pattern. However, after level 8, the amount of overlap error appears to decrease roughly exponentially for each coastline. This is shown in Fig. 7.

In practice it is unlikely that quad levels of less than 8 would be used in an application, so it is the roughly exponential behaviour beyond this level that will

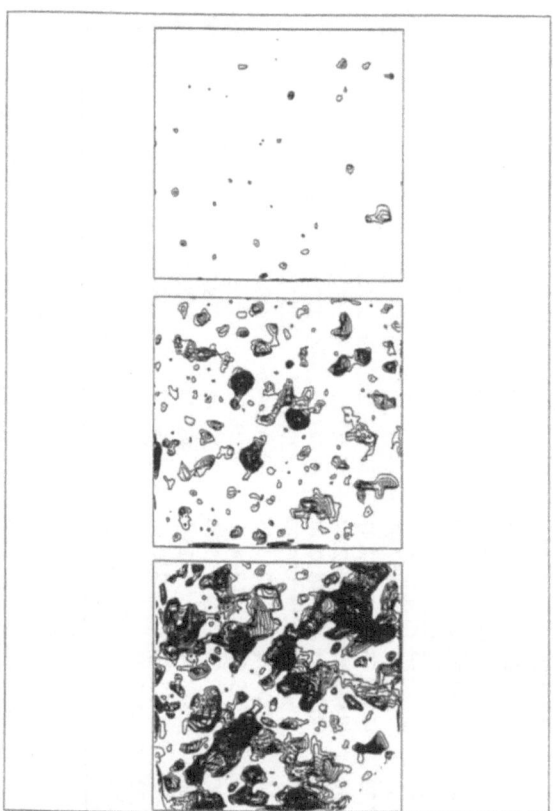

**Fig. 6.** Selected error surface for varying levels of point reduction using VIP algorithm

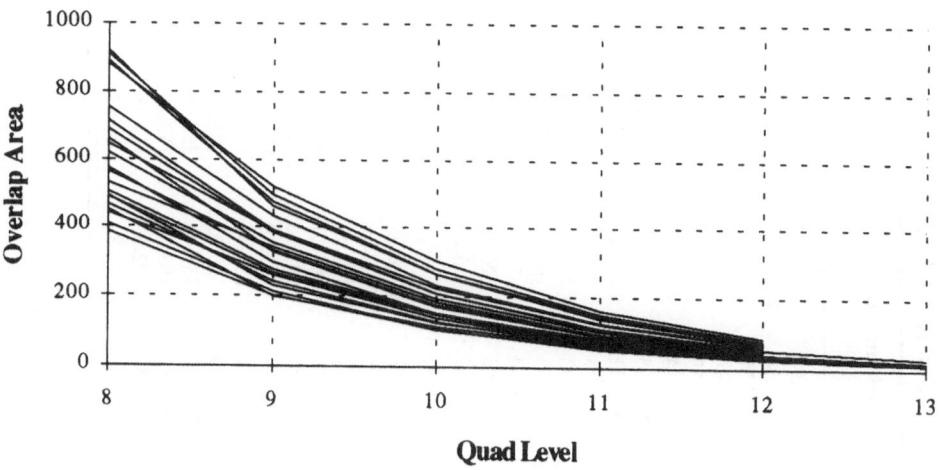

**Fig. 7.** Area of mismatch vs quad level for simulated coastlines of fractal dimension $(f)$ = 1.00 to 1.20

**Table 1.** Area of mismatch by fractal dimension and quad level

Fractal dimension

| Quad level | 1.00 | 1.01 | 1.02 | 1.03 | 1.04 | 1.05 | 1.06 | 1.07 | 1.08 | 1.09 |
|---|---|---|---|---|---|---|---|---|---|---|
| 1 | 29017.50 | 34880.55 | 37234.78 | 38990.45 | 31911.11 | 23133.31 | 22922.69 | 22761.55 | 22615.30 | 22460.09 |
| 2 | 29017.50 | 34880.55 | 37234.78 | 38990.45 | 31911.11 | 23133.31 | 22922.69 | 22761.55 | 22615.30 | 22460.09 |
| 3 | 9677.40 | 5270.19 | 5532.84 | 5990.68 | 3837.73 | 3609.55 | 3888.30 | 4175.94 | 4455.48 | 4724.50 |
| 4 | 7434.89 | 5626.00 | 8601.53 | 8006.54 | 4836.78 | 7414.47 | 7579.35 | 7859.08 | 8205.47 | 8591.23 |
| 5 | 3191.53 | 4190.19 | 6415.78 | 3008.78 | 2839.40 | 2108.65 | 3430.90 | 3016.78 | 2823.54 | 3641.68 |
| 6 | 1534.95 | 2497.73 | 1353.93 | 1517.09 | 1795.68 | 1278.82 | 1316.37 | 1482.67 | 2023.80 | 2171.08 |
| 7 | 814.11 | 2250.38 | 905.92 | 1006.44 | 1052.88 | 893.96 | 690.84 | 722.34 | 1004.63 | 933.54 |
| 8 | 384.65 | 407.41 | 449.89 | 467.36 | 411.83 | 492.42 | 490.38 | 504.42 | 576.00 | 566.12 |
| 9 | 200.33 | 211.44 | 198.90 | 239.88 | 259.03 | 228.92 | 266.22 | 275.40 | 295.55 | 329.87 |
| 10 | 104.85 | 104.41 | 113.98 | 125.99 | 137.80 | 125.40 | 146.55 | 151.24 | 149.82 | 174.12 |
| 11 | 49.94 | 54.81 | 59.89 | 61.77 | 66.74 | 68.66 | 68.61 | 74.54 | 80.10 | 92.04 |
| 12 | 25.88 | 26.29 | 28.46 | 309.85 | 33.53 | 33.48 | 36.48 | 37.78 | 39.68 | 51.91 |
| 13 | 12.32 | 13.24 | 13.99 | 14.63 | 15.54 | 16.69 | 17.46 | 18.74 | 19.38 | 30.10 |

| Quad level | 1.10 | 1.11 | 1.12 | 1.13 | 1.14 | 1.15 | 1.16 | 1.17 | 1.18 | 1.19 | 1.20 |
|---|---|---|---|---|---|---|---|---|---|---|---|
| 1 | 22307.99 | 19520.39 | 19072.61 | 18640.08 | 18236.38 | 17886.69 | 17602.97 | 45717.23 | 46551.32 | 47406.52 | 48273.20 |
| 2 | 22307.99 | 19520.39 | 19072.61 | 18640.08 | 18236.38 | 17886.69 | 17602.97 | 45717.23 | 46551.32 | 47406.52 | 48273.20 |
| 3 | 4984.49 | 19520.39 | 19072.61 | 18640.08 | 18236.38 | 17886.69 | 17602.97 | 45717.23 | 46551.32 | 47406.52 | 48273.20 |
| 4 | 7029.00 | 4985.17 | 5179.60 | 3557.70 | 3803.66 | 4051.95 | 4312.14 | 11443.75 | 11273.91 | 13150.15 | 4968.91 |
| 5 | 3377.87 | 2097.36 | 2182.66 | 2389.00 | 2239.38 | 1626.55 | 1905.90 | 11205.15 | 4537.09 | 4338.16 | 3858.27 |
| 6 | 2008.34 | 1663.40 | 1845.51 | 1819.96 | 1663.36 | 1335.70 | 1242.55 | 9443.94 | 9848.04 | 10242.03 | 1133.91 |
| 7 | 1006.74 | 1067.41 | 1119.54 | 1118.64 | 1088.99 | 1249.57 | 1348.02 | 1256.27 | 1315.70 | 1382.97 | 1791.41 |
| 8 | 662.79 | 530.97 | 621.56 | 691.14 | 715.70 | 648.14 | 755.00 | 918.85 | 889.74 | 910.34 | 880.79 |
| 9 | 327.30 | 339.45 | 352.06 | 383.75 | 388.86 | 388.50 | 432.61 | 461.71 | 475.41 | 494.64 | 520.37 |
| 10 | 170.69 | 183.59 | 193.20 | 208.98 | 207.48 | 223.50 | 230.66 | 264.29 | 262.84 | 278.41 | 302.01 |
| 11 | 87.60 | 92.56 | 101.19 | 108.71 | 108.32 | 122.07 | 124.05 | 137.41 | 145.57 | 148.63 | 160.14 |
| 12 | 44.01 | 47.39 | 50.43 | 54.16 | 56.77 | 62.30 | 63.44 | 69.50 | 74.97 | 77.84 | 84.85 |

**Table 2.** Logarithmic regression coefficients

| Fractal dimension | A | B |
|---|---|---|
| 1.00 | 97648 | −0.6882 |
| 1.01 | 100731 | −0.6865 |
| 1.02 | 100543 | −0.6808 |
| 1.03 | 121188 | −0.6910 |
| 1.04 | 107255 | −0.6741 |
| 1.05 | 97782 | −0.6655 |
| 1.06 | 108544 | −0.6685 |
| 1.07 | 104944 | −0.6608 |
| 1.08 | 128433 | −0.6745 |
| 1.09 | 67454 | −0.5959 |
| 1.10 | 144853 | −0.6747 |
| 1.11 | 83654 | −0.6201 |
| 1.12 | 102741 | −0.6326 |
| 1.13 | 120642 | −0.6403 |
| 1.14 | 117923 | −0.6360 |
| 1.15 | 77757 | −0.5912 |
| 1.16 | 115730 | −0.6241 |
| 1.17 | 153559 | −0.6405 |
| 1.18 | 123034 | −0.6158 |
| 1.19 | 126187 | −0.6148 |
| 1.20 | 104368 | −0.5908 |

be considered from here onwards. For each coastline, a function of the following form is fitted:

$$E = A \exp(BQ) \ ,$$

where $E$ denotes error level and $Q$ = Quad level. This is achieved by ordinary least squares regression of Log(Error) against Quad level. The results for each coastline are tabulated in Table 2. A typical fitted curve (for the coastline of fractal dimension 1.08) is shown in Fig. 8. As can be seen, the correspondence between the observed values and the fitted curve is strong.

When all of the exponential coefficients are inspected in Table 2, it is clear that they all have very similar rates of decay (B coefficient), but that the overall scale factor for each of the error curves increases monotonically with fractal dimension (Fig. 9). It may therefore be reasonable to fit a more general model, covering *all* curves, rather than separate models for each coastline. Such a model will take the form:

$$E_i = A_i \exp(BQ) \ ,$$

where the subscript $i$ denotes which coastline is being referred to. Note that although the scale parameter $A$ is subscripted, the decay rate $B$ is not, since this is common to all coastlines. The results of fitting this model to the data are given in Table 3.

Now that a model has been developed for all of the experimentally simulated coastlines, a logical and useful extension would be to generalise this result so that

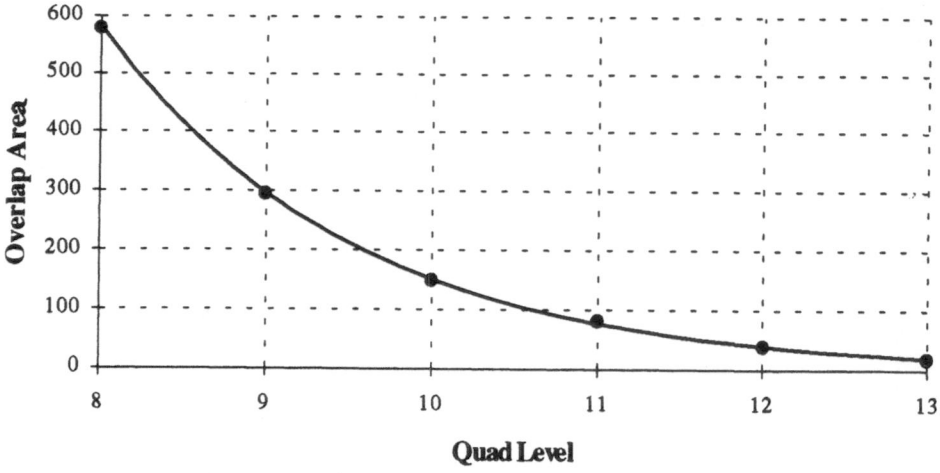

**Fig. 8.** Typical fitted curve for coastline ($f = 1.08$)

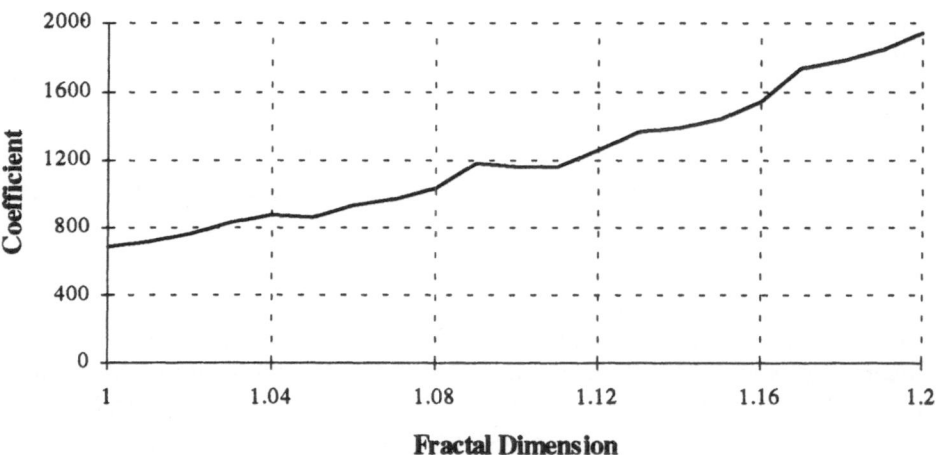

**Fig. 9.** Monotonic increase of B coefficient with fractal dimension

it could be used to estimate error levels given actual coastline or similar cartographic data. For the tabulated $A_i$ it can be seen that increasing the fractal
dimension generally increases the scale factor of the error curve. Thus $A$ could
be modelled as a monotonically increasing function of $f$, the fractal dimension.
The general model would then take the form:

$$E = A(f) \, \mathrm{Exp}\,(BQ) \; .$$

Then given a new polygon, if its fractal dimension is known or can be measured,
the relationship between error and quad level may be estimated. However, the
fractal dimension of a line is not particularly easy to measure. Given the usual
'divider walking' technique (Shelberg et al., 1982), and the somewhat arbitrary

**Table 3.** Regression model with common decay rate

| Fractal dimension | $A_i$ |
|---|---|
| 1.00 | 68 761 |
| 1.01 | 71 794 |
| 1.02 | 76 137 |
| 1.03 | 83 092 |
| 1.04 | 87 723 |
| 1.05 | 86 281 |
| 1.06 | 93 242 |
| 1.07 | 97 271 |
| 1.08 | 103 594 |
| 1.09 | 117 983 |
| 1.10 | 116 005 |
| 1.11 | 116 391 |
| 1.12 | 125 991 |
| 1.13 | 136 864 |
| 1.14 | 139 195 |
| 1.15 | 144 494 |
| 1.16 | 154 223 |
| 1.17 | 173 576 |
| 1.18 | 177 971 |
| 1.19 | 184 441 |
| 1.20 | 194 280 |
| Common decay rate | − 0.6398 |

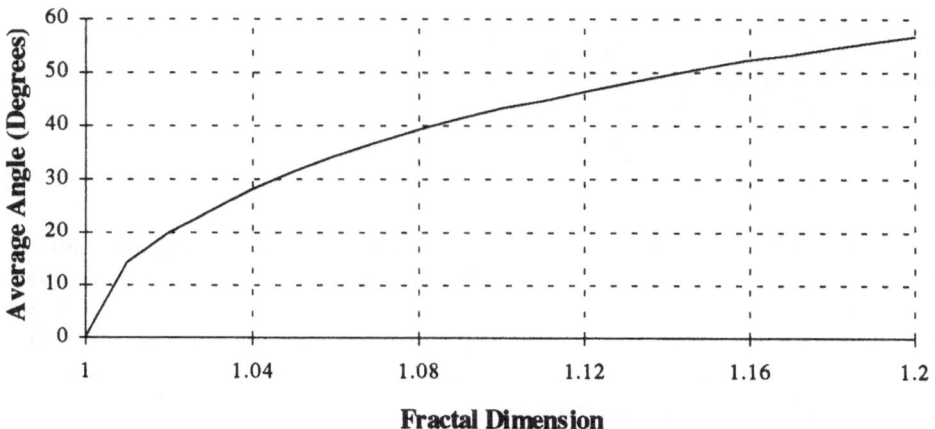

**Fig. 10.** Relationship between average angle ($t$) between digitised segments and fractal dimension

nature of some aspects of the method (such as the initial divider width), it may be simpler to consider some other form of line complexity measure. One possibility may be $t$, the average angle between digitised line segments. There is a strong correspondence between this quantity and the fractal dimension (Eastman, 1985), but the former is considerably easier to measure. For the experimentally generated lines, a plot of $t$ against $f$ is given in Fig. 10. Clearly a strong monotonic relationship exists. In this case, the final model will be specified by:

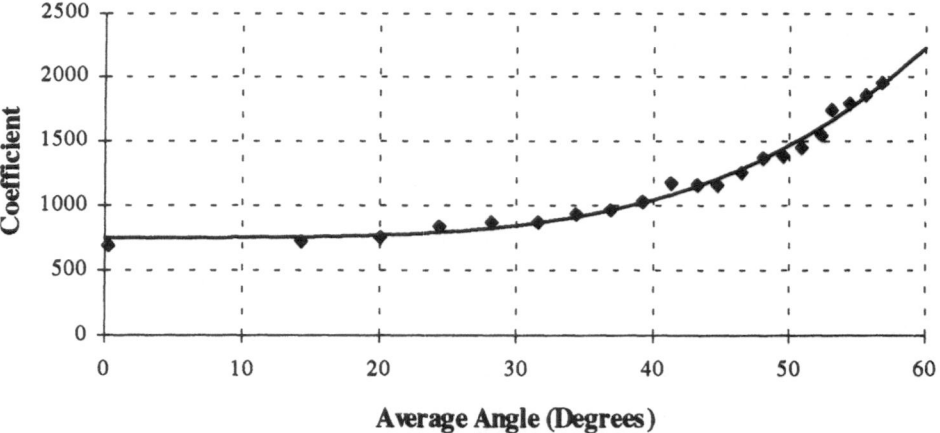

Average Angle (Degrees)

**Fig. 11.** Relationship between average angle ($t$) and scale coefficient ($A$)

$$E = A(t) \exp (BQ) \ .$$

Plotting $A$ against $t$ for the simulated polygons gives the relationship shown in Fig. 11. This may be reasonably fitted by the quartic function shown below:

$$A = Ct^4 + D \ .$$

Ordinary least squares regression returns the values $C = 1.137 \cdot 10^{-9}$ and $D = 753.8$. The fitted curve is also shown in Fig. 11. Thus the final error model for the two dimensional case is arrived at, as:

$$E = (Ct^4 + D) \exp (BQ)$$

The suggested methodology for its use is as follows. Given a polygon, evaluate the average angle between its line segments. From this a function linking quad level and error is obtained. If the user then specifies what (s)he considers to be an acceptable level of error, the inverse of the above function will give a suitable quad level. Clearly, this result may not be an integer, so rounding the answer up to the next integer is recommended.

### 3.2 3D results

For the three dimensional data, the rms error of terrain specification is plotted against percentage of points remaining after the VIP algorithm has been used. This result takes the form of the following curve:

$$E = A((1-P)/P)^B$$

where $P$ denotes percentage of points remaining with $A = 43.29$ and $B = 0.547$.

At this stage, since only one level of terrain complexity has been examined it is not possible to extend the model in the way done for the two dimensional case, however this is planned for future studies.

What is striking about visually examining the results of VIP reduction is the relatively low level of error generally when the number of (x, y, z) reference points is reduced by as much as 94% (see Fig. 6a). This would suggest that without a great loss of accuracy, the storage requirements and processing time required to manipulate and analyse terrain data could be greatly reduced by judicious use of the VIP algorithm.

## 4. Conclusions

It is noted that the accuracy of digital representations of 2D and 3D geographical landscapes rely very much on the sampling strategy and the tolerances used in the processing of data. The choice between vector based or raster based (quadtree) systems, the storage requirements and processing tolerances can have an important effect on accuracy. These choices are left very much to the user but can make the difference between efficient and inefficient use of storage space and processing time.

The effects of sampling strategy and processing tolerances are investigated on the quadding of vector polygon data and the VIPing of DTMs yielding interesting and potentially useful relationships between key variables. In the case of quadding of vector coastlines a clear relationship between line complexity and the chosen quad level (where the quad level is greater than 7) is noted in respect to the level of accuracy required. In respect to the usage of the VIP algorithm to reduce the number of significant points required to represent a 3D terrain surface an interesting relationship between the percentage sample of points retained from the original total and the resultant rms error of the new surface is observed. The work on the VIP algorithm presented here is limited to only one simulated terrain. Further research is required to test this relationship over a range of terrain complexities.

## References

Baugh IDH, Boreham JR (1976) Measuring the coastline from maps: a study of the Scottish mainland. Cartogr J 13:167–171

Buttenfield BP (1989) Scale dependence and self-similarity in cartographic lines. Cartographica 26:79–100

Carpenter LC (1980) Computer rendering of fractal curves and surfaces. Unpublished research paper, Boeing Computer Services, Seattle, Washington

Chrisman N (1982) A theory of cartographic error and its measurement in digital databases. In: Foreman J (ed) Proceedings of the Fifth International Symposium on Computer-Assisted Cartography. Washington DC, pp 159–168

Chrisman N (1989) Modelling error in overlaid categorical maps. In: Goodchild M, Gopal S (eds) Accuracy of Spatial Databases, Taylor and Francis, London, pp 21–34

Eastman JR (1985) Single pass measurement of the fractal dimension of digitised cartographic lines. Paper presented at the Annual Meeting of the Canadian Cartographic Association, New Brunswick

Goodchild MF, Mark DM (1987) The fractal nature of geographic phenomena. Ann Assoc Am Geogr 77:265–278

Jasinski MJ (1990) The comparison of complexity measures for cartographic lines. National Centre for Geographic Information and Analysis, Technical Paper 90–91

Mandelbrot B (1967) How long is the coast of Britain: statistical self-similarity and fractional dimension. Science 156:636–638

Mandelbrot B (1977) Fractals: form, change and dimension. Freeman, San Francisco

Muller JC (1986) Fractal dimension and inconsistencies in cartographic line representations. Cartogr J 23:123–130

Samet H (1984) The quadtree and related hierarchical data structures. Comput Surv 16:187–260

Schalkoff RG (1990) Digital image processing and computer vision. Wiley, London

Shelberg MC, Moellering H, Lam N (1982) Measuring the fractal dimensions of empiricial cartographic curves. In: Foreman J (ed) Proceedings of the Fifth International Symposium on Computer-Assisted Cartography. Washington DC, pp 481–490

# Towards the development of an intelligent spatial decision support system

**Yee Leung**

Department of Geography, The Chinese University of Hong Kong, Shatin NT, Hong Kong

**Abstract.** This contribution deals with the efficient and effective integration of databases, knowledge, intuition, inference, and graphics in spatial decision support systems (SDSS) through fuzzy logic and expert system technology. The purpose is to develop a powerful software environment for building SDSS for spatial decisionmaking with declarative (rule-based) and procedural (algorithms and models) knowledge. The fuzzy-logic-based expert system shell, the nerve center of the SDSS, is first discussed. The general architecture of the SDSS is then proposed and examined. It is apparent that knowledge-based SDSS is a must for a successful development of the field. Systems with little intelligence will not withstand the test of time.

## 1. Introduction

Regional classification, land use planning, urban and regional development, facility location, resource exploration and allocation, and environmental management are examples of spatial decisionmaking problems whose solutions are ordinarily derived through the processing and analysis of voluminous geo-referenced information by policymakers or experts equipped with domain specific knowledge and expertise. To have an effective and efficient solution to these problems, integrated decision support systems with database management, friendly user interface, and high level of intelligence are necessary.

Current spatial decision support systems (SDSS) are more or less information systems taking the form of geographic information systems (GIS) in general or land information systems (LIS) in particular. These systems are predominantly based on Boolean logic which gives no room for imprecision in information, human cognition, perception, and thought processes. The all-or-nothing system imposes artificial precision on intrinsically imprecise database, spatial phenomena and processes. It is inappropriate for informative and truthful data retrieval triggered by complex and imprecise queries. It fails to determine and communicate the extent of imprecision and error to users. Therefore, results of overlay unravel more questions about data quality and boundary mismatch than they solve (McAlpine and Cook, 1971; Burrough, 1986; and Leung, 1990a). Similar problems also exist in other data analysis and cartographic modeling exercises (Bouillie, 1982; Muller, 1987). It is thus imperative to have a more appropriate logical system for the design of SDSS. Such a system should be able to handle and to communicate precise and imprecise information. It should also be able to model natural language and human thought process. Since information and knowledge representations, inferences, input and output controls, system and user interfaces all involve the use of natural languages which are intrinsically imprecise, fuzzy logic is then instrumental in SDSS design.

Arising from the necessity in using the fuzzy set approach in spatial analysis and planning in general (Leung, 1984, 1985, 1987, 1988) and using fuzzy logic in GIS in particular (Leung, 1989a, b, c, d, e, f, 1990a, b), a handful of researchers have attempted to incorporate some limited sort of capabilities to handle imprecision of information and knowledge in GIS. The use of fuzzy sets in database design (Robinson and Strahler, 1984), representation of fuzzy spatial concepts, data, and relationships (Robinson, 1984; Robinove, 1986), query of spatial information (Robinson et al., 1985), and cartography (Bouillie, 1982; Muller, 1987) have been proposed. All these studies are however conceptual with no solid grounding of the mathematical and logical structures necessary for such a line of development. A powerful software environment for entertaining fuzzy reasoning in SDSS is still non-existent.

In addition to an inappropriate logic, human knowledge and expertise have not been effectively integrated into current GIS or SDSS. Low level of intelligence has thus been experienced in data input, retrieval, analysis, and output. Existing systems are severely limited in entertaining complex queries and in integrating rule-based and procedural knowledge (Clarke, 1990; Leung, 1990b; Openshaw, 1990; Fischer and Nijkamp, 1991; Goodchild, 1991). Artificial intelligence, particularly expert system technology, is thus instrumental in developing intelligent SDSS. Together with fuzzy logic, we would have a very useful tool for spatial decision analysis.

The objective of this chapter is to introduce in brief an ongoing research aiming at the development of a powerful software environment for efficient and effective decisionmaking using fuzzy logic, GIS, and expert system technologies. Though the ultimate goal is to develop a fully integrated SDSS within which declarative (rule-based) and procedural knowledge (e.g. algorithms, mathematical and statistical models) can be used at any time throughout the decisionmaking process, the following discussion places more emphasis on the development of an knowledge-based SDSS, an important step towards the construction of the fully integrated system.

In what follows, the architecture of the knowledge-based SDSS is first examined. A framework for the design of a fully integrated intelligent SDSS is then discussed.

## 2. System architecture of the knowledge-based SDSS

To account for imprecision in our reasoning process and decisionmaking environment, fuzzy logic is employed as the mathematical foundation in the design of the knowledge-based SDSS. The system is constructed in such a way that databases, knowledge, inferential mechanism, and graphics can be integrated into a unified whole. Natural language interface is employed for integration and user-friendliness design.

The overall system architecture comprises a fuzzy-logic-based expert system shell (with a maps display module), a fuzzy information retrieval module, and a database management system (Fig. 1). The system is developed in C under a DOS operating system. There is also a LISP version developed under VMS (Leung and Leung, 1990 a, b). Characteristics and functions of the above components are described in the following subsections.

### 2.1 The fuzzy-logic-based expert system shell

The shell is the key component of the overall architecture. It directs control flows and data flows. It is the development tool which facilitates the construction of rule-based SDSS with a high level of intelligence. Fuzzy logic is employed to handle approximate reasoning so that fuzzy and non-fuzzy terms can be employed to make inferences. Furthermore, uncertainties in rules and facts can be specified through certainty factors.

The shell consists of three basic subsystems (Fig. 2). They are the knowledge acquisition subsystem, the consultation driver, and the fuzzy knowledge base. The knowledge acquisition subsystem consists of management modules for objects,

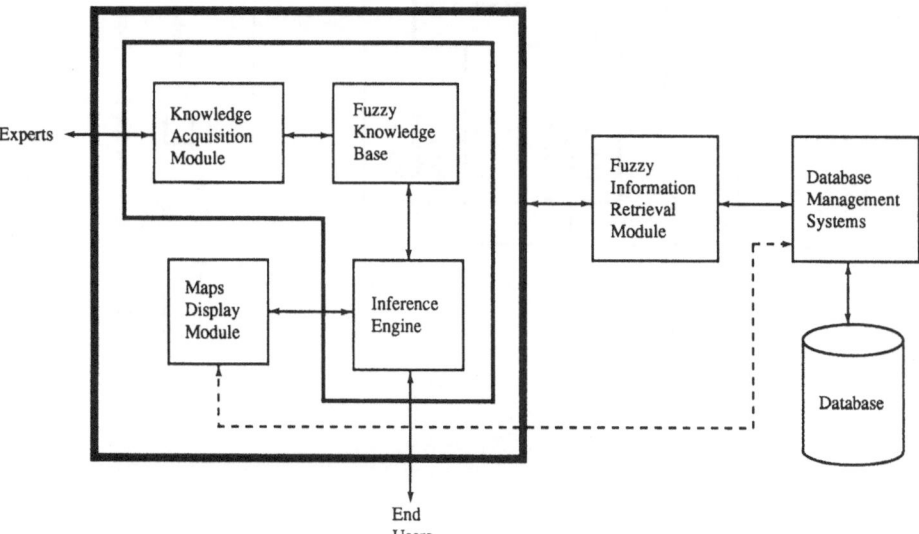

**Fig. 1.** Overall architecture of the knowledge-based spatial decision support system

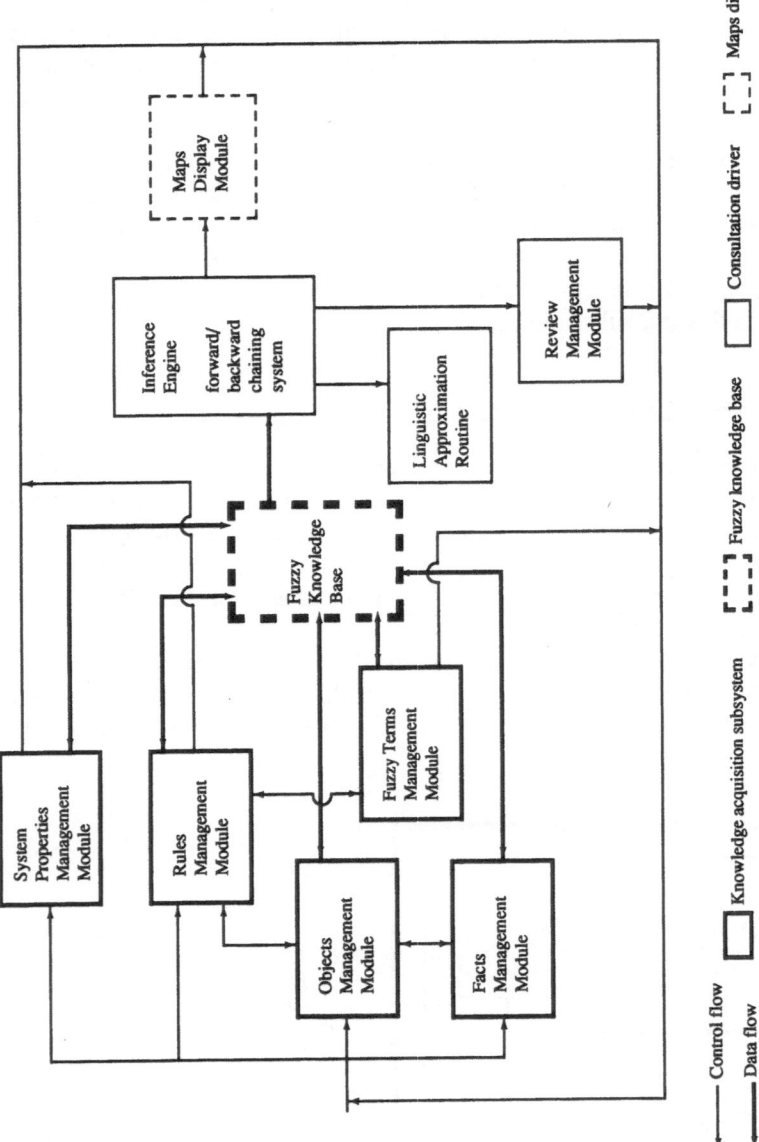

**Fig. 2.** Structure of the fuzzy-logic-based expert system shell (adopted from Leung and Leung, 1990a)

facts, fuzzy terms, rules, and system properties. These modules are responsible for acquiring and managing knowledge in rules and facts. The objects management module creates, modifies, or deletes objects in the system. An object is a basic entity in the system. For instance, "precipitation", "temperature", "slope", and "population density" are all objects which can be contained in any facts about a spatial phenomenon or process. Objects can take on six formats: numeric, binary, single-valued, multi-valued, fuzzy, and mathematical expression.

The fuzzy terms management module provides routines to define fuzzy subsets for corresponding fuzzy terms. They are generally stored as finite dimensional vectors with default values. For example, the fuzzy subset *hot* can be stored as a vector (0 0.125 0.250 0.375 0.547 0.625 0.828 0.875 1 1) whose values are determined with respect to points, such as 24, 25, 26, 27, 28, 29, 30, 31, 32, 33 (°C), in a temperature continuum. The determination can be based on subjective feeling or an objective function. To make it more flexible, mathematical functions defining fuzzy subsets can also be provided by users. The module supports operations such as CREATE, DELETE, VIEW, EDIT, and PRINT.

In the facts management module, there are two alternative ways to obtain facts. One method is to ask the user interactively for the facts, i.e. the values of the objects and the certainty about them. The other is to obtain the values directly from databases or predefined files. A fact is actually a data proposition of the following form:

⟨OBJECT⟩ is ⟨VALUE⟩     (fuzzy/non-fuzzy certainty factor)

For example

Temperature is 30 °C     $(CF = 0.9)$

and

Temperature is *very high* $(CF = approximately\ 0.9)$

are respectively non-fuzzy and fuzzy facts with non-fuzzy and fuzzy certainty factors, CF.

A rule in the rule management module is defined as an implication statement expressing the relationship between a set of antecedent propositions and a set of consequent propositions. Attached to each rule is a fuzzy/non-fuzzy certainty factor indicating the degree of confidence in the rule. The general structure of the rule is

```
(rule ⟨rule-name⟩
if ⟨object 1⟩⟨operator 1⟩⟨value 1⟩ and/or
   ⟨object 2⟩⟨operator 2⟩⟨value 2⟩ and/or
       ⋮
then ⟨object n⟩ is ⟨value n⟩)
certainty is ⟨certainty factor⟩ .
```

The operators can be ordinary inequalities ($>$, $<$, $=$, $\geq$, $\leq$) or fuzzy inequalities ($\gtrsim$, $\lesssim$, $\tilde{=}$, $\gtrapprox$, $\lessapprox$, where "$\sim$" means approximately). The certainty factor can be a precise value in $[-1, +1]$, a fuzzy number, or a linguistic probability (e.g. *highly probable*).

The system properties management module allows users to input or change system properties such as goal objects, inference mechanism, and output devices.

The fuzzy knowledge base stores all knowledge entities such as objects, facts, fuzzy terms, rules, and system properties. Knowledge entities are used by the inference engine to perform inference and consultation. They can easily be modified or updated.

The consultation driver contains three modules referred to as the inference engine, the linguistic approximation routine, and the review management module. The inference engine extracts knowledge from the fuzzy knowledge base and makes inferences via rules and facts. It supports forward (data-driven) and backward (goal-directed) reasonings. Evaluations of rules are based on fuzzy implications and inferences are based on fuzzy logic. A typical spatial inference involving fuzzy terms may be expressed in schematic form as follows:

Rule: If distance to city center ($X$) is *short* ($A$)
      then rent ($Y$) is *high* ($B$)                                   $(CF_1)$

Fact: Distance to city Center ($X$) of polygon $K$ is *very short* ($A_1$)       $(CF_2)$

---

Approximate conclusion: Rent ($Y$) in polygon $K$ is *very high* ($B_1$)        $(CF_3)$

Here, the rule and the fact contain linguistic variables "distance" ($X$) and "rent" ($Y$) whose values are fuzzy terms *short, high,* and *very short* which can be specified as fuzzy subsets $A$, $B$, and $A_1$ respectively (see Leung, 1982, 1985 for a discussion on fuzzy concepts in spatial analysis). Since *very short* is not equal to *short*, then by classical logic the rule cannot be fired and no conclusion can be derived. This is however counter-intuitive. By common sense, we would infer from the fact *very short* ($A_1$) an approximate conclusion *very high* ($B_1$) via the specified rule (see Leung 1989d for further discussion).

Based on fuzzy logic, the fuzzy term $B_1$ in the conclusion can in fact be derived through approximate reasoning by

$$B_1 = A_1 \circ (A \rightarrow B)$$

with certainty factor ($CF_3$) obtained by

$$CF_3 = CF_1 * CF_2 \ ,$$

where

   $A \rightarrow B$ is a fuzzy relation defining the fuzzy implication $\rightarrow$;
   $\circ$ is a compositional rule of inference; and
   $*$ is a fuzzy operation, e.g. fuzzy multiplication, on the certainty factors (fuzzy numbers) $CF_1$ and $CF_2$.

Let $\circ$ be the max-min ($\vee - \wedge$) composition, then the conclusion $B_1$ above can be obtained by

$$\mu_{B_1}(v) = \bigvee_u \{\mu_{A_1}(u) \wedge [\mu_A(u) \wedge \mu_B(v)]\} , \quad u \in \underline{U}, \, v \in \underline{V} .$$

Such an inferential schema supported by the inference engine makes decision analyses more natural and powerful.

The linguistic approximation routine maps fuzzy subsets (mathematical expressions) onto linguistic expressions (natural-language expressions) for any conclusions drawn. The review management module handles various reviews requested by end-users. They can ask "why" a fact is requested by the system and "how" a fact is established. They can also ask about changes in conclusions when certain facts are changed, i.e. "what if".

In addition to the three standard subsystems, a maps display module is also incorporated. It can be used as an independent component or an integral part of the shell (see discussions in 2.3).

## 2.2 The fuzzy information retrieval module and database management systems

To enhance the power of the knowledge-based SDSS, the expert system shell is integrated with database management systems (e.g. Rdb, GIS) through a fuzzy information retrieval module. A flexible interface is provided in the module so that changes of database management systems can be accomplished with minimal effort. Such an integration establishes a communication between expert systems and database management systems and enhances both of their capabilities as a result (Leung and Leung, 1990).

A fuzzy query (Leung, 1990a) in general can be expressed as

$$R_1 + \ldots + R_n + F_1 + \ldots + F_r + L_1 + \ldots + L_q .$$

The term $R_i$, $i = 1, \ldots, n$, is a relational formula taking the form of a relation $R_i(t_1, \ldots, t_m)$ with $t_j$, $j = 1, \ldots, m$, being a tagged term indicating the condition of the attribute $A_j$ which can take on the following forms:

(a) $A_j$ = variable name,
(b) $A_j$ = numerical or linguistic value, or
(c) $A_j \neq (\geq, \leq, >, <)$ a numerical value 0.

The term $F_k$, $k = 1, \ldots, r$, is a fuzzy formula (a formula with fuzzy predicates). The term $L_l$, $l = 1, \ldots, q$, is a precise formula (a formula with precise predicates). The symbol $+$ means the union.

The fuzzy information retrieval module supports a simple fuzzy query language employed by the expert system shell for retrieving precise and imprecise information from the database in GIS. The syntax of the fuzzy query language used by the expert system is defined in extended BNF Grammar as follows:

| QUERY | ::= (⟨relation_name⟩ SYMBOL_LIST NON_FUZZY_CONDITION FUZZY_CONDITION) |
|---|---|
| SYMBOL_LIST | ::= STAT_EXP \| (⟨field_name⟩ ⟨field_name⟩*) |
| NON_FUZZY_CONDITION | ::= NON_FUZZY_CON \| nil |
| FUZZY_CONDITION | ::= FUZZY_CON \| nil |
| STAT_EXP | ::= (STAT_OP NON_FUZZY_CONDITION) |
| NON_FUZZY_CON | ::= (OPERATOR TERM TERM*) |
| FUZZY_CON | ::= (FUZ_OPERATOR FUZZY_EXP FUZZY_EXP*) |
| STAT_OP | ::= max ⟨field_name⟩ \| min ⟨field_name⟩ \| total ⟨field_name⟩ \| average ⟨field_name⟩ \| count |
| TERM | ::= NON_FUZZY_CON \| (COMPARATOR ⟨field_name⟩⟨value⟩) \| (COMPARATOR ⟨field_name⟩ STAT_EXP) |
| FUZZY_EXP | ::= FUZZY_CON \| FUZZY_TERM |
| FUZZY_TERM | ::= (⟨field_name⟩ ⟨membership_distribution_table⟩ ⟨weight⟩) |
| OPERATOR | ::= and \| or |
| FUZ_OPERATOR | ::= and \| or \| comb \| poll |
| COMPARATOR | ::= ⟨\|⟩\| = \|⟨⟩\|⟩ = \|⟨ = |

where

⟨field name⟩ is the name of a field in the database;
⟨relation name⟩ is the name of a relation in the database;
⟨value⟩ is a value of the type corresponding to a field;
⟨member_distribution_table⟩ contains two lists. The first list contains domain values for a field in ascending order. The second list contains the corresponding degrees of memberschip;
⟨weight⟩ is a weighting factor (between 0.0 and 1.0) by which the degree of membership of the fuzzy term will be multiplied.

In the example of climatic classification, the query
    "List the precipitation and temperature of
    all pixels with adequate precipitation and
    a temperature greater than or equal to 23 °C"

can be expressed as
    PIXEL(Name = $x$, Precipitation = $u$, Temperature = $v$) +
    ADEQUATE($x$) + $v \geq 23$ °C

where PIXEL(Name = $x$, Precipitation = $u$, Temperature = $v$) is a relational formula; ADEQUATE($x$) is a fuzzy formula; and $v \geq 23\,°C$ is a precise formula.

Under this situation, the system will generate a query
(Pixel_rec(name precipitation temperature)
(and (temperature $\geq 23\,°C$))
(and (precipitation(adequate)))

and find all pixels which satisfy the relational formula: PIXEL(Name = $x$, Precipitation = $u$, Temperature $= v$) from the frame

| PIXEL ‖ Name | Precipitation | Temperature |
|---|---|---|

.

It then finds all pixels whose temperature is above $23\,°C$ and then searches and arranges in order all these pixels from the frame

| ADEQUATE ‖ Precipitation | $\mu$ |
|---|---|

by their degree of belonging $\mu$, to *adequate*.

The fuzzy query language can be used independently. All queries however can also be generated by the expert system without requiring users to know the query language.

Thus, the fuzzy information retrieval module, together with a database management system, can be used as an independent module or an integral part of the expert system shell. That is, any integrated database management system is permitted to function on its own or some of its operations can be downloaded from the expert system through the inference engine.

## 2.3 The maps display module

Though conventional GIS has rather powerful graphic capabilities, they are however not suitable for depicting imprecision. To display inferential results or queries based on fuzzy logic, a module for displaying maps with gradual transitions, intermediate areas, and fuzzy boundaries is developed as a component of the overall system. Methods of dithering (Leung and Leung, 1990 a, c) and bit-mapping (Leung and Leung, 1990 c) have been implemented.

In dithering, colors are mixed in a single resolution unit (data point) consisting of a fixed number of pixels. It makes use of the spatial integration that our eyes perform. A gradual change of color can be produced by changing the proportions of different colors in a data point. By assigning a color to a classified region, the intensity of a color in a data point can then be used to display the gradation of belongingness within a single region. Similarly, various mixtures of colors can be used to display multiple-region belongingness, gradual transition and fuzzy boundary between regions.

In bit-mapping, the R-G-B color model is employed to depict fuzziness. Generally, the value corresponding to a pixel can be divided into three parts with each part governing respectively the intensity of the red, green, and blue components. The mixing of the red, green, and blue components results in a color on the screen. To show fuzziness on the screen, the three components of a pixel can be adjusted to form a continuous color sequence with respect to the pixel's degrees of belonging to various classes.

While both methods have their merits and shortcomings, other computer graphic techniques are being considered and devised. Display of imprecision is a unique feature in our system which current GIS fails to accomplish. Under our graphic scheme, display of precise spatial structures and processes becomes a special case.

## 3. A framework for the development of an integrated intelligent spatial decision support system

The integrated knowledge-based SDSS discussed above is very flexible and is of a much higher level of intelligence than conventional GIS or SDSS. Its rule-based inference enables users to incorporate their knowledge and expertise into the SDSS environment. The system has successfully been employed to build spatial expert systems with raster database (Leung and Leung, 1990b). Expert systems with vector database will be developed when the data become available.

Large-scale geographical problems are nevertheless highly complex. Effective solutions of these problems require an intelligent use of large databases, structured and unstructured knowledge. Over the years, structured (procedural) knowledge taking the format of mathematical and statistical models have been developed to solve a large variety of spatial problems. Though they are formal and precise, practitioners often find them to be too mechanical, too inflexible, and more or less unaligned with human problem-solving practices which are based quite often on IF-THEN rules supported by the rule-based system discussed above. Solving problems purely by rules, on the other hand, tends to be too loose and unstructured. To be effective, solutions of complicated problems require the employment of procedural and rule-based knowledge simultaneously or interchangably. The former enables us to be structured and logically strict while the latter permits us to exercise our expertise, intuition, and values.

To make decision analysis with GIS even more powerful, we are in the process of developing a tool for building integrated intelligent SDSS within which structured and unstructured knowledge can be employed to solve complicated spatial problems under certainty and uncertainty. Advanced artificial intelligence techniques will be employed to build a system whose overall architecture is conceptualized in Fig. 3.

The core of the system is the fuzzy-logic expert system shell (FLESS) discussed in the previous section. It directs control flows and information flows of the whole system. Its knowledge-base can contain domain specific knowledge acquired from experts. It also contains meta-knowledge for inference control, systems and user interface, and external-routine call.

In general, FLESS has an interface with external DBMS which can be any GIS, Rdb, and remotely sensed information systems. The communication be-

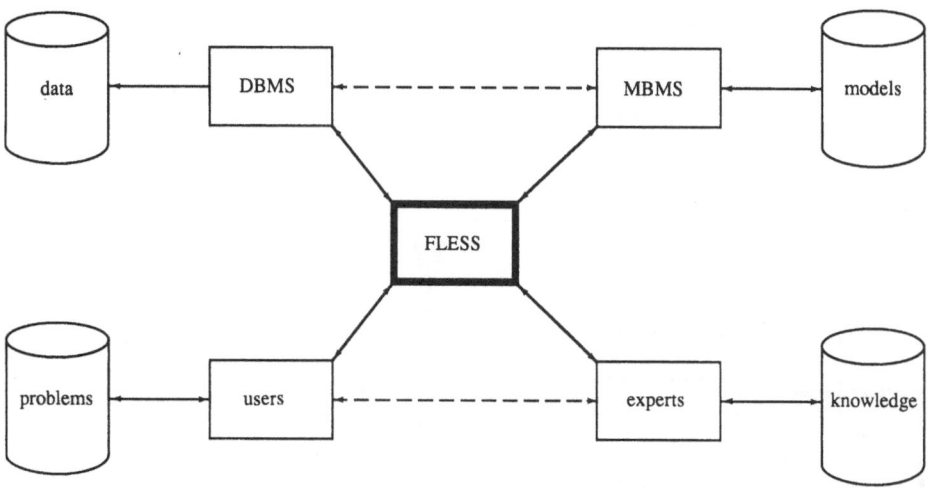

**Fig. 3.** Conceptual design of the integrated intelligent spatial decision support system

tween FLESS and the DBMS is carried out by modules such as the fuzzy information retrieval module previously discussed.

To facilitate communication with external routines such as algorithms, statistical procedures and mathematical models, an interface with a model base management system (MBMS) will be incorporated in FLESS. Parallel to DBMS, MBMS
organizes procedural models into an easy-to-use structure. Calls to MBMS are in-
voked by meta-rules in FLESS. More work needs to be done in this area of
research.

In addition to linkages to DBMS and MBMS, friendly user interface and
knowledge acquisition modules are essential parts of FLESS for man-machine interaction. There are rooms for improvement in their designs and a thorough
discussion is beyond the scope of this contribution.

There are many ways to implement the conceptual framework depicted in
Fig. 3. To effectively use fuzzy and non-fuzzy declarative and procedural knowledge (algorithms, statistical procedures, and mathematical models written in
other languages) in a single system, an attempt to use object-oriented approach
has been made. While there are a lot more works left to be done in design and
implementation, a prototype has been constructed (see Fig. 4) and applied successfully to solve aspatial fuzzy and non-fuzzy optimization problems (Leung et
al., 1989).

The expert system shell contains many objects. All the external interfaces, inference and control mechanisms are implemented by methods in system objects.
The inference engine is used to support the system methods for rule-based inference. The engine can handle any mix of precise and fuzzy reasoning. All system
methods are grouped into the system class "ROOT". Knowledge engineers can use
this class without any declaration.

A knowledge base of an expert system built from the shell is also defined by
a collection of classes, objects and methods. If a user-defined class is declared as

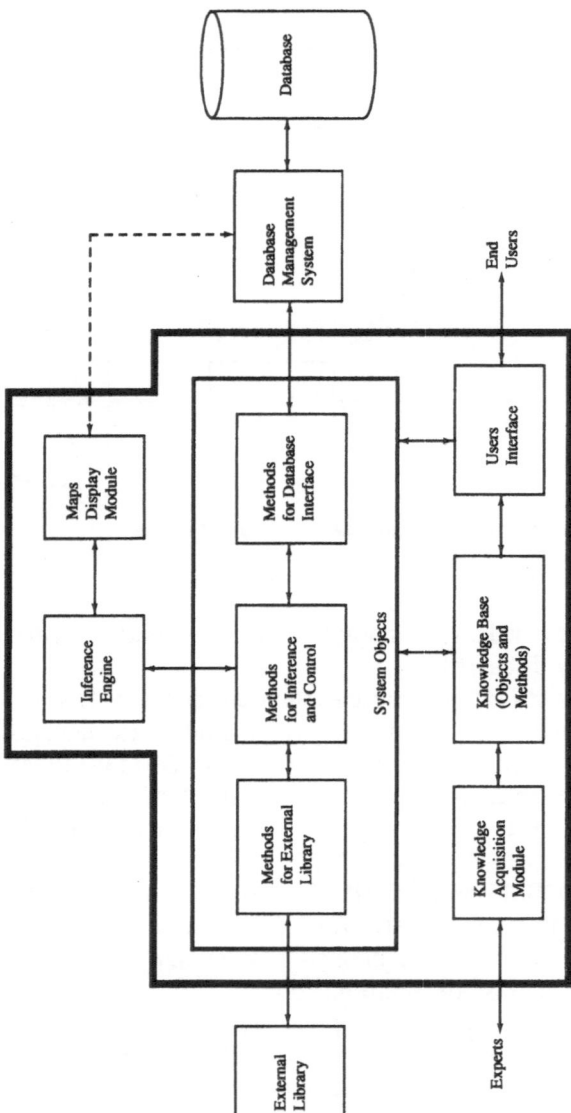

**Fig. 4.** Overall architecture of the object-oriented spatial decision support system

a child of the root class, it can inherit all the capabilities of the system methods. Thus each user-defined object may be treated as an independent entity which inherits inference capabilities from the system class "ROOT".

Rules and procedures are frequently used independently to handle various classic computing and artificial intelligence problems. However, in our expert system shell, these techniques are represented as attributes or methods of objects, and can be intermixed in any proportion to model and solve complex problems. The top level control of an expert system built from the shell is either driven by rules or procedures, though they can be freely mixed at lower levels. The former is similar to a goal-directed system while the latter is procedural driven. A knowledge engineer can model a problem according to his or her method of thinking.

The controls of the system can be classified into intra-objects and inter-objects controls. Intra-objects control is regarded as the self coordination of an object. It is governed by the methods of the object. In other words, intra-objects control depends on the domain knowledge provided by a knowledge engineer. It determines the micro-behavior of the system. On the other hand, inter-object control is employed to coordinate the interactions among objects by inter-object communication. Message passing is the only means for the communication between objects. The macro-behavior of the system is determined by the mechanism of message passing.

End-users can communicate with any object defined by the system or knowledge engineers through the user interface. An end-user usually consults an expert system through a form of conversation. The user interface acts as a bridge between the objects in the system and the users. The user's input is translated by the user interface into messages for invoking appropriate objects in the system. When messages are sent to the user interface, they will be decoded and the resulting information will be shown on the screen.

The knowledge acquisition module is used to capture domain knowledge from knowledge engineers and transform it into internal format and store in a knowledge base. To cater for the need of different levels of users, two modes of inputs are supported. One of the modes is to support knowledge engineers to input knowledge through conventional editors. This mode of input provides a quick means of knowledge input for knowledge engineers who are familiar with the system and knowledge representation. The other mode which supports knowledge input through interrogations is for knowledge engineers or domain experts who do not have a clear concept of knowledge representation.

The development environment of the system is composed of VAX-LISP version 2.1, VAX-Rdb version 2.1 and VAX-PASCAL version 3.4 running under VMS version 4.5 on a VAX-11/780 computer.

## 4. Conclusion

Knowledge-based information systems will play an important role in decision analysis in the future. SDSS without intelligence have very little chance to meet challenges commonly encountered in the highly complex and imprecise decision-making environment. The on-going research described in the present contributon is designed for the purpose of developing powerful tools for spatial decision and policy analyses. Our ability to efficiently and effectively integrate databases, knowledge (declarative and procedural), expertise, intuition, inference, and graphics into an unified system is crucial to human and physical systems engineering. In the years to come, higher level of intelligence through man-machine interaction will be experienced in decisionmakings with integrated databases involving GIS, remote sensing systems, and multimedia systems. Management of uncertainty by probability, fuzzy logic, and/or other non-standard logics will be an essential requirement of all SDSS.

## References

Boullie F (1982) Actual Tools for Cartography Today. Cartographica 19:27–32

Burrough PA (1986) Principles of Geographical Information Systems for Land Resources Assessment, Oxford University Press, Oxford

Clarke M (1990) Geographical Information Systems and Model Based Analysis: Towards Effective Decision Support Systems. In: Scholten MJ, Stillwell JCM (eds) Geographical Information Systems for Urban and Regional Planning. Kluwer, Dordrecht, pp 165–175

Fischer MM, Nijkamp P (1991) Geographic Information Systems and Spatial Analysis, Discussion Paper 14, Vienna, Department of Economic and Social Geography, Vienna University of Economics and Business Administration

Goodchild MF (1991) Progress on the GIS Research Agenda. In: Harts J, Ottens HFL, Scholten HJ (eds) EGIS '91, Proceedings of the Second European Conference of Geographical Information Systems, Vol 1, EGIS Foundation, Utrecht, pp 342–350

Leung Y (1982) Approximate Characterization of Some Fundamental Concepts of Spatial Analysis. Geogr Anal 14:29–40

Leung Y (1984) Towards a Flexible Framework for Regionalization. Environ Plann 16:1613–1632

Leung Y (1985) A Linguistically-based Regional Classification System. In: Nijkamp P, Leitner H, Wrigley N (eds) Measuring the Unmeasurable. Martinus Nijhoff, Dordrecht, pp 451–486

Leung Y (1987) On the Imprecision of Boundaries, Geographical Analysis 19:125–151

Leung Y (1988) Spatial Analysis and Planning under Imprecision. North-Holland, Amsterdam

Leung Y (1989a) Fuzzy Logic and Knowledge-based Geographic Information Systems: 1, A Critique and a Proposal for Intelligent Geographic Information Systems, Occasional Paper No 99, Hong Kong, Department of Geography, The Chinese University of Hong Kong

Leung Y (1989b) Fuzzy Logic and Knowledge-based Geographic Information Systems: 2, Knowledge Representation and Measure of Confidence, Occasional Paper No 100. Hong Kong, Department of Geography, The Chinese University of Hong Kong

Leung Y (1989c) Fuzzy Logic and Knowledge-based Geographic Information Systems: 3, Translation Rules and Truth-values of Fuzzy Proposition, Occasional Paper No 101, Hong Kong, Department of Geography, The Chinese University of Hong Kong

Leung Y (1989d) Fuzzy Logic and Knowledge-based Geographic Information Systems: 4, Rules of Inference, Occasional Paper No 102. Hong Kong, Department of Geography, The Chinese University of Hong Kong

Leung Y (1989e) Fuzzy Logic and Knowledge-based Geographic Information Systems: 5, Acquisition of Knowledge, Occasional Paper No 103. Hong Kong, Department of Geography, The Chinese University of Hong Kong

Leung Y (1989f) Fuzzy Logic and Knowledge-based Geographic Information Systems, A Prospectus, Proceedings of IGARSS '89 12th Canadian Symposium on Remote Sensing, Vol 1, Canada, IGARSS. (See also a revised version: Leung Y (1990a) A Prospectus on Fuzzy Logic and Knowledge-based Geographical Information Systems, Asian Geographer 9:1–9)

Leung Y (1990b) The Development of an Integrated Knowledge-based LIS/GIS, Proceedings of the Annual Conference of the Australasian Urban and Regional Information Systems Association, Canberra, Australia, pp 125–132

Leung Y, Leung KS (1990a) An Intelligent Expert System Shell for Knowledge-based Geographic Information Systems: 1, The Tools (unpublished paper)

Leung Y, Leung KS (1990b) An Intelligent Expert System Shell for Knowledge-based Geographic Information Systems: 2, Some Applications (unpublished paper)

Leung Y, Leung KS (1990c) Analysis and Display of Imprecision in Raster-Base Information Systems (unpublished paper)

Leung KS, Leung Y, Wong MH (1989) The Integration of Rule-based and Procedural Methods to Solve Optimization Problems Through Expert-System Technology. In: Verdegay JL, Delgado M (eds) The Interface Between Artificial Intelligence and Operations Research in Fuzzy Environment. Verlag TÜV Rheinland, Berlin, pp 157–176

McAlpine JR, Cook BG (1971) Data Reliability From Map Overlay, Proceedings of the Australian and New Zealand Association for the Advancement of Science, 43rd Congress

Muller JC (1987) The Concept of Error in Cartography. Cartographica 24:1–15

Openshaw S (1990) Spatial Analysis and Geographical Information Systems: A Review of Progress and Possibilities. In: Scholten MJ, Stillwell JCM (eds) Geographical Information Systems for Urban and Regional Planning. Kluwer, Dordrecht, pp 153–163

Robinove CJ (1986) Principles of Logic and the Use of Digital Geographic Information Systems, U.S. Geological Survey Circular 977

Robinson VB (1984) Modelling Inexactness in Spatial Information Systems, Proceedings of the Pittsburgh Modeling and Simulation Conference, pp 157–161

Robinson VB, Strahler AH (1984) Issues in Designing Geographic Information Systems Under Conditions of Inexactness, Proceedings of the 10th International Symposium Machine Processing of Remotely Sensed Data, pp 198–204

Robinson VB, Thongs D, Blaze M (1985) Machine Acquisition and Representation of Natural Language Concepts for Geographic Information Retrieval, Proceedings of the Pittsburgh Modeling and Simulation Conference, pp 161–166

# Modelling spatial interaction using a neural net

**Stan Openshaw**

School of Geography, Leeds University, Leeds LS2 9Jt, UK

**Abstract.** Neurocomputing has the potential to revolutionise many areas of urban and regional modelling by providing a general purpose systems modelling tool in applications where data exist. This chapter examines the empirical performance of a feedforward neural net as the basis for representing the spatial interaction contained within journey to work data. The performance of the neural net representation is compared with various types of conventional model. It is concluded that there is considerable potential for many more neural net applications in this and related areas.

## 1. Introduction

In recent years there have been major developments in applicable artificial intelligence (AI) technologies that are seemingly very relevant to many areas of geography. Yet geographers have been slow to appreciate both the immediate potential and the likely impact of these new developments on their areas of interest. There are probably several reasons for this. One might be the change of emphasis in geographic research in the 1980's towards non-quantitative methodologies preceeded by a collapse of confidence in the ability of quantitative methods to handle the relevant problems of geography. In addition, there is no strongly computationally orientated research tradition in geography and the term "computational geography" is seldom, as yet, if ever, used. Yet this focus on computationally intensive methods is a key feature of many AI developments. Instead, many geographers continue the traditional emphasis on classical statistical methods that often cannot cope with the complexities of spatial data analysis particularly in a GIS context, with a preference for inferential approaches that are at best over-emphasised and at worst highly misleading. Maybe this reflects the strength of the normal science paradigm and a marked reluctance to embrace strongly inductive and exploratory descriptive approaches; see Openshaw (1990). Whatever the reasons, conventional geographical methodologies effectively preclude many artificial intelligence procedures which are largely based on induction and render invalid the general concept that new knowledge may be created from information by computing machines. Alternatively, it may be that the relevant developments have simply not yet permeated into the geographer's consciousness. Openshaw (1991) provides a general review of some of the opportunities for AI in spatial analysis and modelling. Here attention is focused on one particular form of AI involving the application of what is often termed neurocomputing.

Neurocomputing is a highly contagious subject that has captivated the interest of many scientists and mathematicians. It has been in existence for almost 50 years but it is only since the late 1980's that artificial neural nets (ANNs) have been sufficiently well developed to provide the basis for a practical and general purpose modelling technology with a very wide portfolio of potential applications. Geographers have been slower to appreciate the seemingly wide applicability of ANNs than many other disciplines, yet there are many suitable modelling challenges in geography to which ANNs could be applied; for instance, spatial modelling with noisy and spatially autocorrelated variables, discrete choice modelling, pattern detection, various qualitative decision making models, and space-time series forecasting. This chapter focuses on only one of the possible major application areas for ANNs in quantitative geography which involves the modelling of spatial interaction data.

The modelling of spatial interaction data is important. It is simultaneously one of the most basic and most explicitly geographical of all modelling challenges. What could possibly be more geographical than trying to develop good theoretical, mathematical, and computational representations of the spatial and temporal (when we know how to do it) flows of people, artifacts, etc. amongst and between geographic areas. Recognition of the importance of this modelling task is not new and already there has been an extensive amount of research devoted to this topic. The justification for re-opening this area of modelling as a basis for a preliminary review of what artificial neural nets would seem to offer is as follows:

(1) the modelling of spatial interaction data possesses both theoretical and applied relevancy;
(2) spatial interaction modelling has a special significance in the historical development of mathematical and computer models in geography as the testing ground for new approaches;
(3) performance benchmarks provided by standard spatial interaction models can be used to evaluate the relative efficiency of ANN-based approaches;
(4) it presents a good example of a complex modelling problem that whilst well researched is not yet completely understood;
(5) it constitutes a good test in one of the areas of geography in which neural nets might be used in the future, offering a demonstration of the nature, the strengths, and the weaknesses of neural computing; and
(6) it may help identify some of the fundamental changes in geographic philosophy, in paradigm, and in perspective that could be needed if the ANN type of modelling is to gain a foot-hold in a human geography that is riddled with self-induced and artificial philosophical dellusions about geography.

The paper is in three parts. Section 2 provides a brief introduction to artificial neural nets. Section 3 explores their utility as a basis for modelling spatial interaction data. Section 4 discusses some of the more general aspects of neural nets in a quantitative geographic context. Finally, section 5 offers some thoughts about the future.

## 2. A brief introduction to neural nets

### 2.1 History

McCulloch and Pitts (1943) are usually credited with publishing the first systematic study of what is now known as an artificial neural network. Many other researchers contributed to these developments in the 1940's and early 1950's; for example, von Neumann (1951) made suggestions that the development of brain inspired computers might be interesting. Much of this early work involved the use of very simple neurons; although it was always intended that they would be built into more complex networks. The first neurocomputer (the Snark) was created in 1951 and the first "successful" neurocomputer (the Mark I Perceptron) in 1957 by Rosenblatt and others (see Rosenblatt, 1958). However, these were essential toys developed to demonstrate with very rudimentary computing equipment that electronic brains might well work.

Rosenblatt (1961) showed that a neural net called a perceptron could learn anything it could represent. However, the single layer perceptron was later demonstrated to have severe restrictions on what it could represent. Nevertheless, in the 1960's these perceptrons created considerable interest and resulted in widely publicised predictions that artificial brains were just a few years away. This period of the first successes of neural computing was a time of immense enthusiasm and excessive hype, characterised by a general lack of rigor and increasingly, intellectual exhaustion. According to Hecht-Nielsen (1990) by the mid 1960's it was clear that neurocomputing's era of first successes was drawing to a close. The final episode was a campaign designed to discredit neural network research to the benefit of expert systems approaches. Minsky and Papert (1969) are sometimes regarded as the villains. They proved mathematically that a perceptron could not implement the EXCLUSIVE OR (XOR) logical function. They then used this trivial fact to give the impression that all neural network research had been proven in a rigorous mathematical manner to be a dead-end. As a result the period 1967 – 1982 witnessed little explicit neurocomputing research although it seems that much of the basic foundations were established during this time usually, albeit under other headings.

These "quiet years" were followed in the 1980's by a gradual renewed interest in the subject leading to an explosion of activity from about 1987 onwards. Hopfield (1982, 1984) is often credited with revitalising the field and Rumelhart and McCelland (1986) in popularising its appeal. The Minsky and Papert criticisms were overcome by developing multi-layered perceptrons with non-linear transfer functions that could be trained. The invention of the backpropagation training algorithm was extremely important in establishing the basis for a viable technology. Rumelhart et al. (1986) describe a practical solution although essentially the same technique had existed for a number of years but was not widely used previous to 1986. Nevertheless, this was a most important development and made possible the subsequent surge of interest in this type of neurocomputing. It allows any arbitrarily connected neural network to develop an internal structure that is appropriate for a given task.

The late 1980's will go down in history as the moment in time when this area of study termed variously neuroscience, connectionism, artificial neural networks, neural computing, neurocomputing, artificial neural nets, and parallel distributed processing, has become firmly established and very popular in a wide range of disciplines with both research and commercial applications. The appearance of procedures that make machines learn and remember in ways that bear some resemblance to human mental processes makes it appear possible to actually apply computation to virtually any problem, including those previously restricted to human intelligence. The underlying ideas were and are extremely seductive. As Hecht-Nielson (1990, p. 111) puts it 'The idea of training a system to carry out an information processing function (instead of programming it) has intrinsic appeal, perhaps because of our personal familiarity with training as an easy and natural way to acquire new information processing capabilities. Neurocomputing systems are often endowed with a "look and feel" vaguely reminiscent of animals. Like a pet, this is a technology that is easy to fall in love with.

## 2.2 What are neural nets?

Artificial neural networks are based on an analogy to the supposed workings of the brain that was biologically inspired. They are composed of elements that perform in a manner analogous to the most simple type of brain neuron and they are organised in a way that appears to be related to the anatomy of the brain. Quite simply, the brain is thought to be a large network of highly interconnected neurons, with each neuron being a very basic summation device. However, neither the brain's neuron nor its structure is sufficiently well understood to fully define its functioning. As a result artificial neural nets are based on guesses as to the transfer functions involved and the nature of the adaptive system equations that seem to be used by the brain, sometimes provoking conflict with biologists because of their lack of biological justification. Also the human brain's neurons are thought to be imprecise and slow; with a computing speed of a few milliseconds compared with picoseconds for computers. They compensate by massive parallelism and highly interlinked neurons. It would appear that the brain can solve immensely complex problems in about 100 time steps whilst the most complex AI programs still require millions to solve far simpler problems. Yet despite a superficial resemblance, artificial neural nets exhibit a surprising number of the brain's characteristics. Their most distinguishing feature is their ability to spontaneously learn a desired function from training examples; that is they have a capability to program themselves. They learn from experience, they can generalise from previous examples to new ones, and they can abstract essential characteristics from inputs containing noisy or irrelevant data.

In principle, an artificial neural net can be used to represent and model almost any complex function or system; no matter how dynamic or difficult the task may appear to more conventional approaches. They can cope with fuzzy "qualitative reasoning" by manipulating neural encoding of symbols and also approximate real valued functions. In some ways they are the ultimate in currently available black-box modelling technology but they have to be taught to recognise patterns and analyse data. However, it should also be understood that these artificial neural nets display only minimal levels of real intelligence. They are limited at present by hardware aspects, by lack of knowledge about how the brain really works, and can only be built on a relatively small scale. Yet it would be wrong to over-emphasise these limitations, because the performance of current neural net technology is already adequate for many applications; particularly, in areas with hard problems where the need for brain-like performance is most clearly manifest. Indeed, it is apparent that they may already provide a viable assumption-free alternative to many statistical methods.

Wasserman (1989) summarises the current situation as follows: "We are presented with a field having demonstrated performance, unique potential, many limitations, and a host of unanswered questions. It is a situation calling for optimism tempered with caution" (p. 7). Despite this caution, it is clear that this technology is extremely relevant to many areas of quantitative geography offering a potential for both developing new approaches for dealing with both traditional modelling and quantitative analysis problems and also offering a potentially new strategy for dealing with many other soft and hard problems, previously not amenable to computer modelling.

## 2.3 Basic types and features of artificial neural nets

There are many different types of neural net most of which are variations on the same general neural network model. There are several basic net architectures: non-recurrent with a feedforward process and recurrent nets which have feedback loops. The former has no memory and its representational powers are contained in its weights. The latter is far more dynamic, less stable, but probably capable of a wider range of representational behaviour. Some nets need to be trained to reproduce known results, others will self-organise to capture the structure of data. Table 1 attempts a simple typology of net types; the task is complicated because

**Table 1.** Some neural net variations

---

(1) *Non-recurrent nets*
   single layer perceptrons
   multilayer feedforward perceptrons
   counterpropagation nets
   adaptive resonance theory

(2) *Recurrent or autoassociative nets*
   Hopfield nets
   bidirectional associative memories

---

there are many variations of each type; for example, there are now at least 19 different types of unsupervised learning nets. Introductory texts by Khanna (1990), Wasserman (1989), and Aleksander and Morton (1990) provide further details. The situation is also somewhat confusing because there is no standardisation of terminology with different authors describing the same net in a totally different way.

Here attention is focused on only one particular network type; a generalised semilinear feedforward multilayer perceptron. This net has to be trained in a supervised manner. It is relevant to spatial interaction modelling because it provides the basis for the approximation of virtually any mathematical function or mapping which can be deduced from examples of the mapping's action as seen in observed data. It is essentially what some term a mapping neural network. Hecht-Nielsen (1990, p. 111) summarises this function as follows:

'The problem addressed by mapping neural networks is the approximate implementation of a bounded mapping or function

$$f: A \subset R^n \rightarrow R^m$$

from a bounded subset $A$ of a $n$-dimensional Euclidean space to a bounded subset $f[A]$ of $m$-dimensional space, by means of training on examples $(x_1, y_1)$, $(x_2, y_2)$, $\ldots$, $(x_k, y_k)$, $\ldots$ of the mapping's action, where $y_k = f(x_k)$.

In other words, a mapping net provides a means of function approximation. This is similar to a regression model except that the available functional forms can be infinitely more general and thus probably far more able to represent complicated, nonlinear functions, and discontinuous functions that seem to characterise the real world.

In terms of spatial interaction modelling, a neural net approach brings with it the promise of being able to offer a new view on the nature of the spatial interaction functions that can be deduced from instances of interaction data (rather than theory) as being most relevant. Whether it is possible in this way to rediscover the classical gravitational form of the traditional spatial interaction model is a matter of some interest. More pressing, however, is the need to demonstrate whether or not an ANN can actually represent spatial interaction data.

### 2.4 A multilayered feedforward semilinear perceptron as an example of a backpropagation network

Fig. 1 gives a representation of one important variant of a backpropagation net of the general multilayered feedforward semilinear perceptron type. It shows a basic three layer neural net of the form widely used for modelling applications. There are a series of inputs (i.e., $n$-variables for a training case) which are connected to a series of $n$ fanout neurons which distribute the data to all neurons in a second, hidden, layer. These hidden layer neurons are connected in turn to an output neuron; or to other hidden layers and then to an output neuron. The number of neurons in each hidden layer can vary. Each neuron forms a weighted sum of the inputs from previous layers to which it is connected, adds a threshold value, and applies a nonlinear squashing function to create its output value. This output

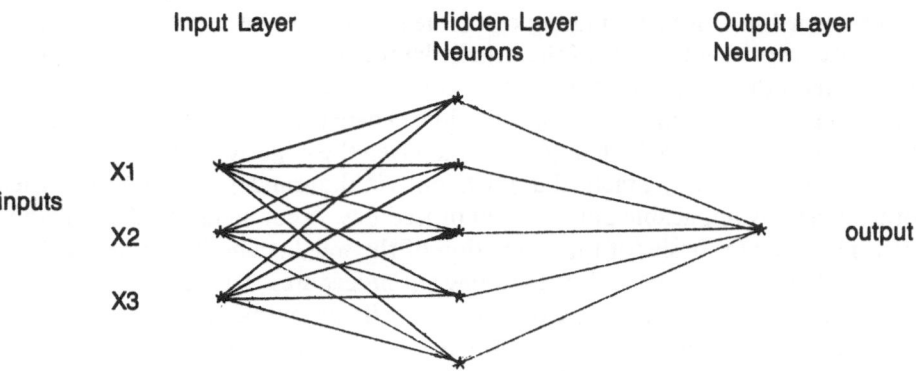

Fig. 1. A simple net with one hidden layer

serves as one input to the next layer to which the neuron is connected, and the process repeated. There are weights associated with each connection which have to be estimated during training and it is by adjusting the weights rather than the pattern of connections that allows the net to learn any relevant functional mapping. Also, each neuron has a sigmoidal activation function which squashes the output; between 0.0 and 1.0 if a logistic function is used. Bipolar functions mapped onto the range $-1$ to $+1$ can also be used. This is important because without this nonlinearity there would be no benefit in having multiple layers since the output from the net would be a simple linear function of the inputs.

In a multilayer net the input to any neuron in layer $k$, $(k)netj$, is the sum of the outputs it receives from all the neurons to which it is connected in the preceding layer

$$(k)inj = \sum_i wji \ (k-1)neti$$

where $wji$ is the weight put on the connection between neuron $j$ in layer $k$ and input neuron $i$ in layer $k-1$. This value is then squashed onto the range 0 to 1 by applying a logistic function, usually with a threshold or bias ($bkj$) so that the inputs do not have to be scaled. The output from neuron $j$ in layer $k$ which is sent as inputs from this neuron to all the neurons in the next layer, is a nonlinear function of the total inputs

$$(k)netj = 1/(1+\exp(-(k)inj+bkj))$$

This process is repeated moving forward from the inputs towards the output layer.

In order to use this net it is necessary to estimate values for the weights and bias parameters so that the net produces the desired outputs. The generalised delta rule formulated by Rumelhart et al. (1986) achieves this objective by regarding training as a nonlinear least squares optimisation problem. They use a steepest descent algorithm to estimate weights that minimise the sum of squares of the errors between the net outputs and known values in the training data set. This requires the calculation of partial derivatives and it is during their computation that back-propagation is used. One major problem here is that the number of iterations needed to complete training of the neural net can be extremely large. This is hardly surprising because of the usually large numbers of weights (or unknown parameters) that have to be estimated. Other problems concern the inefficiency of a steepest descent optimisation procedure. There are many nonlinear optimisation methods which are in theory better but given the large numbers of parameters to be estimated in any practical application (viz thousands) there are only a few other generally applicable optimisation procedures. A conjugate gradient method offers a far better basis for the estimation of the weights and it is somewhat surprising that anyone ever used steepest descent procedures. The conjugate gradient search procedure used here is from the Harwell subroutine library. As is usual, random values are used as initial estimates of the parameters.

Another noteworthy aspects concern the flexibility of the technology. It can handle real value inputs by scaling to a range of either 0 to 1 or $-1$ to $+1$; the output would then be rescaled. It can also cope with integer, qualitative valued data, and bit patterns; for both input and output and it can produce multiple output variables. As a result many problems which are of a non-quantitative nature can be successfully coded for a neural net; for example, the prediction of bit patterns that represent the coding of arbitrary classification codes. Despite their apparent simplicity, feedforward neural nets of this type are extremely powerful. Indeed, applications have been extremely wide-ranging, covering speech recognition, function representation, data compression, and pattern recognition in many different fields. The principal problem is the lack of rules and tools for the design of nets tailored to particular applications. Additionally, if the net is to be simulated in software on a serial computer then there are also maximum size limitations that apply. The average human brain is thought to contain $10^{11}$ neurons but the average neural net looked at in this paper contains no more than $10^3$ and often much less! Even then a typical application may well require 10000 or more evaluations of the net before the nonlinear optimisation process used to estimate the weights converges. If the net is complicated and each evaluation takes only 1 second of CPU time then a single run with 10,000 evaluations would take almost 27 hours of machine time. The need for massively parallel computation is quite apparent.

One general problem with all these methods concerns the risk that only a local minimum will be found to the optimisation problem. Experience seems to suggest that this rarely happens. There are in any case alternative training methods based on simulated annealing that could be applied, albeit at a very great increase in computational costs. A much more serious problem is the very long training times needed. In practice this means several tens of thousands of neural net calculations. The bigger the net, the greater the number of times its outputs and partial derivatives are required by the nonlinear optimisation procedure. Speeding up the net training process is currently the subject of intense research.

There are a few further points of note. First, it is commented that many aspects of the technology are delightfully vague; for instance, the architecture of the net is arbitrary and there are a few different squashing functions that could be used. The power of the method derives from the weights allocated to the connections rather than the pattern of connections, although the latter can also be changed; for example, skipping layers and there is no real necessity for every neuron in layer $k$ to be connected to every neuron in layer $k+1$. The problem is that there is no good theoretical basis for changing the configuration of the linkages. Likewise, the best number of hidden layers and the numbers of neurons in each layer are difficult to determine; indeed, some researchers have embedded their neural nets in an outer optimisation process which is designed to identify the optimal net configuration. It is difficult to know how much it matters for any particular application. Maybe increasing the number of layers and neurons in each layer beyond some value merely delays the parameter estimation process. However, Lapedes and Farber (1988) claim that a neural net seldom needs more than two hidden layers to solve most problems involving real valued input data. There is some evidence to suggest that the accuracy of the mapping is controlled by the number of neurons per layer and not the number of layers. On the hand, whilst two layers of hidden neurons are sufficient they may not be efficient.

Another key uncertainty concerns the size of training sets necessary for valid generalisation and the dangers of over-training in which the representational powers of the net is mainly used to model the random noise contained in the data rather than the more important systematic factors which are essential for generalisation and prediction. Like many areas of neural computing these and related problems have to be investigated by empirical means.

## 2.5 Other types of net that may be useful for modelling spatial interaction

Self-organising maps was one of the most important neural networks discovered during the 1980's; see Kohonen (1984). A self-organising map discovers a continuous topological mapping of a function from one $n$-dimensional space to another $m$-dimensional space by a means of self-organisation based on data examples. It has been shown to be capable of representing highly complex and nonlinear functions. This type of net are often used as pattern detectors but in a spatial interaction context the mapping to be learnt might be considered to involve a spatial interaction field. The only restriction here is the need to ensure that all the variables involved are measured in the same metric space. This may mean that the intervening opportunity type of interaction problem may be more appropriate for this type of mapping.

A recurrent associative network of the type described by Hopfield (1986) may also be relevant to spatial interaction modelling. This type of net starts at an initial state and then converges to one of a finite number of stable states. Associative nets have the ability to remember and retrieve patterns. The interesting point about the Hopfield net is that the energy function it minimises has remarkable resemblance to the entropy-maximising derivation of the spatial interaction model. There is a hint here that maybe a radical alternative route to a net based spatial interaction model is to regard the entire spatial interaction system as a Hopfield net, with each origin-destination zone being a neuron. Whether this is useful is left to others to investigate.

## 3. Applying neurocomputing to spatial interaction modelling

### 3.1 Simple spatial interaction modelling

One of the few geographical applications of a neural net is White (1989). He uses a very simple net to represent a linear region in which there are five zones with consumers and five retail centres, only two of which have shops. This net is trained on 7 cases to reproduce results which are considered to represent the evolution of the city system, capturing its historical dynamics; he clearly has an excellent imagination! Yet it is a example of a feedforward net. A variant is explored here with 10 input neurons and 5 output neurons.

First, let's assume that the consumers are assigned to zones in a random manner. The expected sales in the two shopping centres located in zones 2 and 3 is based on the following simple model

$$S(j) = 0 \text{ if } j = 1, 4, 5$$

$$S(2) = o(1) + o(2)$$

$$S(3) = o(3) + o(4) + o(5)$$

where

$S(j)$ is the sales in centre $j$,
$o(i)$ is the number of consumers in zone $i$.

Training data can be generated by randomly assigning uniformly distributed values to the $o(i)$'s and then scaling them so that they sum to unity. A very simple net is quite able to capture the structure in this data and provide a means of estimating results for other, unseen, data sets sampled from the same generating process.

Table 2 gives the predicted and real sales for the 5 zones, as well as estimates for 10 other data sets. Note that the zero zones are always zero although there is nothing in this net to insist they should be. The net has learned to efficiently represent the algorithm used to generate the training data sets. The training process used a conjugate gradient nonlinear optimisation procedure to identify values for the weights that minimised the sum of squares of the errors. It required 1,279 net evaluations but the final error sum of squares was extremely small; circa 0.000054. By comparison the error in predicting the unseen data was larger; circa 0.001637; but the fit was still very good. It can be improved further by increasing the training set size; a training set of 100 cases reduced it to 0.001024; or by adding additional neurons to the net itself.

A second more complex model is used to investigate the effects of greater complexity in the model generating process. Here a singly constrained spatial shopping model is used. The sales in centres 2 and 3 is now given by

$$S_j = \sum_i O_i D_j A_i \exp(-0.5 \cdot c_{ij})$$

where

$$A_i = 1 / \sum_j D_j \exp(-0.5 \cdot c_{ij}) \ .$$

**Table 2.** Predicted and actual sales in a linear region

| Training case | Sales in | | | | | Predicted sales | | | | |
|---|---|---|---|---|---|---|---|---|---|---|
| | 1 | 2 | 3 | 4 | 5 | 1 | 2 | 3 | 4 | 5 |
| Supervised training | | | | | | | | | | |
| 1 | 0.0 | 0.0 | 0.35 | 0.64 | 0.0 | 0.0 | 0.36 | 0.63 | 0.0 | 0.0 |
| 2 | 0.0 | 0.0 | 0.28 | 0.71 | 0.0 | 0.0 | 0.28 | 0.72 | 0.0 | 0.0 |
| 3 | 0.0 | 0.0 | 0.78 | 0.22 | 0.0 | 0.0 | 0.78 | 0.21 | 0.0 | 0.0 |
| 4 | 0.0 | 0.0 | 0.35 | 0.64 | 0.0 | 0.0 | 0.35 | 0.64 | 0.0 | 0.0 |
| 5 | 0.0 | 0.0 | 0.41 | 0.58 | 0.0 | 0.0 | 0.41 | 0.58 | 0.0 | 0.0 |
| 6 | 0.0 | 0.0 | 0.43 | 0.56 | 0.0 | 0.0 | 0.43 | 0.56 | 0.0 | 0.0 |
| 7 | 0.0 | 0.0 | 0.43 | 0.56 | 0.0 | 0.0 | 0.43 | 0.57 | 0.0 | 0.0 |
| 8 | 0.0 | 0.0 | 0.25 | 0.74 | 0.0 | 0.0 | 0.25 | 0.74 | 0.0 | 0.0 |
| 9 | 0.0 | 0.0 | 0.51 | 0.48 | 0.0 | 0.0 | 0.50 | 0.49 | 0.0 | 0.0 |
| 10 | 0.0 | 0.0 | 0.38 | 0.61 | 0.0 | 0.0 | 0.38 | 0.61 | 0.0 | 0.0 |
| Unseen new data | | | | | | | | | | |
| 1 | 0.0 | 0.0 | 0.23 | 0.76 | 0.0 | 0.0 | 0.21 | 0.78 | 0.0 | 0.0 |
| 2 | 0.0 | 0.0 | 0.17 | 0.82 | 0.0 | 0.0 | 0.14 | 0.85 | 0.0 | 0.0 |
| 3 | 0.0 | 0.0 | 0.27 | 0.73 | 0.0 | 0.0 | 0.26 | 0.73 | 0.0 | 0.0 |
| 4 | 0.0 | 0.0 | 0.25 | 0.74 | 0.0 | 0.0 | 0.25 | 0.74 | 0.0 | 0.0 |
| 5 | 0.0 | 0.0 | 0.45 | 0.55 | 0.0 | 0.0 | 0.45 | 0.54 | 0.0 | 0.0 |
| 6 | 0.0 | 0.0 | 0.38 | 0.61 | 0.0 | 0.0 | 0.38 | 0.61 | 0.0 | 0.0 |
| 7 | 0.0 | 0.0 | 0.30 | 0.69 | 0.0 | 0.0 | 0.30 | 0.69 | 0.0 | 0.0 |
| 8 | 0.0 | 0.0 | 0.51 | 0.48 | 0.0 | 0.0 | 0.51 | 0.48 | 0.0 | 0.0 |
| 9 | 0.0 | 0.0 | 0.40 | 0.59 | 0.0 | 0.0 | 0.40 | 0.59 | 0.0 | 0.0 |
| 10 | 0.0 | 0.0 | 0.25 | 0.74 | 0.0 | 0.0 | 0.25 | 0.75 | 0.0 | 0.0 |

$c_{ij}$ is the distance between zones $i$ and $j$ in a linear region, with intrazonals set at 0.5; $D_j$ is the attraction of centre $j$ and is defined as an uniformly distributed random number between 0 and 100 although for zones 1, 4, and 5 the value is always zero; and $O_i$ is generated as previously.

The net has now to learn how to represent this model. A summary of the results are given in Table 3. A larger net is also investigated. The configuration of these nets are represented in the following manner:

⟨number of input neurons⟩:⟨neurons in layer 1⟩:⟨etc⟩:⟨output neurons⟩

The original net based on White (1989) only contained 10 input neurons and 5 output neurons. Larger nets with an additional hidden layer containing 5 and 10 neurons respectively are also considered. The greater representational power of the larger and more complex nets is offset by the much longer training times. However, all the nets correctly reproduced the right structure with retail sales being concentrated in those zones with retail attractions and with sales values that were fairly close to the expected values. The ANN works well, at least on these small *artificial data* sets.

## 3.2 A simple real world data set

An attempt is now made to model some real interaction data. Tobler (1983) provides a 9 by 9 inter-regional migration matrix for the USA. This is input into a

**Table 3.** Representing sales in a more complex linear region

| Size of training set | Number of net evaluations | Errors: training** | on unseen data** |
|---|---|---|---|
| Original net (15:5) | | | |
| 10 | 2551 | 0.0096 | 0.0957 |
| 100 | 1243 | 0.5209 | 0.0579 |
| 200 | 1189 | 0.9553 | 0.0519 |
| 500 | 1111 | 2.6494 | 0.0472 |
| 1000 | 864 | 5.3905 | 0.0442 |
| Larger net (15:5:5) | | | |
| 10 | 10000* | 0.00000002 | 0.2131 |
| 100 | 10000* | 0.2339 | 0.2383 |
| 200 | 10000* | 0.6486 | 0.0657 |
| 500 | 20000* | 2.4648 | 0.0984 |
| 1000 | 20000* | 5.2869 | 0.0488 |
| Larger net (15:10:5) | | | |
| 10 | 10000* | 0.00000986 | 0.1119 |
| 100 | 10000* | 0.3373 | 0.0841 |
| 200 | 10000* | 0.7528 | 0.0591 |
| 500 | 20000* | 2.3336 | 0.0669 |
| 1000 | 10000* | 5.0770 | 0.0458 |

Notes:  * no convergence after 10000 net evaluations
         ** magnitude of error sum of squares reflects size of training data set

**Table 4.** Performance of various nets on a 9 by 9 interaction matrix

| Net configuration | Number of parameters | Net goodness of fit | Standard deviation of errors | CPU time (seconds) |
|---|---|---|---|---|
| 4:5:1 | 36 | 0.615e-5 | 40.7 | 125 |
| 4:10:1 | 71 | 0.217e-5 | 24.2 | 384 |
| 4:15:1 | 106 | 0.202e-5 | 23.3 | 523 |
| 4:20:1 | 141 | 0.217e-5 | 24.2 | 504 |
| 4:30:1 | 211 | 0.213e-5 | 24.0 | 694 |
| 4:40:1 | 281 | 0.190e-5 | 22.6 | 1108 |
| 4:50:1 | 351 | 0.194e-5 | 22.9 | 1557 |
| 4:2:2:1 | 21 | 0.174e-4 | 69.4 | 14 |
| 4:3:2:1 | 29 | 0.160e-4 | 65.7 | 18 |
| 4:3:3:1 | 34 | 0.686e-5 | 43.0 | 85 |
| 4:4:2:1 | 37 | 0.173e-3 | 132.0 | 1 |
| 4:4:3:1 | 43 | 0.552e-5 | 38.5 | 124 |
| 4:4:4:1 | 49 | 0.433e-5 | 34.1 | 128 |
| 4:5:2:1 | 45 | 0.574e-4 | 39.3 | 51 |
| 4:5:3:1 | 52 | 0.274e-5 | 27.2 | 205 |
| 4:5:5:1 | 66 | 0.279e-5 | 27.4 | 159 |
| Conventional model (singly constrained) | | | 36.0 | 1 |

neural net and its performance compared with a conventional doubly constrained spatial interaction model. The data input to the net consists of individual $T_{ij}$ records, that is the trip matrix is unrolled to leave 72 records consisting of: $T_{ij}$

(dependent variable) with $O_i$, $D_j$, $C_{ij}$, $O_j$, and $D_i$ (as predictors). It should be noted that this is an unconstrained model. The objective is to predict the $T_{ij}$ values. The data are scaled to the interval 0.0 to 1.0 and input into the net and the outputs rescaled back again. Table 4 shows the performance for a number of different net configurations. Many of these nets offer a better goodness of fit than a conventional doubly constrained spatial interaction models, even though the ANN version is unconstrained. Table 5 gives a comparison of a systematic sample of the net's outputs compared with a conventional model.

Note that as the nets gain layers so they take longer to train. Note also that there is no explicit need to consider the problems of spatial dependency. If spatial autocorrelation is important then it will be present in the training data and the net will learn to represent it. In a similar manner it is possible to take into account data errors and different measurement scales. If the dependent variable is binominal or poisson then the net will deduce this from the training data, there is no need to be explicit. This greatly simplifies the spatial response modelling task.

## 3.3 A more complex data set

Finally, an attempt is made to model a 73 by 73 interaction matrix which had been extensively studied using conventional spatial interaction models; see Openshaw (1976, 1979). The data relate to a 73 zone system of journey to work flows obtained from the 1966 census together with historically accurate generalised travel costs. The benefits of using these data was that the performance of a number of conventional spatial interaction models was known and these could be used as benchmarks for evaluating the net based alternatives. The trip matrix is unrolled to yield 5229 cases (100 internal to external zone interactions are ignored) with

**Table 5.** Extracts from a neural net spatial interaction model

| $T_{ij}$ pair | Observed data | Conventional model | Neural net |
|---|---|---|---|
| 1 | 180 | 209 | 180 |
| 5 | 17 | 25 | 19 |
| 10 | 300 | 263 | 300 |
| 15 | 87 | 108 | 86 |
| 20 | 551 | 538 | 551 |
| 25 | 29 | 31 | 32 |
| 30 | 185 | 177 | 181 |
| 35 | 346 | 351 | 346 |
| 40 | 279 | 282 | 279 |
| 45 | 316 | 384 | 316 |
| 50 | 64 | 69 | 62 |
| 55 | 134 | 107 | 136 |
| 60 | 113 | 98 | 117 |
| 65 | 72 | 57 | 71 |
| 70 | 66 | 87 | 66 |

**Table 6.** Performance of various nets on a 73 by 73 interaction matrix

| Net configuration | Number of parameters | Net goodness of fit | Standard deviation of errors |
|---|---|---|---|
| 4:4:4:1 | 45 | 0.811e-1 | 13.2 |
| 4:8:8:1 | 121 | 0.764e-1 | 12.8 |
| 4:8:16:1 | 201 | 0.983e-1 | 14.5 |
| 4:16:8:1 | 225 | 0.983e-1 | 14.5 |
| 4:16:16:1 | 369 | 0.104e-1 | 15.0 |
| 4:4:2:1 | 33 | 0.894e-1 | 13.9 |
| 4:8:2:1 | 61 | 0.894e-1 | 12.6 |
| Singly constrained model | | | 18.2 |
| Doubly constraind model | | | 13.5 |

intrazonal trips being retained. The results are shown in Table 6. All the nets produced levels of performance better than the conventional singly constrained entropy maximising model and many are better than the doubly constrained variant. Remember that the nets are unconstrained and that the inclusion of origin or/and destination constraints greatly boosts the performance of standard interaction models. Clearly the nets work extremely well. The only drawback is the amount of computing time involved. In the runs used for Table 6, each net was run for 30 hours CPU time on a 12 mips/1 megaflop processor; compared with a few seconds for the conventional models. Whether the training process could have been either stopped earlier or else speeded up by changing either the optimisation routine used or the form of the sigmoidal function are matters for further research.

A final test was to investigate whether the interaction pattern captured by the nets in Table 6 might be entirely data specific or contained some generally relevant spatial interaction structure. Table 7 shows what happens when these nets trained on all trips are applied to car only data. In general, these results are surprisingly good and compare reasonably favourably with the levels of performance that can be achieved on the same data by a standard model, see Openshaw (1979).

**Table 7.** Performance of various nets trained on all trips in representing car only trips

| Net configuration | Number of parameters | Standard deviation of errors |
|---|---|---|
| 4:4:4:1 | 45 | 4.4 |
| 4:8:8:1 | 121 | 5.9 |
| 4:8:16:1 | 201 | 4.0 |
| 4:16:8:1 | 225 | 4.1 |
| 4:16:16:1 | 369 | 4.0 |
| 4:4:2:1 | 33 | 5.0 |
| 4:8:2:1 | 61 | 4.4 |
| Singly constrained model | 1 | 3.5 (calibration) |
| Doubly constrained model | 1 | 2.4 (calibration) |

## 3.4 Further developments

The neural net modelling of spatial interaction data offers the modeller a greatly increased freedom to worry more about the application of the model and much less about its specification. The new flexibility might also be used to experiment with different model designs. For instance, interaction data exist as integer counts yet all the models operate in a real number space. A neural net can be built to model interaction as an integer bit pattern. Instead of having a single output neuron which is mapped on to a positive real number, it is possible to have 16 output neurons to represent the bit pattern of a 16 bit integer that represents the trips. An attempt was made to test such a model using a (5 : 40 : 32 : 16) net configuration. After 200 hours of CPU time the 2,040 parameters still had not been adequately trained and the error for the 73 by 73 data was 43 (cf the values in Table 6). However, it is quite clear that the net was slowly starting to provide a reasonable representation. An extract of the results are shown in Table 8.

There are other coding schemes that might well be better than a binary bit pattern in which order is critical. In particular, it might be worthwhile considering grouping the trips into size bands and then using a separate output neuron for each. This would also give an integer output representation.

## 4. ANNs for spatial interaction modelling

The results have demonstrated that a neural net can represent spatial interaction data, although questions might be raised about the amount of computing time involved. This is partly a function of trying to simulate parallel processes on a serial machine and mainly the inefficiency of current training algorithms. It is likely that both problems will soon be solved. From a spatial interaction modeller's point of view there are a number of possible benefits to be gained:

(1) better performing models;
(2) greater representational flexibility and freedom from current model design constraints;
(3) an opportunity to handle explicitly noisy data;
(4) incorporation of spatial dependency in the net representation, currently it is ignored; and
(5) an opportunity to exploit future ANN developments in other fields and import them into geography.

**Table 8.** Extract of an integer interaction net

| Observed trips | Predicted trips | Predicted bits | Observed bits |
|---|---|---|---|
| 0 | 1 | 0000000000000000 | 1000000000000000 |
| 19 | 63 | 1100100000000000 | 1111110000000000 |
| 137 | 31 | 1001000100000000 | 1111100000000000 |
| 158 | 63 | 0111100100000000 | 1111110000000000 |
| 75 | 127 | 1101001000000000 | 1111111000000000 |
| 5 | 7 | 1010000000000000 | 1110000000000000 |
| 73 | 63 | 1001001000000000 | 1111110000000000 |

It also raises a number of interesting questions concerning the generality of the spatial interaction process and the possible viability of seeking an associative memory based approach. The latter type of net might well provide the basis for an universally applicable, pre-calibrated, spatial interaction modelling-net. There is some prospect here of an entirely new approach to modelling spatial interaction based on the recognition that there are only so many different patterns of spatial interaction that can exist. These might be identified, stored in a neural net containing memory, and then used as the basis for intelligent modelling of the world's spatial interaction patterns. The powerful representational properties of neural nets also offers some prospect that zoning system scale and aggregation effects might be taken into explicit consideration, together with time; provided an adequate and broadly representative sets of spatial interaction data sets are available for training. Indeed, this highlights a major problem in that "libraries" of training data largely do not exist and that generations of quantitative geographers generally have been very careless with their data. If the full potential of exploiting ANNs that can "learn" or be taught spatial patterns is to be realised, then some systematic attempt needs to be made to systematically store and catalogue spatial data sets as training examples.

## 5. Conclusions

Feedforward multilayer artificial neural nets can be applied to almost any problem for which training data sets are available. There is no longer any need to assume that model building has to proceed via a theory based specification and there are seemingly many areas in human and physical geography where suitable applications exist if adequate training data were available. Indeed, particular attention might be focused on hard areas such as spatial interaction modelling but also on the modelling and prediction of categorical data; an area where existing statistical methods are highly complex and dependent on the acceptability of various key assumptions that may not hold. Neural nets can also be developed that will replicate the descriptive and predictive functions of current statistical methods, often with an improved level of performance and accuracy. The main drawback is the computer time involved. It is clear, however, that this is no going to be a problem much longer as computer workstations continue to speed-up and cheap multiprocessor technology becomes widespread.

Another major application area is the more general question of spatial data pattern and relationship finding. The GIS revolution has given geographers access to large quantities of map related data. There is great need for computer based methods able to do something "useful" with this rapidly increasing heap of data. Never before in human history have geographers had access to so much geographically referenced information and have so few ideas as to what to do with most of it, partly because the available spatial analytical toolbox is so inadequate. There is a growing need for new methods of analysis able to compress data by spotting patterns that may be of further interest but without being told in advance what to look for or where to look for it. Neural nets are clearly relevant technology that can be used to filter out the rubbish and leave behind information that may be worthy of further attention. It also provides a means of developing computational models and representations in areas of research where previous only qualitative approaches and soft methodologies could be used.

Neural computing is very much in its infancy. However, it is abundantly clear that already there are many potential applications in geography. There are many-many problems which are hard to solve with conventional methods. Indeed, one cause of the end of the first quantitative revolution in geography was the failure of standard statistical methods and models to cope with the analysis tasks they were presented with. Artificial neural net technology offers a means of progressing beyond what is possible with classical analysis technology. It can cope with soft data, it can offer a degree of brainlike performance and intelligence not found anywhere else, and it is becoming easier to apply. There is, however, a cost associated with these developments that may well hinder its diffusion. Namely, the black box nature of the technology and the difficulties associated with understanding what the neural net is representing, which may be widely used as a justification for non-use and the continuation of current styles of research. If this happens on a large scale, then it would represent a major failure of geography and geographers to exploit the full potential of neurocomputing in particular and AI in general.

Neural nets are a revolutionary technology with the potential to provide a new and different focus to quantitative geography during the 1990's. However, there are two key sets of problems that have to be faced: first, there are the technical aspects of developing and demonstrating appropriate ANN technology in a geographical context; and second, there are attitudinal problems in that geographers will have to learn to become more flexible regarding the methodologies that are perceived to be useful in geographic research. Maybe this will involve a degree of unlearning what was previously considered to be the accepted wisdom. All revolutions must have been like this and bring with them a complex mix of potential benefits and threats. Neurocomputing is no different and it will be very interesting to follow the diffusion and uptake of neural technology and the speed by which these methods become part of mainstream geography during the 1990's. It is certainly possible that the 1990's will witness a major revolution in quantitative geography and see a great expansion of those parts of geography where computational methods might be used. ANNs are likely to be an important component of these developments.

## References

Aleksander I, Morton H (1990) An introduction to neural computing. Chapman and Hall, London
Fukushima K (1975) Cognitron: A self-organising multilayered neural network. Biol Cyber 20:121–136
Grossberg S (1976) Adaptive pattern recognition and universal recoding: part 1, parallel development and coding of neural feature detectors. Biol Cyber 23:121–134
Khanna T (1990) Foundations of neural networks. Addison-Wesley, Reading Mass
Kohonen T (1984) Self-organisation and associative memory. Springer, Berlin
Hecht-Nielsen R (1990) Neurocomputing. Addison-Wesley, New York

Hopfield JJ (1982) Neural networks and physical systems with emergent collective computational abilities. Proceedings of the National Academy of Science 79:2254–2258

Hopfield JJ (1984) Neurons with graded response have collective computational properties like those of two-state neurons. Proceedings of the National Academy of Science 81:3088–3092

Lapedes A, Farber R (1988) How neural nets work. In: Anderson ZA (ed) Neural information processing systems. American Institute of Physics, New York, pp 442–456

McCulloch WS, Pitts WH (1943) A logical calculus of the ideas immanent in nervous activity. Bull Math Biophys 5:115–133

Minsky M, Papert S (1969) Perceptrons: an introduction to computational geometry. The MIT Press, Cambridge Mass

Neuman J von (1951) The general and logical theory of automata. In: Jeffress LA (ed) Cerebral Mechanisms in Behaviour. Wiley, New York, pp 1–41

Openshaw S (1976) An empirical study of some spatial interaction models. Environ Plann A 8:23–42

Openshaw S (1979) Alternative methods of estimating spatial interaction models and their performance in short-term forecasting. In: Bartels CPA, Ketellapper RK (eds) Exploratory and explanatory statistical analysis of spatial data. Martinus Nijhoff, Boston, pp 201–226

Openshaw S (1990) Spatial analysis and GIS: A review of progress and possibilities. In: Stillwell J, Scholten H (eds) Principles of geographic information systems. Elsevier Applied Science, Amsterdam, pp 153–63

Openshaw S (1991) Some suggestions for the development of AI tools for spatial analysis and modelling in GIS. Ann Reg Sci 26:35–51

Rosenblatt F (1958) The perceptron: A probabilistic model for information storage and organisation in the brain. Psych Rev 65:386–408

Rosenblatt F (1961) Principles of neurodynamics. Spartan Books, New York

Rumelhart DE, Hinton GE, Williams RJ (1986) Learning representations by back-propagating errors. Nature 323:533–536

Rumelhart DE, McCelland JL (1986) Parallel distributed processing, vols 1, 2. The MIT Press, Cambridge Mass

Tobler W (1983) An alternative formulation for spatial interaction modelling. Environ Plann A 15:693–704

Wasserman PD (1989) Neural computing: Theory and practice. Van Nostrand Reinhold, New York

White RW (1989) The artificial intelligence of urban dynamics: Neural net modelling of urban structure. Papers Reg Sci Assoc 67:43–53

# Part C
# Applications

# Geographical information systems and dynamic modelling

## Potentials of a new approach

**Wolf Dieter Grossmann and Sigrid Eberhardt**

Institution for Ecosystems and Environmental Studies, Austrian Academy of Sciences, Kegelgasse 27, A-1030 Vienna, Austria

**Abstract.** Combinations of dynamic models and Geographical Information Systems (GIS) have a vast potential to solve problems. Deficiencies and advantages of GIS and dynamic models are described. A multifaceted description of complex systems allows three different types of combining dynamic models and GIS. Different classes of dynamic models are used within these combinations. These are: − complex aggregated dynamic feedback models (e.g. like those of Odum, Forrester or of the AEAM work [1]), − simple generic dynamic models (in particular object oriented models) − models of physics based on partial differential equations (e.g. those for heat conduction or dispersal of noise or transport of gaseous pollutants). The model dynamics are combined with GIS held "base maps" to produce time series of maps, so called "Dynamic Maps". Base maps combine spatial features which are locally important for the dynamic process and are used to either modify or even form the dynamics. Different types of models need different types of GIS-held base maps and are adequate for different types of problems. This paper is based on a number of actual applications of one of these combinations; an overview is provided on potential applications of the whole new approach.

## 1. Geographical information systems

Geographical information systems (GIS) have become very important for storing evaluating, depicting, updating and processing of spatial data.

There are two types of representation of spatial data in the computer: the raster and the polygon format. Maps usually have the form of many different shapes, that is, of polygons. Most remote sensing data have the form of raster data. Many GIS's now offer the possibility to process both, raster and polygon data and to convert one format into the other. A good GIS allows to store the

---

[1]   e.g. Odum 1982; Forrester 1968; Essa 1982.

topology of spatial relationships, i.e. which polygon is adjacent to which others, adjacent to which line features (line features: rivers, power lines, telephone networks etc.) and so on.

Updating of data is easier with GIS once the initial spatial information has been entered into the GIS. One of the most important features of GIS is the so-called map overlay: here different thematic maps are combined as if they were drawn on transparencies and a new map showing the intersections of all lines is generated. This feature is needed, when for example the ecological suitability of an area for a forest is evaluated. All spatial factors that are relevant for the growth of the forest are combined, as e.g. a map on the soil, a map on orography (altitude, slope, aspect), a map on soil moisture etc. All spatial features are about the same within each polygon of this new map generated through overlay of these thematic maps. Maps generated through overlay show the intersection of the different maps from which they are combined. They have many more polygons than the original maps or more information is available in each raster. Such overlay map can be used to assess the combined information, e.g. with respect to ecological suitability for a species or the suitability of areas in a region for infrastructural construction because all the combined information is additionally offered in a data bank attributed to the overlay map.

The results of evaluations with GIS can be shown in many different formats, e.g. as statistics or as maps. Maps are a very good means of communicating results as people have a unique capability to understand patterns.

GIS have a multitude of other features, which together allow a comprehensive and sophisticated use and processing of spatial data (Ashdown and Schaller 1990).

The details in the map do *not* represent the reality. The data are only a "*model of reality*", what is called the "data model" in GIS terminology.

- The boundaries between polygons are actually transition zones.
- homogeneous polygons are actually heterogeneous. There is an underlying continuous reality which is overlaid with a grid or a system of polygons. (These two examples are stressed by Goodchild 1991.)
- Although the definition where one spatial type ends and another begins is often somewhat arbitrary the resulting map looks unambiguous.

Other disadvantages:

- the collection of digital data is expensive and time consuming
- GIS data are more or less static
- the same numbers often have different meanings in different locations (e.g. in a case study it was found that a mountainous sub species of spruce naturally had downward bend needles whereas this "lametta syndrome" was one of the early symptoms of tree damage in other locations)
- the use of GIS needs specialized long training.

## 2. Dynamic models

Dynamic models depict complex structures and help to evaluate its inherent dynamics. Many different types of dynamic models exist to meet the manifold requirements for dealing with structures and their dynamics. A few examples are: feedback models (e.g. of the type developed by J. Forrester, H. Odum, or the AEAM group), Input/Output models, dynamic general equilibrium models, dynamic linear optimization models etc.

Dynamic models allow a better understanding of complex relationships. Their structure can be communicated to achieve a common understanding within a research group. They may show unexpected behavior which also helps in understanding problems. Some types of models tend to correct mistakes in data, in particular feedback models.

New insights in nonlinear dynamics have demonstrated that even some simple systems (Lorenz 1963; May 1974; Haken 1978; or summarized in Gleick 1988) and almost all more complex systems can exhibit unpredictable behavior. Whereas this inherent unpredictability cannot be overcome, the so to speak normal, practical unpredictability from complexity will be diminished. The situation may be ambiguous in that respect that in some areas the claims for predictability have to be withdrawn but that predictions will become much more powerful and practical in other areas. It is difficult to know which predictions will be corrected and to what degree and which predictions will be wrong.

Dynamic feedback models have difficulties to process numerous details. Manifold software packages exist to run dynamic models. Therefore dynamic models come in all disguises; no common standard exist. Communication of researchers is seriously impeded by this fact.

Education in structural thinking and in model building is often not good or does not even exist (nearly no university offers courses).

Models are often cross disciplinary. Specialized scientists tend to reject the transgression into their area by a model building group. They tend to defend their turf; modelers tend to underestimate the value of the knowledge available in the specific fields.

These factors limit the applicability and usefulness of dynamic models and impede the possibilities for training of systems people. As a consequence, models tend to be not nearly as good and useful as they could be.

Many dynamic models successfully process numerous spatial variables. But these are either structurally simple models, e.g. transport models of physics or they use structurally simple spatial data, e.g. geometrical grids. Models that are suitable to predict the future development of complex systems, in particular feedback models, are not adequate to process numerous spatial details. One reason valid for dynamic models based on difference equations: Here spatial details are depicted as state variables. Hence numerous spatial details lead to large scale systems. But large nonlinear systems of difference equations are theoretically capable of exhibiting whatever dynamics is desired. The general agreement here is that small models are useful and can be validated, large models are useless for prediction and not even good in supporting understanding of the system which was modelled. Already in 1973 Lee has written his "Requiem for Large Scale Models" (Lee 1973) where he as a user and developer utters his disappointment and frustration about the uselessness and total failure of large scale models.

An attempt to include spatial details into models would only be appropriate if the model environment also provides all capabilities of a GIS.

## 3. Reasons for combining geographical information systems and dynamic models

GIS data usually are voluminous. Some of the spatial information is static, e.g. orographical features, but other information is actually dynamic. It takes time and is expensive to update voluminous information. Therefore data in GIS tend to be static, although the dynamic information should be dynamic.

Some of the information, e.g. on soil types or microclimates, has been collected during years or even decades. Some data, e.g. several characteristics of microclimate, are long term average values. By definition it takes time to determine long term averages. Rapid updates are technically all but impossible for such types of information.

Hence GIS data are often crucially deviating from the real situation. Mechanisms or procedures are needed to achieve a faster "tracking" of the real situation by the data in the GIS.

Can remote sensing help in this more frequent update of GIS data? Remote sensing information can be converted into GIS formats. Some satellites submit their information quite frequently. Hence, remote sensing data is less static than information which was terrestically collected. But remote sensing data can only solve some of the problems of GIS data because remote sensing has other shortcomings. Some of these are: 1) not all terrestrial data can be observed with remote sensing, e.g. soil types. 2) Classification of new phenomena or of phenomena in unusual combinations is difficult and sometimes impossible. Terrestrial data collection and remote sensing are often complementary, not mutual substitutions.

Information produced by dynamic models is by its very nature dynamic. Could this information be used to try to anticipate the future development of spatial data? Again two reasons pose difficulties here: 1) multi-loop feedback models are inadequate to handle voluminous spatial data. 2) The availability of spatial data is uneven; in some areas many good data are available, while in others few data exist. Processing of uneven data in models is difficult.

But not only practical reasons limit the use of extensive spatial information in dynamic models. Theoretical considerations show that such an approach is inadequate. Therefore one of the authors has developed a multifaceted method to overcome some of the difficulties described here (e.g. Grossmann 1983 or 1991 a). In this method three different types of problems (or "facets") are seen in complex systems: − change of the structure; − dynamics generated by one particular structure; − and the details that are inherent in the structures and form them. Different methods are most suitable to tackle these different problems. GIS for example are often very good to process details, feedback models are adequate to model structures and to reveal the dynamics that are inherent in these structures. Therefore structures are evaluated with models; spatial data are processed with GIS. Each of these two facets of complex systems (complex structures and details) is processed with adequate methods. This multifaceted method allows to process details separately from the dynamic models and to combine the results of the different methods afterwards. The whole method is iterative. The different types of dynamic maps described here are created by combinations between different types of models and different kinds and levels of details processed with the GIS (first results: Grossmann et al. 1984).

Models have some of the features needed to solve complex problems; GIS have others. Therefore these two tools are combined to allow more adequate solutions to such problems.

Three *different types of combinations of dynamic models with base maps*[2] exist, depending on the class of model. These are:

- complex aggregated dynamic feedback models,
- the so to speak "classical" transport models of physics, depicted as partial differential equations,
- simple generic models.

The resulting three different combinations are adequate for different types of problems.

## 4. Connection of aggregated dynamic models and base assessment maps

Many problems exhibit complex structures and aggregated characteristics. Examples for complex structures are 1): the relationships between demand for land, suitability of land for a purpose and resulting land use change, 2): the manifold factors in the preservation of bio diversity, or 3): the global structures involved in weather, climate and climate change.

Such structures can adequately be depicted with aggregated dynamic feedback models. Often the dynamics are similar through large regions, e.g. due to national laws or in continental build up of ozone during high pressure regimes. In the extreme such spatially extended variables can be global, as for example in the $CO_2$ increase. But locally the dynamics are *modified* by spatially varying factors, e.g. altitude, steepness, soil type, by administrative regulations or by the vegetational changes in the case of the $CO_2$ increase.

Therefore a dynamic model is used to produce time series describing such spatially extended variables. These numeric values are afterwards locally *modified* using a base assessment map generated with a Geographical Information System. The base assessment map (BAM) depicts the spatial distribution of factors important for the local modification. It could e.g. be the suitability for agricultural use or the risk for plants to be affected by ozone or the net primary production. These local factors could either be shown as polygons within a map or as raster data. The spatially extended dynamic variables are modelled only once in this method, using one central model, not within each polygon separately, as would be the case in the method of the "active area dynamics", see below.

The combination between the dynamics and the BAM is for example done as follows: the models specifies that an area of $0.5\,km^2$ will be converted from agriculture to forests in 1992. The BAM would in this case be generated in such a way as to depict the suitability for agricultural use. Those agricultural areas will be converted first that are the least suited for agriculture. Another example: if the conversion is intended for fixing $CO_2$, land which allows a high primary produc-

---

[2] We differentiate between "Base Maps" and "Base Assessment Maps". Base maps are the general format of GIS data for combination with dynamics; base assessment maps are the special form of base maps necessary for the combination of base maps and *aggregated* dynamics.

tivity could be converted to forest use. In this way the definition of the BAM depends on the problem. The BAM usually depicts a polygon- or raster-wise assessment of a suitability or a risk.

The resulting map is plotted with the GIS in all these types of combinations between dynamics and areas. The locally modified values attributed to the polygons are plotted for all points over time, for which maps are desired. A feedback from the resulting maps to the model is possible but difficult.

## Applications

*Forest damage by pollutants*[3]: The dynamic model contains the time series of emissions of important pollutants (precursors of ozone, i.e. volatile organic compounds, carbon monoxide and $NO_x$, sulphur dioxide) and the different scenarios how these will change in the future. The model describes the interactions between amount of biomass and resulting increased deposition caused by the vegetation and resulting concentrations of air pollutants. The BAM is based on the local distribution of the concentrations as modified by factors such as location of sources of pollutants, prevailing wind directions, exposure to pollutants, species and age of trees and resulting sensitivity to different types of pollutants and so on. Such a BAM is a risk map for the forests to be affected by forest damage. It does not show the actual amount of damage. The results of the dynamic model show the amount of damage, but not its location. This amount is distributed to the forest areas according to their respective risk to be affected. As models and BAMs differ with the hypothesis on the causes of the damage this method also allows the testing of hypotheses on forest damage. Different hypotheses lead to different maps which can be compared with the actual distribution of damage.

*Tourism and its interactions with the regional economy and new agricultural strategies* (see Haber 1985; Grossmann and Clemens-Schwartz 1985): Tourism needs specialized infrastructure. Therefore a region which would be attractive for tourists (due to climate, natural environment and location) has many more visitors, if such an infrastructure exists. However, selective tourists tend to avoid an area, if the density of tourists exceeds what they tolerate. The dynamic model describes such a mechanism and the resulting deterioration of an area. Several different BAMs will be used: the first one depicts the suitability for construction of infrastructure. The second one assesses the attractiveness for tourists. Combination of these two depicts those areas where infrastructure is likely to be built and tourists are likely to reside. From this the first candidates for destruction of areas by tourism can be derived. The resulting spatial pattern reminds of an attack of locusts on an area.

## Spatial classes

Spatial classes are an important concept in land-use change. Land area within one class but within different categories can be converted into each other. Two examples are provided:

---

[3]    Reported in Grossmann 1991 b–d and Kopcsa and Grossmann 1991 or Grossmann and Schaller 1986.

*a) Forest die-back:* Areas with healthy forest can be transformed into areas with forest affected by damage. There exist in all five damage categories in the German classification scheme: category 1: healthy forest; category 2: slightly damaged forest (loss of needles between 10% and 25%), category 3: damaged forest (loss of needles between 25% and 60%), category 4: severely damaged forest (loss of needles higher than 60%) and category 5: dead or dying forest. Forest of each category could change into each other category (category 5 by either regrowth or afforestation). Therefore only one spatial class exists in this problem area.

*b) Study on agricultural land-use changes:* Different categories of land were differentiated in an agricultural study which could not be transformed into each other. The suitability for agricultural use depends on several criteria: – steepness of the area, – quality of the soil, – exposure of the area, – content of rocks, – elevation of the area. Agricultural area of the suitability class 5 is very poor, elevated rockly land with a low quality of the soils, suitability class 4 is somewhat less rocky, less elevated, more favorably exposed etc., suitability class 1 is flat lowland area with a good quality of the soils. These five types of land belong to different spatial classes because steep, elevated land cannot be converted into flat area with good soil.

## Shortcomings

The method could use very detailed spatial information. But usually the preciseness of the available different variables and their scale differ considerably. The actual measurement may only be valid within a few or just one polygon. The changes calculated by the method in the individual polygons are not stored and cannot be used in the model because the model only deals with aggregates. Therefore after some time the results become imprecise or even wrong. Also, changing environmental factors can have very different local effects. This can locally change properties or even structures of subsystems. But again the data of separate polygons cannot feed back on the aggregated model. Therefore the results of the aggregated model can stray from the actual development.

Changes in the different polygons happen in different ways. Therefore average values are only correct for a short time span. Also average values usually are not appropriate for very heterogeneous systems. But the method is based on average behavior which is locally modified. A specific version of another model would be needed to predict the development in each specific polygons if polygons are different. This is described in Sect. 5. A summary of advantages and disadvantages of the different methods is given in Table 1.

## Software tools

The DYS-ARC software allows to evaluate models and translate their dynamics into dynamic maps using a BAM. This software asks, which BAM to choose and which time series of data to read. Afterwards it asks for the points in time for which to make the maps, checks the times and then starts a GIS to produce the corresponding time series of maps. DYS-ARC can also process dynamics from

**Table 1.** Characteristics of methods to link space and time (Grossmann 1990)

| Area of application | Typical applications |
| --- | --- |
| 1. Anticipation/ forecasting | Planning, perception of dangers, early warning, test of scientific hypotheses. |
| 2. Environmental monitoring | Improved understanding of remote sensing data. Environmental early wearning. |
| 3. Economic early warning | Economic planning, support for political decision-making. |
| 4. Land-use planning | Land-use planning: Cities, villages, regions, federal states. Planning of landscapes and green areas. Planning routes for high speed trains. Planning, also of investments, in the area of tourism, recreation and spare time. |
| 5. Use of resources | Management of agricultural and forest areas. Development of pollution of aquatic systems over space and time. Calculation of production of drinking water and its quality. Development of forest damage. Management of non-renewable resources. Management of renewable resources (e.g. water). |
| 6. Planning of investments (Planning of sites) | Changes of markets, regions, accessibility for traffic, connectivity within traffic net works. |
| 7. Risk assessments | Routing of dangerous goods. Siting of waste disposal sites. |
| 2. Environmental planning | Tasks in protection of nature, rehabilitation, clean-up, area-related planning. |
| 9. Environmental impact assessment | Problems with waste dumps. Planning of construction. Planning of infrastructural investments. |
| 10. Transport process | Maps on emissions and deposition. Monitoring and updating of these maps ("dynamic Kataster"). Spatial and temporal changes of water pollution. Environmental impact assessment. |
| 11. Strategic management | Dynamic strategic management. Economic early warning. |
| 12. Research | All of these applications and basic research. |

sources other than models. DYS-ARC allows to handle problems with up to 99 different spatial classes.

## 5. Connection of models of physics and GIS

Many transport processes of physics can be described with partial differential equations, which are solved numerically. Examples are: − electromagnetic waves, − conduction of heat, − dispersal of pollutants, − flows of substances and gases, e.g. of groundwater, − propagation of noise.

One example are models on the transport of pollutants. Here so to speak "wave fronts" on the dispersal of a substance are calculated. The calculation proceeds along a grid which resembles a fishing net. Based on the values from two grid points a new value in the direction of the transport is calculated. In addition values from the boundary of the area are included for the calculation. This procedure gives a spatio-temporal development. This development is *modified* by geographical details from the GIS, e.g. hills. Hills become preferred areas of deposition of the pollutant, if they are located downwind.

Mathematically the same equations are used for each grid point but with the specific data from this grid point. Therefore the same model has to be applied up to millions of times, depending on the size of the grid, to derive spatio-temporal developments. This contrasts with aggregated models, which calculate all values for past and future points of time in one model run. The models used here are mathematically simple compared to aggregated models. Complexity enters as many grid points are connected.

Züger from the Austrian research center Seibersdorf has coupled a transport model of air pollutants with the Geographical Information System ARC/INFO. W. Flake from ESRI/Munich has connected a dispersal model of noise with ARC/INFO[4]. No software exists, which solves the problem of connection of transport models and GIS in general in the way DYS/ARC does for the connection of aggregated models and BAMs.

For advantages and disadvantages of this type of spatio-temporal modelling see Table 2.

## 6. Active area dynamics: The connection of generic models and GIS

Generic models describe simple situations in a prototypical way. They are applied to a specific situation by being provided with data from this application, usually stored in a data base. They are used in connection with maps, if the dynamics, which have to be calculated for the polygons and a base map are more complex than models of physics.

Many informations and attributes are valid throughout the whole area. But the limitations of this approach were mentioned in Sect. 2: locally changed properties of separate polygons cannot be evaluated by aggregated model. In such cases information exists which can only be exploited, if individualized models are used in each polygon separately.

*Example:* In the modelling of the phases of forest development the same generic model of forest development processes the data of each polygon of a base map. The base map will have comparatively few polygons, as usually not many different phases of forest development exist in an area. A base assessment map usually has many more polygons. (In a case study on forest damage the BAM of the Rosalia area of 600 ha consisted of 6000 polygons. But the Rosalia area has only 33 departments consisting of between one to five different age classes each, giving a total of 89 sub-departments. Therefore the base map for the active area dynam-

---

[4]    Züger: Austrian Research Center Seibersdorf, A-2444 Seibersdorf, Austria. Flake: ESRI Germany, D-8051 Kranzberg, Ringstrasse.

ics would have 89 polygons). If the model is run for a period of 100 years, depicting the development of a forest area with mixed ages, then an average of 40% of the forest areas are cut within the first 40 years. It is no longer feasible to apply the method of aggregated dynamics to an area that has changed so considerably. But the development of these areas can be calculated individually with a generic model which runs in each polygon separately and is based on the individual data of each polygon.

As generic models in such applications calculate the development of each polygon individually, the development of each polygon is depicted by its own model object, so that its own dynamic becomes visible. The name "active area dynamics" derives from this method.

Adjacent polygons often are connected to each other by transport processes. Therefore some applications of active area dynamics periodically need an interim balancing calculation. The changes caused by altered vicinity conditions are calculated and the updated data for each polygon are stored back in the data base. These data will be used for the next step.

The query language has to meet tough requirement: It must be possible to regard each polygon as complete in itself, because there are as many individualized model objects as polygons. Therefore each polygon should be handled as one separate element. The query language must support abstractions and object behavior to meet these requirements. In addition query possibilities like in a relational data base language are needed: e.g. it must be possible to ask the software to depict all model objects that represent polygons between specified age classes or polygons with a specified degree of forest damage or any combination between such queries.

### Possible applications

*Study of the relationship between forests and precipitation.* (*a*) Microclimatic processes directly affect only small areas, although they are able to influence extended regions. This may in particular be true for the evapotranspiration by forests and the resulting influence on rain fall. Some forests store water and release it during dryer periods. Also they evapotranspirate so that larger forest areas could increase precipitation downwind.

(*b*) Now growth of the vegetation is simulated with the method of active area dynamics. An overlay map depicting the ecological conditions of different sites in a larger area is used as a base map. An object oriented simple forest growth model is put on each polygon of the base map to simulate plant growth and evapotranspiration. The resulting evapotranspiration is passed on to a climate model and the change of precipitation, most likely an increase in rain, due to the development of the vegetation is calculated. The new rainfall patterns are in turn entered into the GIS to modify the base map of the ecological conditions of sites. As forests upwind grow or are planted, additional areas may become suitable for growth of forests.

(*c*) Forest growth is modelled on all areas including those just made suitable for forest growth. The new or modified forest areas in turn could increase precipitation still further downwind. By iteratively repeating this process it can be deter-

**Table 2.** Applications of dynamic maps (Grossmann 1990)

|  | Aggregated models | Transport models | Active area dynamics |
|---|---|---|---|
| Area of application | Depiction of spatially extended phenomena which are locally modified with Base Assessment Maps. | Transport processes which can be described with simple mathematical equations and which are modified by local GIS data. | Forecasting of somewhat more complex systems based on polygons where spatially extended phenomena can be neglected. |
| Effort | Effort was recently drastically reduced due to new modelling tools (STELLA, PC DYNAMO, SIMCON, DYS/ARC). Computational capacity of a larger PC suffices. Considerable effort for the determination of connections within models. Comparatively low demand for data. | Standard models for the about 11 types of transport processes. High to extremely high computational demand. High demands on data collection. | No standardized models. Suitable languages for modelling not well known. High demands on data collection of non-standard data. For many applications extremely high computational demand. Overview difficult which model brings which results. |
| Advantages | Flexible overall optimization. Model results often are very similar to past actual development. Structural equivalence of models and reality is possible. | Precise, often unique solution. Well established method. Characteristics of data are well known. Little difficulty in determining suitable model because standard procedure. | Supports research and management on diversity over space and time. May allow new form of land-use (complex land-use). Very flexible, but simple models. |
| Disadvantages | Requires specific knowledge and learning of connected thinking. | Models are extremely inflexible. Problems due to isolated particular areas. | Overall views usually cannot be derived with bottom-up procedures. |

mined where forests are possible and of which type given the initial rainfall regimes in that area. This process might well influence climate in larger areas of regions or even continents (NRC 1990). The method of active area dynamics can help to determine the final potential vegetation patterns if rainfall is evaluated beginning at the coasts and repeating the calculation ever further downwind.

*GIS supported agriculture for decreased environmental impact.* The threat of leaching of nitrates into the ground water depends on several factors, e.g. soil type, nitrate content of the soil, soil structure, nutrient uptake by the plants or location of ground water layers. In this application a risk map is produced to depict the risk that nitrates can leach to the groundwater. This is overlaid with a map showing the availability of nitrates. An object oriented model on plant growth is combined with a map showing the ecological conditions of the site ("suitability map"). This model is additionally fed with climate data. The model is used to show the predicted growth of an agricultural species on the suitability map. The uptake and hence availability of nitrates is depending on the growth of the plants and the initial availability of nitrates. The object oriented model is put onto each specific polygon that differs from its neighbors with respect to ecological conditions or availability of nitrates. With this combination of an object oriented model and a suitability map the development of nitrate availability can be predicted and monitored with a few actual measurements.

This map on availability of nitrate is overlaid with the risk map on leaching of nitrate. The resulting map shows the actual, temporal risk of leaching of nitrates. This latter map could even be used to spatially control the application of nitrates.

In a similar fashion other dangerous substances could be modelled.

### Shortcomings of the method of active area dynamics

Active area dynamics are much slower than aggregated dynamic models, as a small model is run for each polygon. If not only local, but in addition also spatially extended variables exist, these in addition have to be calculated for each polygon.

The generic models need more detailed information to achieve precise results than the aggregated models. It is possible that no sufficiently precise and complete information is available. As a consequence, the generic models would not be more precise than an aggregated dynamic model, they just would be slower.

In general the two types of dynamic maps cannot be compared because they use different kinds of information and are applied for different purposes. E.g. the BAM in the Rosalia area (mentioned above) would be far more precise than the map for the active area dynamics with 89 polygons, but the dynamics produced individually for these 89 polygons by the active area dynamics would be more correct than those from the aggregated model – as long, as no spatially extended factors enter. These factors could not be entered into the generic model, because it would make them fairly complex.

### Software tools for active area dynamics

The connection of generic models and GIS is methodologically difficult, because the bookkeeping is complex, where the model has to get which data and where it achieves which results. The development of a suitable software is urgently necessary but the expenditure of work is very high. But this new software is now under development. This software has to some extent, as mentioned before, to link capabilities of a relational data base and a modelling environment.

### 7. Summary and outlook

As different types of dynamic models have different areas of applications and different advantages and disadvantages they can for some applications be linked. This further extends the already vast applicability of dynamic maps.

The different types of dynamic maps allow new possibilities:

- Improved classification of remote imagery (a remote image is taken at a specific time). Dynamic maps can be produced to show the same area at the same time as the remote imagery. The dynamic maps can be based on data from terrestrial observation. In that case the two descriptions of the area are based on different sources of information. Therefore a comparison of differences can help to classify the remote imagery. Very often the classification of remote imagery is a major problem, in particular if new or unexpected developments occur.
- Support for environmental monitoring. The expected development (from any vehicle of anticipation) is translated into dynamic maps and compared with the actual ongoing development (from terrestrial observation, false-color infrared photographs or remote imagery). Differences may reveal unexpected developments at a very early time. The differences may be evaluated with statistics and comparison of patterns.
- Maps for guiding and scheduling management actions in ecosystems (e.g. where to plant or harvest what at a specific time). In the most elaborated form, a complex multispecies plant-model is translated into maps for a specific plantation. These maps will support schemes of complex land-use (Haber 1979, 1985; Haber et al. 1984).
- Assessing the applicability of models. Finding errors in structures and data.
- Calibrate, (in-)validate or improve models. One data base (the data used for the dynamic model) is used for calibration. The dynamic model is then applied in the process of generating dynamic maps. The resulting maps can in turn be compared with reality. This allows additional procedures for validation or "invalidation".
- Update information in the GIS. This update is partially based on model predictions which will not always be valid but will in each case demonstrate where the fastest or most drastic changes are either expected or are likely to occur and where hence efforts for updating by actual observation should have priority. This allows to partially overcome the limitations with the nearly static character of spatial information in GIS.

The coupling between dynamics and spatial details is a translation process that exists in reality. Therefore it is much more than a mere trick to circumvent the limits of dynamic models or of GIS'. As the bulk of the information is kept in data banks the dynamic models are kept small. As a necessary and adequate consequence this coupling prevents large scale models from coming into existence. Each type of data is processed with those methods that are most adequate for this processing. The result, the dynamic maps, allow many new possibilities for validation.

# References

Ashdown M, Schaller J (1990) Geographic information systems and their application in MAB-projects, Ecosystem Research and Environmental Monitoring. MAB Mitteilungen Nr. 34. Deutsches Nationalkomitee Bonn 250 pages (German and English)

ESSA (1982) Review and evaluation of adaptive environmental assessment and management. Environment Canada, Vancouver

Forrester JW (1968) Urban Dynamics. MIT-Press, Cambridge

Gleick J (1988) Chaos. Bantam Books, New York

Goodchild MR (1991) Presentation (by other speaker) in the 1991 "Regional Science Association" 31st RSA European Congress

Grossmann WD (1983) Systems approaches towards complex systems. In: Messerli P, Stucki E (eds) Fachbeiträge der schweizerischen MAB-Information, vol 19, Bundesamt für Umweltschutz, Bern

Grossmann WD (1990) Erfahrungen mit der hierarchischen Systemmethodik und dem Einsatz dynamischer Modelle im MAB 6-Projekt Berchtesgaden. (Experiences with the hierarchical sytems method and with the application of dynamic models in the MAB6 Project Berchtesgaden). Institution for Ecosystems and Environmental Studies. Austrian Academy of Sciences. Kegelgasse 27, A-1030 Vienna

Grossmann WD (1991 a) Model- and strategy-driven geographical maps for ecological research and management. In: Risser PG, Mellilo J (eds) Long term ecological research. (SCOPE Band). Wiley, New York

Grossmann WD (1991 b) Einsatz von Risikokarten in der Waldschadensproblematik: Konzept, Probleme und Ergebnisse im Projekt Lehrforst Rosalia (Assessing hypotheses on forest decline with geographical maps of risk: specifications and problems). Zbl ges Forstwesen 108:3 – 13

Grossmann WD (1991 c) Einsatz dynamischer Modelle in der Waldschadensproblematik: Anwendungsfelder, Probleme und Ergebnisse im Projekt Lehrforst Rosalia. (Application of dynamic models to the problem of forest damage: possibilities, problems and results). Zbl ges Forstwesen 108:15 – 35

Grossmann WD (1991 d) Ergebnisse der Anwendung von dynamischen Karten ("Zeitkarten") im Lehrforst Rosalia. (Results of the application of dynamic maps in the demonstration forest Rosalia). Zbl ges Forstwesen 108:215 – 235

Grossmann WD, Clemens-Schwartz (1985) OLIMP (Olympic Impacts) – Ein Modell über die Auswirkungen geplanter Olympischer Winterspiele in Berchtesgaden 1992, (OLIMP-Olympic Impacts – A Model on the Effects of intended Olympic Winter Games in Berchtesgaden 1992). In: German National Committee MAB (ed) MAB, vol 21. German National Committee MAB, Bonn, pp 225 – 242

Grossmann WD, Schaller J, Sittard M (1984) "Zeitkarten": eine neue Methode zum Test von Hypothesen und Gegenmaßnahmen beim Waldsterben. (Dynamic maps: a new method to test hypotheses and counter policies in the problem area of forest (dieback). Allgemeine Forstzeitschrift, München

Grossmann WD, Schaller J (1986) Geographical maps on forest die-off, driven by dynamic models. Ecol Modell 31:341 – 353

Haber W (1979) Raumordnungskonzepte aus der Sicht der Ökosystemforschung. Forschungs- und Sitzungsberichte der Akademie für Raumforschung und Landesplanung (Hannover) 131:12 – 24

Haber W (ed) (1985) Mögliche Auswirkungen der geplanten Olympischen Winterspiele 1992 auf das Regionale System Berchtesgaden. Technische Universität, Freising

Haber W, Grossmann WD, Schaller J (1984) Integrated evaluation and synthesis of data by connection dynamic feedback models with a geographic information system. In: Brandt J, Agger P (eds) Methodology in landscape ecological research and planning. Proc 1st International Seminar. Int. Assoc. of Landscape Ecology. Roskilde University Center, Roskilde

Haken H (1978) Synergetics. 2nd edn. Springer, Berlin Heidelberg New York

Kopcsa A, Grossmann WD (1991) Erstellen der Risikokarten für den Lehrforst Rosalia. Zbl ges Forstwesen 108 (in press)

Lee DB Jr (1973) Requiem for large-scale models. Am J Planners 34

Lorenz EN (1963) The predictability of hydrodynamic flow. Trans NY Acad Sci Ser 2, 25:409 – 432

May RM (1974) Biological populations with nonoverlapping generations. Science 186

Odum H (1982) Systems ecology. Wiley, New York

NRC (1990) National Research Council: Research strategies for the US global change research program. National Academy Press, Washington, DC

# Computer-aided regional planning

## Applications for the Perth and Helsinki regions

**Geoffrey G. Roy**[1] **and Folke Snickars**[2]

[1] Department of Computer Science, University of Western Australia, Perth, Australia
[2] Department of Regional Planning, Royal Institute of Technology, S-10044 Stockholm, Sweden

**Abstract.** The ISP computer system for interactive spatial planning was originally developed as a part of a project concerning land-use planning in the Malmö region of Sweden. In its first versions it was set up for use on mainframe computers. Land-use models were developed using this version of the computer system for Stockholm and Perth. During the last few years the system has been redesigned for a Sun workstation environment. The background for this redevelopment has been the use of the system for the preparation of a land-use plan for the metropolitan region of Helsinki. The philosophy of the ISP system is to develop an interactive work environment for the generation and evaluation of land-use plan alternatives. The focus is on man-machine interaction and flexibility of model development. The current paper outlines the properties of the current version of the ISP system and the traits of the models designed for Perth and Helsinki planners. The paper uses the Perth application to demonstrate the modelling capabilities within the system. The Helsinki application is used as a means to illustrate some traits of the graphics interface.

## 1. Introduction

Modelling exercises for forecasting and scenario construction in regional land-use and transport planning have had varying levels of popularity over the years. The early efforts were encouraged by the rapid developments in computing technology which enabled the solution of large sets of equations and the use of iterative solution techniques that were not previously available. In most instances, the underlying forecasting and planning models were relatively crude in their theoretical outlook, or so narrowly framed, that their relevance to practical planning problems was limited.

Our experience with planners and planning agencies has led us to a number of conclusions concerning the use of analytical techniques in plan-making situations. The first conclusion is that detailed knowledge of the future of specific parts of the regional system, generated by mathematical models, can contribute valuable input into the planning process. On the other hand, the final integration of this information into the decision-making processes is still very much a human endeavour. The second observation is that there is a necessary balance to be achieved between the computer-based analyses and the human decision-making processes if the performance of both aspects are to be optimized. A further conclusion is that there is still considerable scepticism amongst planners as to the value of sophisticated modelling systems in forecasting and scenario construction, particularly where the planners and their political clients cannot fully understand the market processes involved and the way that the analyst creates model abstractions of these. The fourth conclusion is that there is a growing need for planners to be able to better document the state of their study regions to be able to effectively develop policy, and to apply alternative planning scenarios to study the likely impacts of these policies. There is an urgent need for planners to place claims on modelling and computing to accommodate the specific problems that pertain to regional land-use and transport planning.

These conclusions do not indicate that any particular forecasting techniques or models are likely to be more acceptable than any other. What is indicated, is a need to develop flexible and theoretically well-founded computer systems which provide the necessary knowledge about restrictions and options for the planner. Such systems might be considered as decision support systems or expert systems depending on the level of scientific or experience-based knowledge included, see also Newton et al. (1987), Kim et al. (1989) and Kim and Han (1991). We prefer to use a more general notion of CARP (Computer-Aided Regional Planning) systems to express the bias towards improving the decision-making abilities of the user. In this way we presume, at the outset, that the user should be firmly in control of the computer system and be required to make most of the important decisions in design and manipulation of the forecasting techniques and models. The tools should be adapted to the context of the decision-making at hand.

Our objective in designing CARP systems is to provide the planner with a set of tools to support practical work in the planning process. The man-machine relationships must therefore be carefully designed and implemented in a way that is appropriate for the level of skill and knowledge of the prospective users.

One attempt to formalize these ideas into an operational system has been under development by the authors for some time and is currently being used by the State Planning Commission, Perth, Western Australia, the Office for Regional Planning and Urban Transportation, Stockholm County Council, Sweden and the Regional Planning Authority, Helsinki, Finland. The operational aspects of the ISP system are described elsewhere, see Snickars et al. (1981), Roy (1982), Roy and Snickars (1983) and Roy (1986). For the current paper, the intention is to discuss the formulation of the Perth and Helsinki models and to show how they are integrated into the ISP system, see also Roy (1986) and Eerola et al. (1989).

## 2. Background to the Perth application

The development of the Perth metropolitan region is probably not substantially different from most other large Australian cities. Perth is characterized by low residential densities, low levels of public transport utilization and a concentration of workplaces in the CBD area. The planning strategies available for Perth planners are various and will not be explored here. It is sufficient to say that these strategies will have impacts on a whole range of social, economic and political aspects of the Perth region. The planners' task is not necessarily to solve these problems but rather to propose feasible alternatives which will be further selected by a range of social and political decision processes.

Our direct concern is therefore not so much with the solutions obtained, or the decisions made, but rather to ensure that the best possible information has been generated and is available at the right level within the planning process.

The efforts to formalize models and methods for forecasting in the Perth region have been largely in response to the specific needs of the planning agency and the availability of data. In some areas there are gaps in the framework developed, which could be overcome given more effort in data collection and model design. The effort has been largely oriented to the development of a comprehensive framework for the regional modelling activity under the condition that this framework should comply with the capacity of the agency to absorb innovations in planning methodology.

The scope of the Perth model was selected to encompass two major aspects of the regional development. The application should deal with population development over space and time with specific interest on household formation and preferences for different classes of dwellings. The economic structural change should be forecasted in terms of employment development over space and time within a set of industry categories.

These variables would form the basis for the evaluating of a range of indices which would give the planners some idea of the impacts of alternative planning strategies in the areas of housing and residential-land demand, workplace location, commuting and other measures of the state of the regional land-use and transport system. While some parts of the modelling process are weaker than others, the framework is well defined, customized to the agency, and operational within the ISP system.

The forecasting processes are intended to project the development of the region into the future. From the user's perspective, the changes occurring over time will be central to the analysis of impacts of planning policy. The forecasting processes are designed to forecast to 1991 and then on to 2021 in five yearly increments. The base year being 1986, that is, using the most recent census data in the Perth metropolitan region.

## 3. System structure for the Perth model

The ISP system imposes certain restrictions on the formalization of a CARP system. While these restrictions might not be inherent to CARP systems in general, they describe the framework for our work with the application in Perth. In general we view the CARP system of consisting of four elements.

There is a regional description in terms of the basic data variables which describe the state of the region, historically and currently, and by which the future states can also be described. This forms a basis for an information system which the user can report from, or extract the required indices or indicators for evaluation purposes.

There is a set of models, forecasting techniques and any other analytical tools which enable the projection of the historical development of the region into the future. These models can be of varying complexity, some affecting small sets of variables, others of a more global nature. In essence we attempt to choose the best available procedures which will be of operational use for the planning agency.

There is a specification of planning policies designed to control, modify or influence the development of the region. These policies may be derived from social or political origin, or from what might be considered good planning practice. In practical terms these policies may influence the relationships among variables, between variables and scenario parameters or any other control system imposed by the user as a means of influencing the development of the region.

There is a specification of planning scenarios which contain those parameters exogenous to the processes built into the CARP system. We would expect such parameters to include items as region-wide population development, total employment within the chosen subsets of economic sectors, income development in the region, or future paths for the preservation of the natural environment. The idea is to create scenarios for those external factors which are essentially driving the development of the region.

For the Perth application, we are particularly interested in the spatial development of the region within given growth scenarios. The spatial development is constrained by the availability of land within various land-use categories and the spatially defined planning policies imposed by the user. These elements are integrated into the ISP system, which allows, through a natural language interface, the user to manipulate (add, change or delete) elements of policy regarding land-use and transportation which can be imposed on the forecasting processes.

The formal mathematical description of the model is in two parts. Firstly, there are those parts of the models, i.e. sets of variables, equations or constraints which are essential to the description of the regional system. These parts are fixed at the outset of an application and cannot normally be altered by the user. Secondly, there are user-imposed policies, normally interpreted as linear, or ratios of linear, equations or constraints within the computer system which are added to the models for forecasting purposes. The model building processes are automatically controlled with the help of the system, thus allowing the user to build many different models to reflect planning policy, and changes in policy, over a period of work with the system.

Currently the ISP system allows the user to impose linear functions in an interactive fashion. While this may be theoretically limiting, it has not yet proven to be a practical constraint in the application of the CARP methodology.

Some elements of the models implemented for a given region must be formally coded into the programs of the ISP system. This is necessary where complex evaluation functions are required and where some of the basic structural relationships in the models must be described. In general, the forecasting processes can be chosen to suit the nature of the variables included and the context of the decision-making at hand. We have often found that the knowledge concerning the mechanisms of change is ill structured and little region-specific knowledge exists to formally define the necessary models.

The ISP system is designed to accommodate three levels of forecasting, in an effort to develop a strategy to cope with the sequential nature of decision-making in planning:

(1) Prior models are developed for those sets of variables where future states can be described or estimated without overall knowledge of the state of the region being available. These models are used for first order estimates of future states for subsets of variables.

(2) A comprehensive forecasting module where rationed and free submarkets are cleared simultaneously in a so-called core model. There is a default forecasting method for this stage, based on minimum-information theory. This method is used to give an estimate of the most likely future state of the region in the absence of better information. If better information is available it should be used already in the prior models. However, an economic market-clearing model could also replace parts of the standard core model.

(3) Posterior models are employed to refine forecasts in selected areas of the system, given that there is an estimate of the overall regional state. These posterior models can be used to disaggregate information via the application of conditional forecasting. They can also be used to compute performance indicators, the values of which will affect the conditions for the three models steps in the next time period.

Table 1 gives a list of variables used in the Perth application. Each of these variables is defined for the 65 zones into which the Perth metropolitan region is divided. The variables fall into two broad categories which may be called core and side variables. Core variables are those which are the independent variables in the simultaneous forecasting step. This means that they play an interdependent role in determining the forecasting result among various submarkets which mutually affect one another. Core variables can be included in user-defined policies as a means of influencing the state of the region.

Side variables are dependent variables which may be determined from the core variables or from forecasting processes over which the user has no interactive control. In some instances, side variables are introduced for computational convenience so that certain attributes are readily available to the user within the data bases without the need to recompute them each time they are required in an analysis. In the Perth application, population by family and dwelling type and zone is a core variable whereas commuting flows are side variables. The latter are forecasted conditional on the former.

In spatial terms, the Perth region has been divided into zones in an attempt to approximate the intraregional activity variation in the region. Each zone is considered homogenous for modelling purposes. In this application, land-use in each zone is classified into three classes, urban, industrial and CBD. This allows us to more accurately describe the type of activities present in each zone even if the modelling process does not keep track of where in the zone the activities may occur.

**Table 1.** Variable definitions for the Perth model including a four letter mnemonic used in the man-machine interaction

| Number | Mnemonic | Description |
|--------|----------|-------------|
| *Core variables* | | |
| 1 | pald | Population of households with no dependents in low density dwellings |
| 2 | pfld | Population of households with dependents in low density dwellings |
| 3 | pamd | Population of households with no dependents in medium density dwellings |
| 4 | pfmd | Population of households with dependents in medium density dwellings |
| 5 | pnpd | Population in non-private dwellings |
| 6 | dald | Dwellings of low density having households with no dependents |
| 7 | dfld | Dwellings of low density having households with dependents |
| 8 | damd | Dwellings of medium density having households with no dependents |
| 9 | dfmd | Dwellings of medium density having households with dependents |
| 10 | mliv | Manufacturing workforce living |
| 11 | rliv | Retail workforce living |
| 12 | sliv | Service and primary production workforce living |
| 13 | cliv | Community service workforce living |
| 14 | bliv | Building and construction workforce living |
| 15 | uliv | Unemployment living |
| 16 | nliv | Non-workforce living |
| 17 | murb | Manufacturing employment in urban areas |
| 18 | mind | Manufacturing employment in industrial areas |
| 19 | mcbd | Manufacturing employment in cbd areas |
| 20 | rurb | Retail employment in urban areas |
| 21 | rind | Retail employment in industrial areas |
| 22 | rcbd | Retail employment in cbd areas |
| 23 | surb | Service and primary production employment in urban areas |
| 24 | sind | Service and primary production employment in industrial areas |
| 25 | scbd | Service and primary production employment in cbd areas |
| 26 | curb | Community service employment in urban areas |
| 27 | cind | Community service employment in industrial areas |
| 28 | ccbd | Community service employment in cbd areas |
| 29 | burb | Building and construction employment in urban areas |
| 30 | bind | Building and construction employment in industrial areas |
| 31 | bcbd | Building and construction employment in cbd areas |
| *Side variables* | | |
| 32 | dlow | Total number of low density dwellings |
| 33 | dmed | Total number of medium density dwellings |
| 34 | prdo | Total distance travelled by private mode from zone |
| 35 | pudo | Total distance travelled by public mode from zone |
| 36 | prdd | Total distance travelled by private mode to zone |
| 37 | pudd | Total distance travelled by public mode to zone |
| 38 | prto | Total time of travelling by private mode from zone |
| 39 | puto | Total time of travelling by public mode from zone |
| 40 | prtd | Total time of travelling by private mode to zone |
| 41 | putd | Total time of travelling by public mode to zone |
| 42 | lurb | Area of land classed as urban |
| 43 | lind | Area of land classed as industrial |
| 44 | lcbd | Area of land classed as CBD |
| 45 | etot | Total employment |
| 46 | clpr | Commuting trips leaving zone by private car |
| 47 | clpu | Commuting trips leaving zone by public transport |
| 48 | capr | Commuting trips arriving at zone by private car |
| 49 | capu | Commuting trips arriving at zone by public transport |
| 50 | free | Not yet used |

For a formal description of the Perth model we can let $X_{ij}(t)$ represent the state of the region at time $t$ for variable $i$ in zone $j$, $Y_{ij}(t)$ the prior guess of the state of the region at time $t$ for variable $i$ in zone $j$, and $Z_{ij}(t)$ the posterior calculation of the state of the region at time $t$ for variable $i$ in zone $j$. Then, the state at time $(t+1)$ is determined from the sequence of three operations:

$$Y_{ij}(t+1) = A\{X_{ij}(t), Z_{ij}(t)\} ,$$

$$X_{ij}(t+1) = E\{Y_{ij}(t+1)\} , \tag{1}$$

$$Z_{ij}(t+1) = P\{X_{ij}(t), X_{ij}(t+1)\} ,$$

where $A\{.,.\}$ represents a set of a priori functions applied to the known historical development of the region to obtain a first estimate of the state at time $(t+1)$, i.e. $X_{ij}(t+1)$, $E\{.\}$ is a forecasting function for the core set of variables, and $P\{.,.\}$ is a function to transform the core forecasts to side variables and to inputs to the prior step of the next time period. The latter functions will usually only provide estimates for part of the variable set and thus may not always ensure that the estimated state is consistent. There may be econometric forecasts as a part of both the prior and the posterior variables. The trajectory over time developing through this prior mechanism is adjusted to the path taken in the earlier time-stages via the core forecast.

The application of the default minimum-information model, in the core forecasting step allows all variables to be estimated, or adjusted, to ensure both internal consistency as well as compliance with the specified scenario parameters. The $E$-function at time $t$ is the reduced form of the optimal solution to a mini-mum-information problem, the objective function, $I$, of which is given by (2):

$$I\{X_{ij}(t), Y_{ij}(t)\} = \sum_i \sum_j [X_{ij}(t) \cdot \ln\{X_{ij}(t)/Y_{ij}(t)\}] . \tag{2}$$

The minimum-information problem implies the minimization of the non-linear function (2) subject to a set of linear constraints, as proposed by Snickars and Weibull (1977). This type of default forecasting process has three important prop-erties for use in CARP systems. Firstly, the forecasts are conservative, that is, the models will not generate changes in system states unless there are external or inter-nal driving forces at work, either in the form of scenario definitions or stemming from user applied policy. Secondly, through this simple default method, we are able to cope with varying qualities of information across the system without fun-damental solution problems. Thirdly, we can solve large problems quite quickly with efficient algorithms as the gradient technique developed by Eriksson (1980). It is absolutely necessary that the algorithm is efficient and reliable in con-vergence, since the model must be solved for a number of time periods during each forecasting process.

The posterior functions, $P\{.,.\}$, complete the estimation process, allowing the estimation of side variables, preparing for the computation of performance in-dicators and making adjustments for variables to be used in the next round of prior model calculations.

Repetition of this process in a sequential fashion provides the mechanism for forecasting over several time periods. At each time period, the state from the most recent forecast will form the basis for the next forecast period. The precise form and content of the prior and posterior models is not constrained to any specific model class. We would choose to use those model structures that seem most appropriate for the given application and data availability.

## 4. Models used in the Perth application

Due to the requirement from the planning agency to forecast for a period of several decades we expect to see substantial changes to the regional structure, especially in the possible outward growth of the Perth urban region. Forecasting to such distant points in time is of doubtful validity. It is still necessary for the planning agency to do so and they must act on other information in the current method is not used. It is clear that under the limited resources available, we will not be able to devise adequate models to produce generally reliable results. In this situation we accept to devise procedures that incorporate the expert knowledge of the Perth planners. The system is there to increase the quality of their decision-making and the quality of the ISP system must be compared to their best forecasting alternative. This strategy is the basis for development of CARP systems as the ISP. Investment decisions have to be made with or without decision support with a scientific base. The use of the ISP system will not only allow the testing of many policy alternatives but will also allow the planner to adapt the model building to local circumstances.

The submodels discussed below are not intended to represent a definitive statement of the dynamics of the Perth urban region. We include them here as a demonstration of the type of simple models which can be used in an attempt to represent the expert knowledge as provided by the planners. It is their qualitative model of how residential choice affects future land demand which is to be transformed into a quantitative model with the help of the system.

For the Perth region, the choice of residential location is considered to be a principle determinant of the form of the urban system in the foreseeable future. We have made efforts, therefore, to model the demand for urban land by making an a priori allocation of population according to a simple model which preferentially allocates population to land in zones that are empty when adjacent to zones are already heavily exploited. This strategy attempts to follow the observed development of the region as a consequence of a complex function of economic and demographic factors, and historical policy instruments. Although it would be possible to devise micro-economic models of residential demand, this commitment has not yet been made by the planners in the Perth metropolitan planning authority. The formulation of their qualitative principle as a prior forecasting mechanism is, in fact, equivalent to setting up an expert system on the basis of the planners' rules-of-thumb. This simple expert model is thus designed to substitute for a more elaborate residential choice model which could better estimate population location according to factors such as income, social fabric, land prices, and accessibility to services and employment.

**Table 2.** Example of performance indicators from the Perth application

| | |
|---|---|
| 1 | Employment density on Urban zoned land (persons/ha) |
| 2 | Employment density on Industrial zones land (persons/ha) |
| 3 | Employment density on CDB zoned land (persons/ha) |
| 4 | Utilization of Urban zoned land (%) |
| 5 | Utilization of Industrial zoned land (%) |
| 6 | Utilization of CBD zoned land (%) |
| 7 | Proportion of medium density dwellings (%) |
| 8 | Public transport trips from zone (%) |
| 9 | Public transport trips to zone (%) |
| 10 | Average travel distance from zone as origin (km) |
| 11 | Average travel distance to zone as destination (km) |
| 12 | Average household size (persons/household) |
| 13 | Percentage of population not in workforce (%) |

The results of this initial allocation of population, and other a priori models which might be implemented, form the starting point for the minimum-information forecasting procedures as described by (2). Adjustments to all core variables are made to find the most probable state of the region that is consistent with all imposed constraints and policies. All or some subgroup of core variables may be included into this part of the forecasting procedure as the users require the ability to express policy.

Following these adjustments, we enter the posterior models for further forecasting activity. In the Perth application, the major development is a simple commuting model which is designed to allow the user to extract a range of indices associated with changing commuting patterns. The commuting model is an attempt to provide some information on labour market behaviour which is determined as a consequence of changing land-use patterns. The development of accessibility properties of the land-use is also determined sequentially over time by using results from the core forecast to influence the posterior model step.

The posterior models, like the one above, are designed to add further information to the forecasting process given that we have produced a consistent estimate for the complete set of core variables. There are no essential restrictions as regards the form of posterior calculations, and any further computations based on the core variables can be made.

In a more general context, the evaluation of the forecast results can be aided by performance indicators which allow the user to measure the impacts of scenarios and policies along a number of dimensions. Table 2 shows an example list of indicators in use in the Perth application.

The main distinction between performance indicators and side variables is that the latter are not stored in the solution data bases but are computed as necessary. This means that their formulation can be altered without having to recreate complete data bases for all previously generated solutions. Because of their general nature, performance indicators must be formally programmed and linked into the ISP system. The ISP system also allows users to interactively create functions for evaluation purposes. These are limited to linear combinations and ratios of model variables and performance indicators.

For the Perth application we have three types of constraints, viz scenario constraints, market-clearing constraints and constraints arising from user-defined policies. The scenario constraints provide the primary relationships between scenario parameters and model variables. Typically, the scenario parameters provide the region-wide control totals for the various population classes and the employment levels in the economic sectors. The forecasting model, therefore, aims to estimate the regional distribution of activity subject to the region-wide scenarios. Market-clearing constraints are designed to ensure that the model variables remain internally consistent. For example, there is a claim for balance between workforce and employment at the region-wide level. Also a balance between population and workforce living within each zone needs to be maintained for the forecasts to be plausible and consistent statements about future conditions.

We also have sets of constraints to ensure that the limits of land availability are satisfied. In this context, we have provided two options for the users. The first one is to allow any activity to be located on any type of land type within a zone. The second is to restrict land-use to stay within the designated land-use classification. The first of these options could allow urban land activities to spill over onto industrial zoned land if the demand for that type of land was sufficiently large. The second option would ensure that this could not occur. The choice between these two approaches is a matter of planning policy.

We invariably will come across sets of variables that are poorly determined either via a scenario constraint or through some specific consistency requirement. These occur when the knowledge of the underlying mechanism for regional change, and the dynamics of urban and regional growth, is insufficient to correctly model the system behaviour. In such circumstances we must rely on the expert knowledge of the users to provide additional information to calibrate the forecasting processes. Typically, this calibration is based on the user's knowledge and understanding of the region, including the complexities and subtleties which cannot be included in our formal model specification. These considerations may give rise to sets of model constraints which will ensure that the forecast results are within the expected bounds.

The first two constraint sets presented above constitute what we might describe as the basic model structure, including the necessary mechanisms to correctly model system behaviour given some projected scenario parameters. The results generated by such a model might be viewed as a non-intervention forecast. Planners, and planning agencies, are of course often required to devise intervention schemes which will satisfy a range of social and economic objectives. These can be achieved in a number of ways, including the imposition of planning policy which directly effects the spatial distribution of activities.

The ISP system in fact encourages the user to test out such policies as a part of the process of developing and evaluating a range of planning strategies. The user may describe such policy in terms of constraints which are built into the forecasting processes.

## 5. The Helsinki application

We have made use of the Perth application to demonstrate some of the modelling capabilities of the ISP system. The Helsinki implementation is of more recent origin and is currently under further development. There are some peculiarities of the Helsinki model which stem from the way that Helsinki planners perceive their main problems in land-use and transport planning. The Helsinki model is more sophisticated than the Perth one because of the way that the software system has been developed for the application. We will not go into detail concerning the Helsinki application but mainly use it to demonstrate in graphical form how the ISP system operates. The work environment for the analyst is a SUN work station, preferrably with a high-resolution colour monitor.

The metropolitan region of Helsinki consists of ten municipalities, the City of Helsinki being the biggest one in population and employment. The City is located along the coast and the rest of the municipalities surround the central region on both eastern and western sides as well as extending the metropolitan region inland to the north. In the ISP system application for Helsinki there are 66 forecasting regions which can be aggregated to the ten municipalities for modelling and evaluation purposes.

Helsinki is a growing metropolitan region with some large-scale suburban developments to the north of the core region. The ownership of land is concentrated to large developer companies in several parts of the region. There is a conflict between the aims of these companies to transform their land to urban use and the goals of the public to create a region which is well served by technical infrastructure, rapid transit and other public services. There is also farming land surrounding the metropolitan region and the recent tendencies for urban expansion to spill over onto non-planned rural land is another worry for the Helsinki planners. Helsinki is surrounded by a beautiful coastline which is under heavy pressure to be further exploited for summer housing. The Helsinki region has several ring roads and public transport relies heavily on trams and buses. The economic specialisation of Helsinki is clearly towards service production and commerce. The high-technology firms seem to cluster in the municipality housing the technical university to the west of the City.

The specific situation for land-use transport planning in Finland has led the Helsinki planners to develop a model with the help of the ISP system which is very land-use oriented indeed. This can be gathered from the summary of model variables in Table 3.

The central variables of the Helsinki application are floor areas for sets of land-use activities. They were the first ones to be introduced. The population and employment variables were originally computed from the floor space ones. This meant, for instance, that there were no region-wide scenarios for population and employment. Instead, the development of the region in this regard was derived from the decisions to build floor space to be occupied at predetermined densities.

Two graphs have been included to demonstrate the capabilities of the ISP system as a tool for the creation and evaluation of land-use policy in the Helsinki region. They contain both historical data and forecasts as well as examples of performance indicators.

**Table 3.** Variables in the prototype version of the ISP model for Helsinki

| Number | Mnemonic | Description |
|--------|----------|-------------|
| *Core variables* | | |
| 1 | lflo | Floor area of low density dwellings |
| 2 | hflo | Floor area of high density dwellings |
| 3 | oflo | Floor area of office employment |
| 4 | iflo | Floor area of industrial employment |
| 5 | sflo | Floor area of service employment |
| 6 | refl | Floor area of rest of employment |
| *Side variables* | | |
| 7 | pdfl | Floor space potential for dwellings |
| 8 | pefl | Floor space potential for employment |
| 9 | plnd | Available land area |
| 10 | dlnd | Floor space density |
| 11 | blnd | Built land area |
| 12 | lpop | Population in labour force |
| 13 | hpop | Population outside of labour force |
| 14 | tpop | Total population |
| 15 | oemp | Office employment |
| 16 | iemp | Industrial employment |
| 17 | semp | Service employment |
| 18 | remp | Rest of employment |
| 19 | temp | Total employment |
| 20 | dvar | Adjustment in floor space density required to satisfy available potential |

The first part of Fig. 1 shows the basic structure of the ISP system plus a window where the sectoral and regional definitions have been collected. It may be noted that the ISP system is both menu and command driven. The icons on top of the main window show that four out of five potentially available windows are still not in use. These windows may be used to display results of forecasts for comparison. The bottom part of the figure shows the use of two windows to display the results of a forecast. A quick overview indicates that there will be a decentralisation of the population from centre to periphery in relative terms. The two maps have been scaled in such a way that the class limits are the same.

The ISP system contains the capability of interactively constructing any regional planning and forecasting model. In the leftmost part of Fig. 1, there is a list of menus showing that policies, constraints ans scenarios can be handled under separate menus. The user also has control over commands to bring in new regional subdivisions for result presentation or new subgroups of result variables to handle. In these contexts, there is always a simultaneous treatment of the historical data and the forecasts.

Figure 2 shows the use of the ISP system for displaying the results of computations of performance indicators from a forecasted solution. The accessibility conditions have been computed using travel times between subregions by car. The measure is the employment potential using a distance decay function weighted by the number of jobs in the destination regions.

The top part of the figure shows the time trajectories of the accessibility in two different ways. In the left part, temporal prisms are given for each subregion.

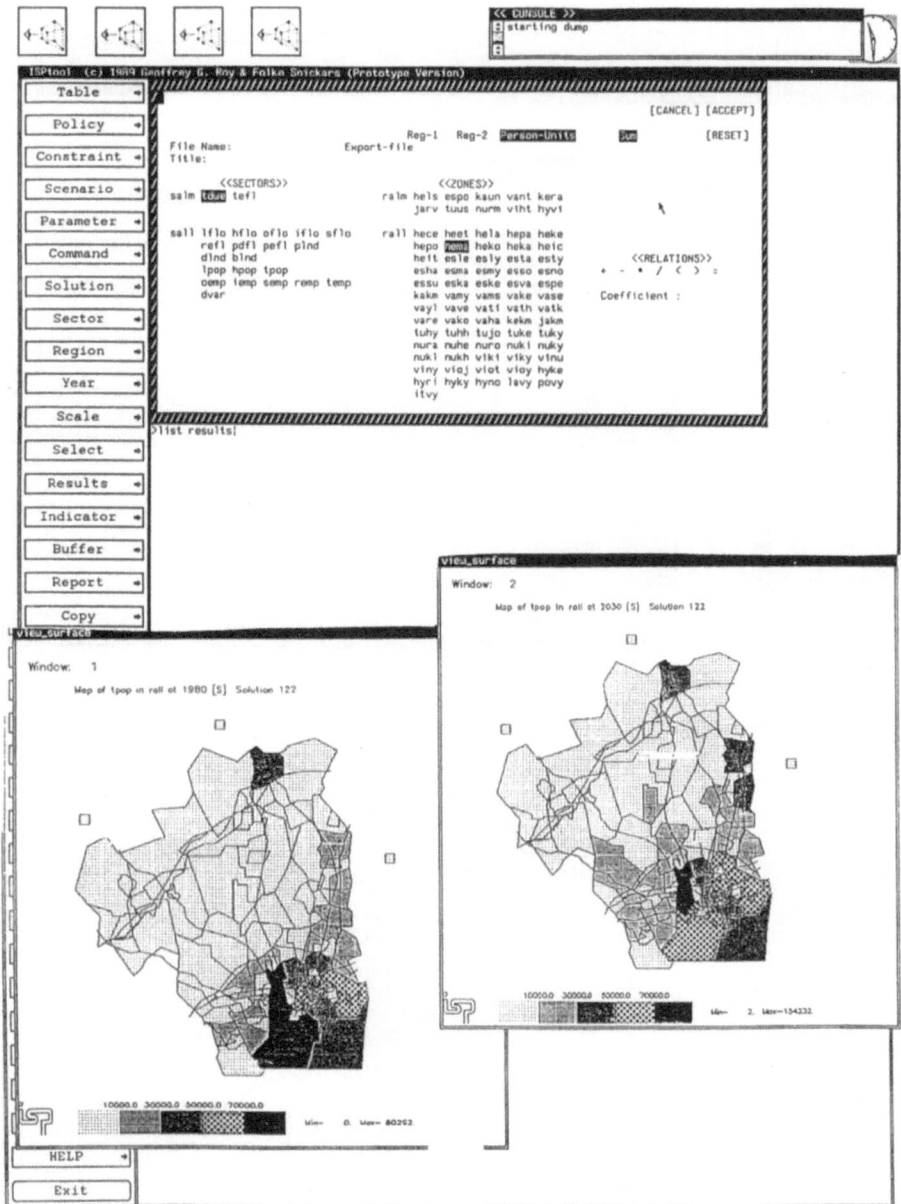

**Fig. 1.** Example of graphical display capabilities of the ISP system as regards menu options, window work and map presentations

The map has been rotated in such a way that there is maximum overview of the conditions in the different subregions. In the right part of the figure, simple time trajectories are used to depict the same indicator. It should be noted that the indicator is shown for an aggregation of the 66 subregions into ten municipalities in both parts of the figure. It is possible to present results for any predetermined

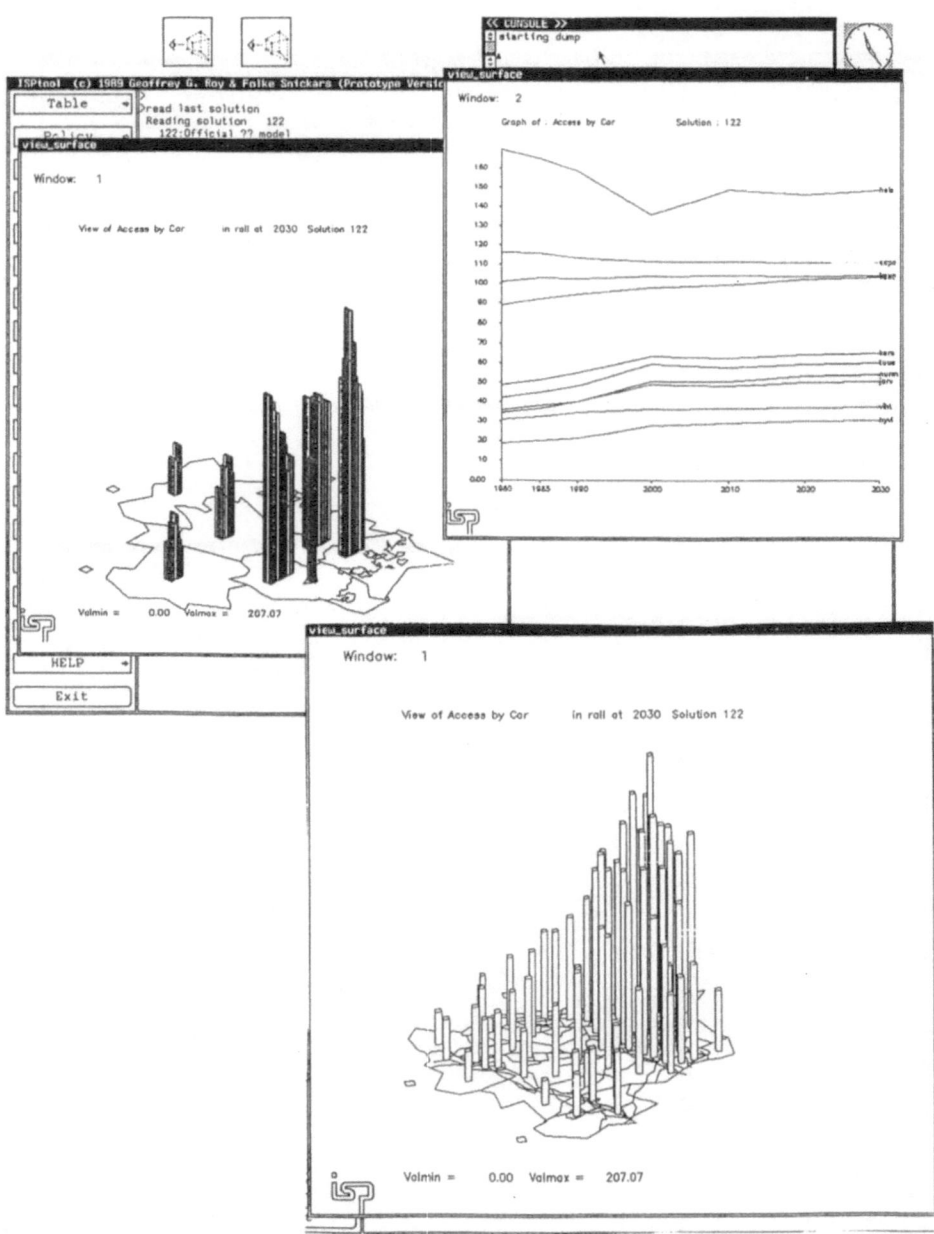

**Fig. 2.** Examples of results of forecasting with the ISP system in the Helsinki application. Accessibility to work from different zones by private car

aggregation of subregions for these graphical representations as with all the other ones which are available.

The bottom part of Fig. 2 shows the accessibility conditions by micro subregions for the forecast year 2030. The map has been rotated in such a way that the regions with the highest accessibility are placed at the back. It may be noted that graphs, prisms and maps can be employed to represent also information showing the difference between time periods or between forecast solutions. In fact, any combination of the results which can be computed using the simple arithmetic operations shown in the right part of the upper segment of Fig. 1, including the use of brackets to change their order of precedence, may also be represented in graphical form.

The Helsinki application is still under way. The strategy has been to develop a framework for land-use modelling with very simple variables and models at the outset. During the period of development, further complexity can be added to the basic ISP structure, and as further planning issues are to be dealt with also new graphical ways of interactively presenting the results can be provided. The strength of the ISP system in this process is its flexibility and the comprehensiveness of its historical and forecast data bases. Any forecast solution is stored together with all of the information necessary to reconstruct it if the need will arise, e.g. as a result of model reformulation.

## 6. The ISP system environment

The ISP system provides an integrated set of tools to assist in model development and testing as well as forecasting, analysis and reporting of results. It is specifically designed to be used by planners and as such employs a user-friendly interface which allows the user to communicate in a natural style of language. The user interface depends largely on computer graphic displays to report results from the data bases, and allows the user to make comparisons between several solutions. Alternative planning policies can be interactively described and the consequences evaluated from within the system.

The value of the system is therefore closely related to the efficiency in which a user can formulate a policy, use it for forecasting a new set of results, and study the consequences. Given that all forecast results are consistent both internally and with the scenarios, this provides a substantial improvement over manual methods and probably an increase in the reliability of the results. It is unusual for a planning agency to fully develop more than two or three plan alternatives using manual methods, even for a major review of a land-use and transport planning scheme. The ISP system, once set up and the application specific models implemented, provides a substantial increase in the planners' ability to evaluate many alternative planning strategies.

This paper has presented a brief review of the application of the ISP system to the development of planning models in the Perth and Helsinki regions. These applications are one of the first attempts to fully implement a CARP system that can be used by planners as a part of their data-to-day planning activity. CARP systems do not, in themselves, guarantee good quality models, forecasts or optimised solutions. They do, however, provide a tool-box for planners to use and adapt to their own specific needs.

The use of CARP systems will therefore place greater demands on the skills of planners as they will be required to be more explicit with the expression of policy issues and the evaluation of impacts. Systems which will externalize these issues will also encourage planners to spend more effort in developing and evaluating policy and less time on traditional number-crunching activities.

There are three main conclusions which can be drawn from our experience in working with regional planning agencies in introducing computer systems for decision support in their organisations.

The first one relates to the process of modelling the urban systems to accommodate for the local pecularities. It is first of all important to understand the local administrative structures and politicts, e.g. the balance between decision power and resources. This involves the identification of the players in the planning game and their respective agendas for development work. The knowledge and skills of the persons within the top and middle management hierarchy will be important for room to be made for efforts in development work.

There is a tendency in many organisations to listen more to proponents for the establishment of general-purpose information data bases than computer systems for specific planning purposes. The strategies in this regard will determine the expectations that the client will have on the support system, the readiness of the personnel to accept new technology and thinking, and the technical support that will be made available within the organisation.

The second conclusion relates to the balancing of model complexity with planning objectives and the learning strategies for the personnel involved. The introduction of complex models at the outset of an application of a decision support project will place high demands on the staff's skill and experience, and demand that substantial time is allocated to learning. In this mode of use, it might be advisable not to enter the decision support system into the organisation itself, but keep it external, selling the services from it to the organisation. It will not be possible to make use of expert judgement among the planners to any substantial extent if this mode is chosen. Another alternative is to develop learning strategies which will allow model complexity to increase over time as experience grows. There is still a need to trade off model validity and reliability with the costs for equipment and manpower involved in setting up the decision support system.

The third conclusion is that it is important to agree with the planning agency client that the decision support system is accepted for long-term use. If the system is accepted, and open to development, it can change the organisational structure of the agency, alter the competence needs, and enhance work productivity. If the system is accepted, the quality of it can be successively enhanced and the competence of the organisation continually increased. If this attitude is not taken, the application of computer-based decision support systems will not be successful in the planning organisation. This will reduce the pace of technological innovation in planning and may weaken the position of the planning profession.

## References

eerola E, Roy GG, Snickars F (1989) The ISP system and its application to land-use planning in the Helsinki region. Research Report, Helsinki Regional Planning Authority, Helsinki

Eriksson J (1980) On solving linearly constrained maximum entropy problems. Lith-Math-1980-14, Department of Mathematics, University of Linköping, Sweden

Kim TJ, Han S-Y (1991) Machine learning, expert systems and an integer programming model: Application to facility management. Working Paper, Department of Urban Planning, University of Illinois, Urbana-Champaign

Kim TJ, Wiggins L, Wright JR (1989) Expert systems: Applications to urban planning. Springer, Berlin Heidelberg New York

Newton P, Sharpe R, Taylor M (1987) Desktop planning. Croom-Helm, London

Roy GG (1982) An introduction to the ISP system for land-use planning. WP-82-70, IIASA, Laxenburg, Austria

Roy GG (1986) Urban modelling and decision-support systems. Proceedings 21st IAG Conference, Perth

Roy GG, Snickars F (1983) CARP: Computer-aided regional planning. Proceedings 1st Australian Conference on Computer Graphics, Sydney, pp 1-6

Snickars F, Andersson Å E, Holmberg I, Roy GG, Schultz J, Strömquist U (1981) Economic development and land-use in southwest Skåne. WP-81-00, IIASA, Laxenburg, Austria

Snickars F, Weibull J (1977) A minimum information principle: Theory and practice. Reg Sci Urban Econ 7:137-168

# Residential quality assessment

## Alternative approaches using GIS

Ayse Can

Department of Geography, Syracuse University, Syracuse, NY 13244, USA

**Abstract.** This paper focuses on the construction of residential quality scores as a preliminary step towards defining neighborhoods in urban areas. Two issues have been addressed: (1) the effect of spatial scale in the delineation of boundaries, and (2) variations in the assessment of residential quality when alternative methods are used. The spatial analysis undertaken in this research is integrated into a vector-based GIS environment to facilitate information exchange as well as the generation of topological information. The City of Syracuse (NY) is selected as the study area.

*Key words:* Residential quality, generalized concordance – discordance analysis, spatial autocorrelation, GIS.

## 1. Introduction

In the last two decades, the neighborhood has become the major spatial scale used in urban policy design and evaluation at the local government level. This is mostly an outcome of one of the requirements – citizen participation – of the Community Development Block Grants (CDBGs) established by the 1974 US Housing and Community Development Act. This act not only marked the beginning of increased local government responsibility to allocate federal funds but also required that local governments initiate neighborhood planning programs in order to increase citizen participation. This has resulted in planning which uses "the neighborhood" as the major entity in developing and implementing improvement projects in the areas of housing, social services, and economic development.

An essential component of neighborhood-based planning programs is the definition of officially recognized neighborhoods. Although methods for delineating neighborhood boundaries vary considerably depending on program objectives, typically an initial stage has involved the identification by the planning agency of discrete geographic entities that are relatively homogeneous residential environments. There is a consensus in literature that residential quality is a multidimensional concept encompassing socioeconomic conditions of residents, accessibility to employment centers and other urban activities, environmental characteristics, and the quality of local public services. However as reported by Rohe and Gates (1985, p. 73) the majority of the neighborhood planning programs have utilized only socioeconomic characteristics since such information is widely available through the US Census and Population and Housing. Although not a common practice, another approach has been the involvement of citizens and community leaders in the final designation of neighborhood boundaries.

This paper focuses on the construction of operational measures of residential quality based on socioeconomic indicators to identify homogeneous areas which will assist in the delineation of neighborhood boundaries in urban areas. Specifically the effect of spatial scale and the method used in the measurement of quality scores is addressed. With respect to the former, two major socioeconomic reporting zones of the US Census Bureau, namely tract and block group are employed. With respect to the latter, factor analysis and generalized concordance analysis are selected as representative methods from multivariate statistics and the multicriteria evaluation framework respectively. The spatial distribution of quality scores based on different scales and methods are examined using spatial autocorrelation statistics in order to discriminate between tract and block group as the spatial unit that should be used in the formation of neighborhoods. To illustrate these concepts the City of Syracuse in New York state has been selected as the study area.

The spatial analysis undertaken in this research is integrated into a vector-based GIS environment via ARC/INFO. This facilitates both the integration of attribute information with locational information and the generation of intermediate information needed in the construction and evaluation of quality indices. US Census Bureau's TIGER files are used to create geographic base maps, whereas the Census of Population and Housing is used in the generation of the attribute database. Algorithms developed in the $C$ programming language for the generalized concordance analysis and spatial autocorrelation statistics are integrated into ARC/INFO using macros. This not only adds to spatial analytical capabilities of a vector-based GIS but also illustrates the benefits obtained in the application of GIS to decision making in urban applications. The spatial query and analytical capabilities of a GIS will definitely facilitate the incorporation of criteria based on topological characteristics in addition to enhancing simple visual examination.

Next section presents a discussion of issues, namely the spatial scale and the method, involved in the construction of residential quality scores. In Sect. 3 a formal examination of the spatial distribution and structure of residential quality scores computed via alternative methods and scales is provided. A specific concern of this section is the quantification of the extent of spatial clustering among geographical units with respect to residential quality using spatial autocorrelation statistics. In the final section some concluding remarks are presented.

## 2. Construction of residential quality scores

With respect to the construction of composite residential quality indices, two interrelated issues deserve particular attention: (i) the spatial scale used to measure socioeconomic indicators, and (ii) the method employed to construct an operational measure of quality.

## 2.1 Spatial scale

Census of Population and Housing (Census Bureau 1980) provides the major source for socioeconomic data in planning applications. The three major geographic reporting zones for urban areas, block, block group, and tract, form a nested hierarchy. Blocks constitute the lowest level in the geographical hierarchy with block groups as aggregations of contiguous blocks and tracts as aggregations of contiguous block groups. Both block groups and tracts are modifiable geographical units since their delineations are arbitrarily done. The arbitrary nature of aggregation might lead to two interrelated problems – the scale and the aggregation problem (Arbia 1989). The scale problem refers to the varying nature of statistical measures for a given phenomenon depending on the geographic scale (level of resolution) used, whereas the aggregation problem refers to variations in statistical measures that might result when data is used for different aggregations of areal units. It has been shown that both spatial patterns and variable associations exhibit considerable differences at different scales and aggregation levels.

Two levels of spatial scale are used in constructing residential quality scores using socioeconomic data from the 1980 Census of Population and Housing Summary Tape File 1 (STF1) which contains complete count data[1]: (1) block group, which is composed of groups of census blocks and averages about 1100 people, and (2) tract which is an aggregation of block group and averages about 4000 people. Although it would be desirable to include block level in the analysis, information at this level is suppressed (except for total population and housing unit counts) for a considerable portion of the urban area due to confidentiality purposes if the total number of people are less than 15 and/or total number of housing units are less than 5. For the study area under consideration, over 30% of the blocks contained suppressed values, hence this study focuses on the block group level and tract level only.

The following are the selected variables that are typically used to capture the demographic and housing dimensions of socioeconomic variations: percentage of non-white persons (NONWHITE), percentage of female-headed single parent households with children under 18 present (SNG_FEM), percentage of occupied housing units that lack complete plumbing facilities (PLUMB), percentage of occupied housing units that have 1.01 or more persons per room (CROWD), percentage of vacant housing units (VACANT), the median value of specified owner-occupied housing units (VAL_OWN), the median contract rent of specified renter-occupied housing units (VAL_RENT). The first variable is a measure of ethnic homogeneity, whereas the second one is an indicator of poverty. The

---

[1]    STF 3 which is based on a sample count of households, provides a much more diverse set of socioeconomic variables for tracts. However since the purpose of the present investigation is to provide a comparison of results obtained at different scales, it is more important to use the same variables instead of using a more exhaustive set of variables.

general housing conditions are measured by the plumbing conditions, density of residence, and vacancy rate. The last two variables are indicators of housing value.

Table 1 provides the means and standard deviations for the selected variables as well as the correlation coefficients at both spatial scales. In accordance with the findings in literature, the spatial variation in the distribution of variables as exhibited by the standard deviations is much lower indicating the presence of a higher degree of homogeneity when the tract level is used. Also the higher values of correlation coefficients leads to the indication of stronger variable associations at the tract level.

The attribute information constructed for tracts and block groups is merged with locational information created in ARC/INFO for the study area. The precensus TIGER/Line files constitute the database in the construction of the tract and block group level maps[2]. TIGER files are digital street network files that have become one of the standard spatial data exchange formats for GIS. They have provided an inexpensive and relatively easy way of constructing geographic base maps in many GIS implementations that do not require a high degree of map accuracy. A technical documentation is available from the author (Can 1991) outlining the use of TIGER for the purposes of polygonization and address geocoding with ARC/INFO and with stand-alone $C$ algorithms.

## 2.2 Methodology

Two alternative approaches are taken for the development of an overall measure of residential quality using the individual socioeconomic indicators. The first involves factor analysis from multivariate statistics, which has constituted a commonly used approach to data reduction in evaluation research, e.g., Kain and Quigley (1970) and Smith (1973)[3]. The seven socioeconomic indicators are subjected to factor analysis using the correlation matrix. Standardized component scores on the first component are treated as composite residential quality scores. As can be seen in Table 2, the first component accounts for over 75% of total variation at the tract level and 68% at the block group level.

The second approach, on the other hand, looks at the assessment of residential quality as a discrete multicriteria evaluation problem and generates an overall performance score for geographical units based on their socioeconomic attributes. Although several procedures from multi-objective decision analysis exist to carry out multiple criteria ranking, here generalized concordance-discordance analysis is used (for a review of these methods, see Rietveld 1980). The use of this technique in residential quality assessment is not considered to be a standard approach and will be treated separately.

---

[2]    The precensus TIGER/Line files are readily available on IBM-standard computer tape and CD-ROMs (ISO 9660 standard). For a listing of reference materials on TIGER, please refer to pp. 2–3 in Census Bureau (1990).

[3]    Factor analysis is a method based on correlations to reduce the overall variation among variables into a smaller number of mutually independent variables, called factors. For a review see Johnston (1978).

**Table 1.** Summary statistics and correlation coefficients

*Tract level*

Means and standard deviations from 56 census tracts

|  | NONWHITE | SNG_FEM | PLUMB | CROWD | VACANT | VAL_OWN | VAL_RENT |
|---|---|---|---|---|---|---|---|
| Mean | 0.19260213 | 0.10730679 | 0.09947804 | 0.02332804 | 0.08165679 | -27066.071 | -175.75754 |
| Std Dev | 0.25443237 | 0.0778222 | 0.07028208 | 0.01854771 | 0.06290699 | 9664.03411 | 34.0078303 |

Correlations

|  | NONWHITE | SNG_FEM | PLUMB | CROWD | VACANT | VAL_OWN | VAL_RENT |
|---|---|---|---|---|---|---|---|
| NONWHITE | 1.00000 |  |  |  |  |  |  |
| SNG_FEM | 0.87192 | 1.00000 |  |  |  |  |  |
| PLUMB | 0.69014 | 0.79717 | 1.00000 |  |  |  |  |
| CROWD | 0.82572 | 0.92092 | 0.84867 | 1.00000 |  |  |  |
| VACANT | 0.76126 | 0.83417 | 0.97541 | 0.85745 | 1.00000 |  |  |
| VAL_OWN | 0.46661 | 0.68818 | 0.71238 | 0.66151 | 0.68312 | 1.00000 |  |
| VAL_RENT | 0.31424 | 0.63132 | 0.54427 | 0.58222 | 0.50432 | 0.76131 | 1.00000 |

*Block group level*

Means and standard deviations from 182 block groups

|  | NONWHITE | SNG_FEM | PLUMB | CROWD | VACANT | VAL_OWN | VAL_RENT |
|---|---|---|---|---|---|---|---|
| Mean | 0.1828769 | 0.10393934 | 0.09852988 | 0.02313857 | 0.08170555 | -28015.659 | -177.12082 |
| Std Dev | 0.2528489 | 0.08669964 | 0.0894965 | 0.02353282 | 0.07974587 | 11220.1176 | 38.143316 |

Correlations

|  | NONWHITE | SNG_FEM | PLUMB | CROWD | VACANT | VAL_OWN | VAL_RENT |
|---|---|---|---|---|---|---|---|
| NONWHITE | 1.00000 |  |  |  |  |  |  |
| SNG_FEM | 0.83626 | 1.00000 |  |  |  |  |  |
| PLUMB | 0.64035 | 0.67189 | 1.00000 |  |  |  |  |
| CROWD | 0.67588 | 0.77475 | 0.61246 | 1.00000 |  |  |  |
| VACANT | 0.70437 | 0.71432 | 0.97056 | 0.63890 | 1.00000 |  |  |
| VAL_OWN | 0.42019 | 0.58336 | 0.62174 | 0.48855 | 0.61455 | 1.00000 |  |
| VAL_RENT | 0.34384 | 0.55626 | 0.52345 | 0.48186 | 0.49006 | 0.70585 | 1.00000 |

**Table 2.** Results of factor analysis

*Tract level*
Eigenvalues of the correlation matrix

|  | 1 | 2 | 3 | 4 | 5 | 6 | 7 |
|---|---|---|---|---|---|---|---|
| Eigenvalue | 5.3167 | 0.9074 | 0.3996 | 0.2092 | 0.1012 | 0.0496 | 0.0163 |
| Difference | 4.4094 | 0.5078 | 0.1904 | 0.1079 | 0.0517 | 0.0333 |  |
| Proportion | 0.7595 | 0.1296 | 0.0571 | 0.0299 | 0.0145 | 0.0071 | 0.0023 |
| Cumulative | 0.7595 | 0.8892 | 0.9462 | 0.9761 | 0.9906 | 0.9977 | 1.0000 |

|  | First component loading | Squared correlation |
|---|---|---|
| NONWHITE | 0.82162 | 0.675062 |
| SNG_FEM | 0.94789 | 0.898497 |
| PLUMB | 0.92313 | 0.852175 |
| CROWD | 0.94326 | 0.889733 |
| VACANT | 0.93267 | 0.869881 |
| VAL_OWN | 0.80668 | 0.650737 |
| VAL_RENT | 0.69329 | 0.480648 |

*Block group level*
Eigenvalue of the correlation matrix

|  | 1 | 2 | 3 | 4 | 5 | 6 | 7 |
|---|---|---|---|---|---|---|---|
| Eigenvalue | 4.7659 | 0.9083 | 0.5960 | 0.3191 | 0.2693 | 0.1169 | 0.0244 |
| Difference | 3.8576 | 0.3123 | 0.2770 | 0.0497 | 0.1525 | 0.0925 |  |
| Proportion | 0.6808 | 0.1298 | 0.0851 | 0.0456 | 0.0385 | 0.0167 | 0.0035 |
| Cumulative | 0.6808 | 0.8106 | 0.8958 | 0.9413 | 0.9798 | 0.9965 | 1.0000 |

|  | First component loading | Squared correlation |
|---|---|---|
| NONWHITE | 0.81120 | 0.658044 |
| SNG_FEM | 0.89548 | 0.801879 |
| PLUMB | 0.88338 | 0.780355 |
| CROWD | 0.81315 | 0.661205 |
| VACANT | 0.90087 | 0.811571 |
| VAL_OWN | 0.75733 | 0.573545 |
| VAL_RENT | 0.69234 | 0.479338 |

*Generalized concordance-discordance analysis:* Generalized concordance-discordance analysis is a multicriteria evaluation method that is a variant of the general family of ELECTRE methods (Roy 1971). It has been widely used in various urban and regional planning applications in the evaluation of alternative actions according to multiple criteria (for a review of applications, see Van Delft and Nijkamp 1977). This method is based on the quantification of differences in criteria values across alternatives using pairwise comparisons. It is more powerful than the weighted summation technique which is a very commonly used one in evaluation research in discriminating across alternatives (Rogerson et al. 1989). In the context of residential quality assessment, the goal of the decision problem is to rank geographical units according to selected socioeconomic indicators. The spatial units, i.e., tracts and block groups, constitute a finite set of alternatives $[i = 1, \ldots, N]$ and socioeconomic indicators constitute a finite set of criteria $[j = 1, \ldots, K]$.

The first step involves the construction of the impact matrix, $P$, with dimensions $(N \times K)$ based on alternatives and criteria. The elements $P_{ij}$ are the original

data values. The elements of this matrix need to be normalized with values ranging from 0 to 1, in order to make the criterion values comparable. Although there
are various standardization methods, here two linear transformations are selected
depending on the direction of criteria in contributing to the overall objective. For
those criteria that contribute in a positive way, the standardized value, $R_{ij}$, equals
$P_{ij}/P_j^{max}$, where the denominator is the maximum value of the corresponding
criteria. For those attributes that contribute in a negative manner, the standardized value, $R_{ij}$, equals $1-(P_{ij}/P_j^{max})$. This way all normalized impacts contribute
in the same direction to the overall objective.

The second step involves the construction of concordance and discordance
measures based on a pairwise comparison of alternatives. The concordance measure, $c_{ii'}$, is the overall dominance score for alternative $i$ for those criteria over
which it performs better than alternative $i'$. The discordance measure, $d_{ii'}$, on the
other hand, measures the degree of dominance alternative $i'$ has over $i$. These are
defined as follows:

$$c_{ii'} = \sum_{j \in C_{ii'}} w_j \; ,$$

where

$$C_{ii'} = \{j \,|\, R_{ij} > R_{i'j}\} \; . \tag{1}$$

$C_{ii'}$ is the concordance set which includes all criteria for which alternative $i$ is
better than, i.e., has a higher value than alternative $i'$.

$$d_{ii'} = \max_{j \in D_{ii'}} [(w_j \,|\, R_{ij} - R_{i'j}\,|)/d_j^{max}] \; , \tag{2}$$

where

$$d_j^{max} = \max_{ii'} (w_j \,|\, R_{ij} - R_{i'j}\,|)$$
$$D_{ii'} = \{j \,|\, R_{ij} < R_{i'j}\} \; . \tag{2}$$

$D_{ii'}$ is the discordance set which includes all criteria for which alternative $i$ is
worse than, i.e., has a lower value than alternative $i'$.

In these equations, $w_j, j = 1, \ldots, K$, is a vector of weights which determine
the relative importance of criteria. Weights measure the preferences of priorities
attached by the decision maker(s) to each criterion. There are various preference
measurement techniques that can be used in multicriteria decision analysis which
are in principle based on either the ranking or rating of criteria. Within the context of multi-attribute utility theory, utility functions form the basis in the quantification of preferences. Alternative specifications of utility functions as well as
common methods used in applications to determine their parameters are
presented in Rietveld (1980). The specification of preferences is a somewhat subjective issue depending on the nature of the decision problem, the information
available, and the decision makers. There is not exact method for the determination of weights (Voogd 1983, p. 109).

In the present application given the nature of the decision problem, there are no a priori reasons for assuming that criteria contribute disproportionately to the overall evaluation of residential quality. Therefore all criteria are taken to contribute equally, i.e., $w_1 = w_2 = w_K$. If, however, there are any reasons to believe that households might discriminate unevenly among criteria then this can be taken into account by assigning different weights.

Once the concordance and discordance measures are computed the next step involves the formation of overall dominance indices which provide the final ranking of alternatives. These are computed as follows:

Concordance-dominance index:

$$C_i = \sum_{i'=1}^{N} c_{ii'} - \sum_{i'=1}^{N} c_{i'i} \ . \tag{3}$$

Discordance-dominance index:

$$D_i = \sum_{i'=1}^{N} d_{ii'} - \sum_{i'=1}^{N} d_{i'i} \ . \tag{4}$$

In the present application, the lower values of the concordance-dominance index and the higher values of the discordance-dominance index indicate higher residential quality given the way criteria is structured in the impact matrix.

It should be noted that an important consideration when multi-criteria evaluation techniques are used is the potential presence of uncertainties. As reported in Voogd (1983) uncertainties might arise due to incomplete or incorrect information on criteria or alternatives and/or specification of preferences. In addition uncertainties might result from the inherent assumptions of the method used. In the current application, given the somewhat deterministic nature of the decision problem, i.e., the number of alternatives are known and fixed, and the criteria are objective indicators measured on a quantitative scale, the amount of uncertainty is confined to the specification of weights and any measurement errors present in the data. Although not pursued here, it would be desirable to carry out a sensitivity analysis by using alternative ratings of the criteria.

An algorithm in $C$ programming language is developed for the generalized concordance-discordance analysis. The maximum number of 225 alternatives and 20 criteria are allowed. If unequal weights are used, their ratings, which should sum to 1, need to be interactively specified. A listing of this algorithm is available from the author upon request. The algorithm can either be used as a stand-alone program with an ASCII file containing the impact matrix, or can be run from within ARC/INFO using a macro. The macro is written so that the program can access directly the polygon attribute table (PAT) of an ARC/INFO coverage.

Multi-criteria analysis is one of the modules included in the NCGIA Core curriculum under decision making in a GIS context (NCGIA 1990). Indeed concordance-discordance analysis is presented as a goal-programming method when discrete alternatives are present. Therefore the incorporation of this analysis into ARC/INFO provides an enhancement to the current analytical capabilities of GIS in the area of spatial decision analysis.

## 3. Analysis of spatial distribution of residential quality scores

The spatial distribution of residential quality scores that are computed by two methods, i.e., factor analysis and concordance-discordance analysis, are presented in Figs. 1–3 for census tracts and block groups in the city of Syracuse. In these maps four-equal class intervals are used. The shading pattern utilizes a decreasing density of shading to reflect an increasing level of residential affluence. In all maps, the neighborhood boundaries designated by the city of Syracuse are shown

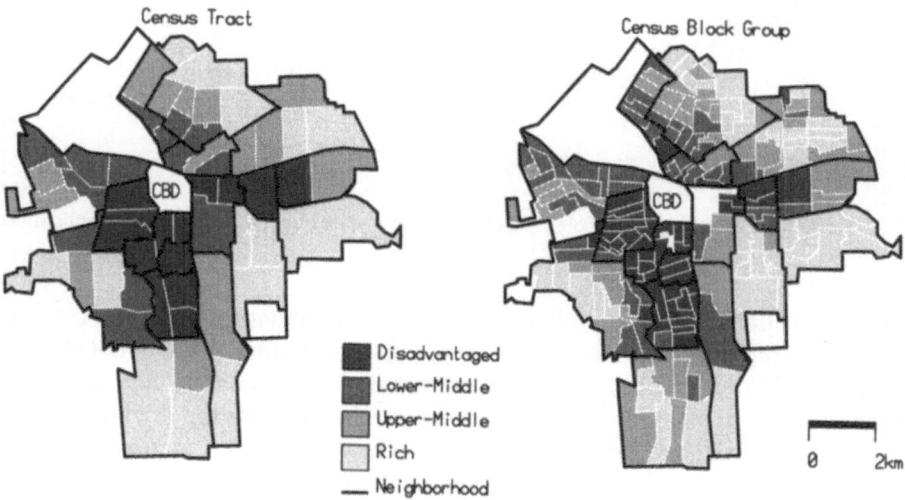

**Fig. 1.** Distribution of principal component scores. (The white lines denote tract and block group boundaries)

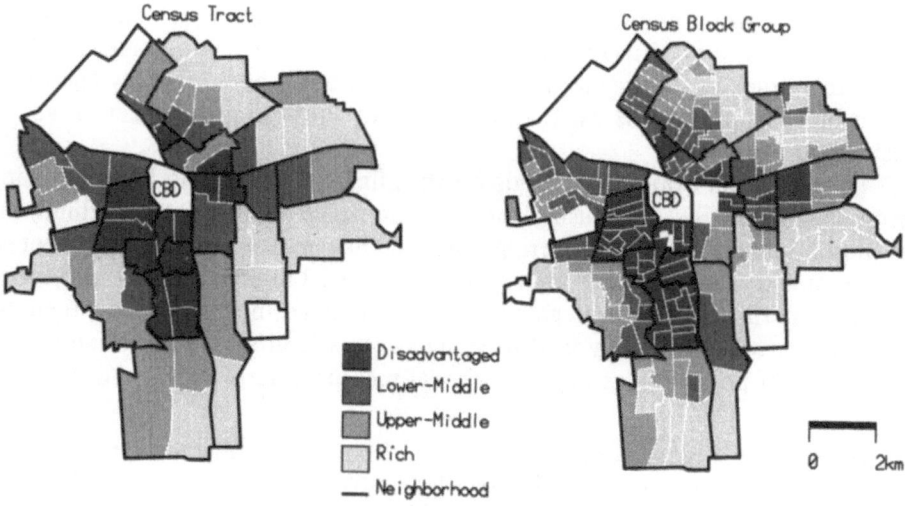

**Fig. 2.** Distribution of concordance scores. (The white lines denote tract and block group boundaries)

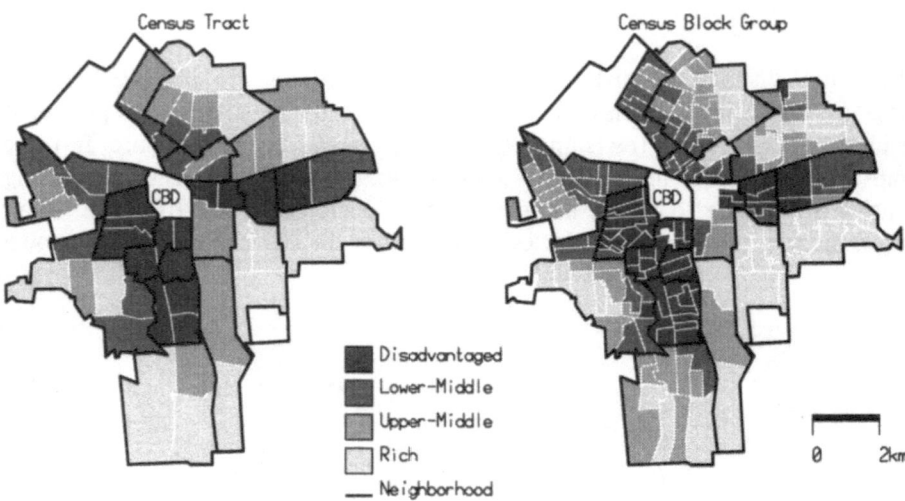

**Fig. 3.** Distribution of discordance scores. (The white lines denote tract and block group boundaries)

by a thick solid line pattern, whereas the white line pattern depicts the tract or block group boundaries[4]. On these maps, geographical units that are not shaded but still within the city boundaries contain suppressed values for total number of persons and/or housing units and therefore not included in the analysis.

The visual examination of these maps indicates a strong similarity in the spatial patterning of residential quality measured according to different methods. There is a clustering of the densest shading stretching out from the city center to the south and east which coincides with the location of poor and transitional neighborhoods. The areas with higher residential quality are found in the outer parts of the city mostly clustering in the eastern part. The extent of heterogeneity with respect to residential quality within the official neighborhoods is exhibited by the block group maps. As commonly done, the definition of neighborhoods by the Syracuse city officials is based on the tract level. These boundaries which seem to border relatively uniform tracts, cut through homogeneous territories when block group level is used especially for the medium and higher ranges of residential affluence.

The definition of neighborhoods for the purposes of neighborhood planning programs should be done using the scale at which quality scores exhibit the greatest amount of spatial clustering. Research has indicated that if the extent of spatial autocorrelation is not high among the spatial units, i.e., tracts and block groups, and the resulting aggregate unit is not small, i.e., neighborhoods, then the aggregation will most likely lead to some systematic grouping of variables and thus result in possible bias in the statistical measures (Arbia 1989). An analogy

---

[4]     These neighborhoods were defined as a part of the program called Neighborhood Statistics initiated by US Census Bureau in response to request by the neighborhood representatives and city officials. The Census Bureau made available the 1980 statistics for these neighborhoods to those localities that participated in the program.

from the world of photography can be made. If one were to consider that a picture is made up of dots, then the size of the dots (level of resolution) is critical in determining whether or not an image is recognizable. If the dots are too large, the image detail will be lost as one area is averaged with other areas to create the dot. This type of effect is the same as that felt in geographical applications when the level of aggregation in data collected obscures the underlying spatial pattern. The selection of appropriate scale to be used in aggregation can be aided using the spatial autocorrelation statistics, such as Moran's $I$ or Geary's $C$ index. In this research, Moran's $I$ is used to measure the extent of spatial clustering among tracts and block groups with respect to residential quality scores based on two methods.

The form of Moran's $I$ is formally given (Cliff and Ord 1981):

$$I = \frac{N}{\sum_i \sum_j w_{ij}} \frac{\sum_i \sum_j w_{ij}(X_i - \bar{X})(X_j - \bar{X})}{\sum_i (X_i - \bar{X})^2} , \tag{5}$$

where

$w_{ij}$ = the element of $W$, the spatial weights matrix, with dimension $N \times N$
$X_i$ = the vector of the attribute values, $i = 1, \ldots, N$ or $j = 1, \ldots, N$
$\bar{X}$ = the mean of the attribute values.

It should be noted that for a row-standardized $W$, i.e., with elements $w_{ij} = w_{ij} / \sum_j w_{ij}$, the first term in (5) drops out since the sum of the elements in $W$ is equal to $N$.

The variance, $V(I)$, and expected value, $E(I)$, of Moran's $I$ under the assumption of normality are:

$$V(I) = \frac{N^2 S_1 - N S_2 + 3(\sum_i \sum_j w_{ij})^2}{(\sum_i \sum_j w_{ij})_2 (N_2 - 1)}$$

$$E(I) = \frac{-1}{N-1} ,$$

where

$$S_1 = 1/2 \ \sum_i \sum_j (w_{ij} + w_{ij})^2$$

$$S_2 = \sum (\sum w_{ij} + w_{ji})^2 .$$

The variance, $V_r(I)$, under randomization is:

$$V_r(I) = \frac{N[(N^2 - 3N + 3)S_1 - N S_2 + 3 S_0^2] - b_2[(N^2 - N)S_1 - 2N S_2 + 6 S_0^2]}{(N-1)(N-2)(N-3)S_0^2} ,$$

where

$S_0 = \sum_i \sum_j w_{ij} = N$ (for a row-standardized matrix)

$$b_2 = \frac{1}{N} \frac{\sum_i (X_i - \bar{X})^4}{(\sum_i (X_i - \bar{X})^2 / N)^2} .$$

Given the variance and the expected value, the standard normal variate, $Z$, can be computed in the usual manner for statistical inference. A positive value of Moran's $I$ if statistically significant will indicate the clustering of similar residential quality scores across spatial units.

There are several ways by which the elements of the spatial weight matrix, $w_{ij}$, can be a priori specified depending on the adjacency assumptions. The commonly used measures include simple contiguity, simple contiguity weighted by the length of the common border, and the distance between the centroids of geographical units. In this application the simple contiguity definition is used. Hence two spatial units are considered to be contiguous, i.e., $w_{ij} = 1$, if they have a common boundary (join). A major difficulty in the construction of the spatial weight matrix is the identification of joins when large data sets are used. One of the ways this can be facilitated is the use of a vector-based GIS, such as ARC/INFO, whose data model is structured so that complex topological information among spatial units can be easily identified. Kehris (1990) and Chou (1989) have already taken advantage of ARC/INFO's data model and have developed algorithms in Fortran to extract the contiguity information in the construction of the spatial weight matrix to compute spatial autocorrelation statistics.

The current study takes a similar approach and develops an algorithm in $C$ language to compute Moran's $I$ test for spatial autocorrelation using ARC/INFO's data files for a coverage. The program uses the information from the Arc Attribute Table (AAT) and Polygon Attribute Table (PAT). The AAT provides the necessary contiguity information in order to form the spatial weight matrix in its two reserved fields, namely (LPOLY # (the polygon number to the left of a join) and RPOLY # (the polygon number to the right of a join). The PAT, on the other hand, contains the attribute on which the autocorrelation statistics want to be computed. A listing of this algorithm is also available from the present author. A macro is written so that this algorithm can be executed from within ARC/INFO.

The results of Moran's $I$ for spatial autocorrelation are presented in Table 3. The values of the test statistic are very similar across residential quality scores computed using factor analysis and generalized concordance-discordance analysis at both spatial scales. As indicated by substantially larger $Z$-values, the extent of clustering is much stronger among block groups than tracts. This leads to the indication that the aggregation using the tract level may lead to bias. Therefore we conclude that the block group should be used in the formation of neighborhoods.

## 4. Conclusions

This study has focused on one of the major components of neighborhood-based planning, namely the computation of residential quality scores as a prerequisite

**Table 3.** Moran's *I* test for spatial autocorrelation

*Tract level*

|  | (1) | (2) | (3) |
|---|---|---|---|
| I | 0.5101 | 0.5018 | 0.4621 |
| Mean | −0.0182 | −0.0182 | −0.0182 |
| (Normality) |  |  |  |
| Variance | 82.205E-04 | 82.205E-04 | 82.205E-04 |
| Z-Value* | 5.8262 | 5.7353 | 5.2970 |
| (Randomization) |  |  |  |
| Variance | 82.049E-04 | 83.735E-04 | 82.383E-04 |
| Z-Value* | 5.8317 | 5.6826 | 5.2913 |

*Block group level*

|  | (1) | (2) | (3) |
|---|---|---|---|
| I | 0.7552 | 0.7678 | 0.7543 |
| Mean | −0.0055 | −0.0055 | −0.0055 |
| (Normality) |  |  |  |
| Variance | 23.878E-04 | 23.878E-04 | 23.878E-04 |
| Z-Value* | 15.5673 | 15.8266 | 15.5495 |
| (Randomization) |  |  |  |
| Variance | 23.818E-04 | 24.023E-044 | 23.833E-04 |
| Z-value* | 15.5868 | 15.7788 | 15.5642 |

(1) Factor scores; (2) Concordance-dominance scores; (3) Discordance-dominance scores
* All significant at 0.0001 level

for defining neighborhoods. It has provided methods and tools that can aid in the formation of neighborhoods by planning authorities and decision makers. It has illustrated that spatial scale and method are important in the designation of neighborhoods when socioeconomic data is used. Although only objective socioeconomic data has been used, the same approach can be applied using qualitative data. In addition the same approach can be taken to the evaluation of neighborhood quality using the current designations.

Although not carried out in this paper, a clustering algorithm is the next logical step in the formation of more homogeneous neighborhoods with respect to residential quality given a size and/or population criteria for clusters, as is commonly done in political redistricting. For a review of currently available GIS redistricting software, see GIS World (1991, vol. 4, #4). This study was primarily concerned with the preliminary formal analysis of the spatial patterns and the organization of residential quality scores at different scales. This is considered to be an essential step prior to the aggregation of geographical units into neighborhoods.

This paper has demonstrated how data input/output as well as analytical capabilities of a GIS (ARC/INFO) when complemented by other tools, enhances the researcher's capabilities in spatial analysis. It has demonstrated how the data model of a vector-based GIS can facilitate the analysis of geographic data. The

incorporation of generalized concordance-discordance analysis and a test statistic
for spatial autocorrelation has added flexibility to the use of GIS.

*Acknowledgements.* An earlier version of this paper was presented at the 31st European Congress of
the Regional Science Association, August 27–30, 1991. The Census of Population and Housing 1980
Summary Tape File 1 and the TIGER files used in this paper were made available by the Syracuse
University Library Data Archive. The GIS work is conducted at the Advanced Graphics Research Laboratory (AGRL) of Syracuse University on a Sun-Sparc workstation using ARC/INFO® v. 5.0.1.,
registered trademark of Environmental Systems Research Institute (ESRI). I would like to thank
Gerald A. Talen for his programming assistance in the development of *C* algorithms.

# References

Arbia G (1989) Spatial data configuration in statistical analysis of regional economic and related
    problems. Kluwer, Dordrecht
Can A (1991) Polygonization and address geocoding using TIGER files. Working Paper (in progress),
    Department of Geography, Syracuse, University, Syracuse, NY
Chou YH (1989) Analyzing the spatial autocorrelation on polygonal data in GIS. URISA Conference
    Proceedings, Vol 4, pp 138–148
Cliff A, Ord K (1981) Spatial processes, models and applications. Pion, London
Delft A van, Nijkamp P (1977) Multicriteria analysis and regional decision-making. Martinus Nijhoff,
    The Hague/Boston
GIS World (1991) A GIS magazine published by GIS World, Inc., CO, Vol 4 (4)
Johnston RJ (1978) Multivariate statistical analysis in geography. Longman, New York
Kain JF, Quigley JM (1970) Evaluating the quality of the residential environment. Environ Plann
    2:23–32
Kehris E (1990) Spatial autocorrelation statistics in ARC/INFO. Research Report No 16, North West
    Regional Research Laboratory, Lancester University
NCGIA (1990) Application issues in GIS. NCGIA Core Curriculum, Vol 3, University of California,
    Santa Barbara
Rietveld P (1980) Multiple objective decision methods and regional planning. North-Holland, New
    York
Rogerson RJ, Findlay AM, Morris AS, Coombes MG (1989) Indicators of quality of life: some meth-
    odological issues. Environ Plann A 21:1655–1666
Rohe WR, Gates LB (1985) Planning with neighborhoods. University of North Carolina Press, Chapel
    Hill London
Roy B (1971) Problems and methods with multiple objective functions. Math Programm 1:239–266
Smith DM (1973) The geography of social well being in the United States. McGraw Hill, New York
US Census (1980) Census of Population and Housing, 1980: Summary Tape File 1 Technical
    Documentation. US Census Bureau, Washington DC
US Census (1990) TIGER/Line Precensus Files, 1990 Technical Documentation. US Census Bureau,
    Washington, DC
Voogd H (1983) Multicriteria evaluation for urban and regional planning. Pion, London

# Information systems for policy evaluation: a prototype GIS for Urban Programme impact appraisal in St. Helens, North West England

Alexander F. G. Hirschfield, Peter J. B. Brown, and John Marsden

The Urban Research and Policy Evaluation Regional Research Laboratory (URPERRL), Department of Civic Design, University of Liverpool, P.O. Box 147, Liverpool L69 3BX, Great Britain

**Abstract.** This chapter describes some key features of an information system designed by URPERRL (Liverpool) for monitoring urban needs and targeting resources in a large metropolitan local authority in North West England. Particular attention is paid to the identification, assembly and integration of the various data sets required for monitoring purposes, and how these are used in a GIS-linked decision support system to provide intelligence and guidance on questions of resource allocation and in planning for the future.

St. Helens Borough Council is a large metropolitan authority in North West England. Every year the authority submits proposals to the Department of the Environment for additional funding through the British Government's Urban Programme to support a range of social, economic and infrastructural projects. Typically, proposals are drawn up in the absence of information about their likely impact. The authority has lacked the data sets and information systems to enable basic questions to be addressed concerning likely local project impacts and their effectiveness in meeting needs.

The authority commissioned URPERRL at the University of Liverpool to fill some of the gaps by designing and creating a decision support system for managing its Urban Programme resources. The system created for St. Helens incorporates a wide range of spatially-referenced data sets covering demography, social conditions and basic infrastructure, handled within an ARC/INFO-based GIS framework.

The database has been used to address a number of policy-relevant questions concerning the operation of the Urban Programme in St. Helens. These include:

- reviewing the existing definition of priority areas;
- targeting projects to specific areas and beneficiary groups; and
- assessing the impact of projects against objectives and output measures.

This chapter discusses some of the steps required to achieve all of this (such as, for example, searching for data, capturing data, designing the information system) and describes some of the ways in which GIS features have proved useful in the production of outputs from the process (such as, for example, in priority area profiling). The chapter concludes with some thoughts on how the system might serve as a model to support strategic planning in other local authorities in the UK and elsewhere.

## 1. Background and introduction to the Urban Programme

In 1990, a brief but innovative pilot study was carried out by the Urban Research and Policy Evaluation Regional Research Laboratory (URPERRL) on behalf of St. Helens Council, one of the five Metropolitan District Councils in the county of Merseyside in North West England. St. Helens is a large industrial town with a population of 189 000. It is the home of the famous glass manufacturing company, Pilkingtons. In common with other declining industrial centres in northern England, the town suffers relatively high levels of unemployment and social deprivation, albeit not on the same scale as neighbouring Liverpool. The project was primarily concerned with building a database which could be used by the authority to monitor the deployment of extra resources that it has received in recent years from the Urban Programme.

Four categories of projects are supported: economic, social, environmental and housing. In order to qualify for resources from the Urban Programme, a project must meet the following criteria:

- It must be a special project which is additional to the Council's main programme (i.e. the Urban Programme cannot be used to fund an activity or service which the Council or its main committees are already committed to provide).
- It must relate to one of the four project categories or areas of action (i.e. social, economic, environmental or housing).
- It must address, either directly or indirectly, problems of deprivation. This means that priority is given to projects located in areas of special need.

St. Helens MBC first became an Urban Programme Authority in 1987. The conferment of this status gave official recognition to the fact that the town faces serious problems of social and economic deprivation which can only be addressed if additional resources are made available. Since 1987, over £8m has been received by St. Helens through the Urban Programme and a total of 116 projects have been implemented.

In common with the Urban Programme nationally, economic and social projects account for the major share of the expenditure (some 80 per cent of the total), with environmental schemes making up 11 per cent and housing projects 9 per cent of total spending (Fig. 1). As is the case in other Urban Programme authorities, St. Helens is very much concerned with monitoring its expenditure. The Authority is eager to establish whether its Urban Programme resources are being spent in the right areas and whether they are benefiting those most in need.

In order to improve its monitoring capability, in November 1990, the Authority commissioned URPERRL to undertake a pilot study in which six key objectives were pursued:

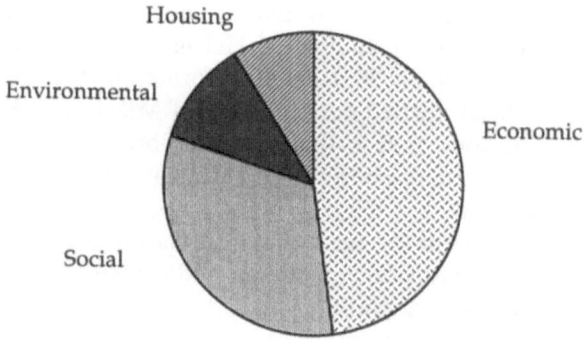

**Fig. 1.** St. Helens Urban Programme: Average annual expenditure by type of project 1987/1992

(1) to review the current definition of the authority's priority needs areas;
(2) to analyse the spatial distribution/impact of Urban Programme projects;
(3) to provide support and information for the Council's Urban Programme submission;
(4) to develop a GIS-based strategic planning tool for use throughout the authority and elsewhere;
(5) to enable the Council to make full use of the data from the 1991 Census; and
(6) to run a short GIS awareness and training course for council officers.

URPERRL was particularly well placed to undertake the project. The laboratory, which was set up as part of the main phase of the Economic and Social Research Council's RRL Initiative, has a specific remit to undertake urban research. In recent years it has built up considerable expertise in designing information systems for monitoring social and economic conditions and evaluating the impacts of urban policy (Hirschfield et al. 1989, 1992; Worrall 1990a, b, 1991; Rao 1991; Worrall and Rao 1991).

The first four objectives were concerned primarily with how to make better use of existing data sets and computer software (particularly GIS software) for monitoring purposes. The Council's fourth objective was to ensure that, as far as possible, the strategic planning tool developed for the Urban Programme in St. Helens was capable of being exported to other Urban Programme authorities. It was also hoped that the end-products of the study could be used as a model for other types of strategic planning in St. Helens and elsewhere.

From the outset, it was recognised that the prospects of being able to export the concepts, methods and other deliverables from the research to other authorities could be enhanced considerably if the database formed part of a geographical information system (GIS). A GIS would be particularly useful because of its ability to integrate disparate forms of data using common spatial referencing systems, such as postcodes or the National Grid. Exportability could be maximised if a leading GIS software package was to be used as an interface for the strategic database.

ARC/INFO is one of the most widely used multi-feature GIS packages currently available. It is the primary GIS software used by URPERRL. At the time of the pilot study it was decided that ARC/INFO would be introduced into a number of local authorities on Merseyside, including St. Helens, as part of a county-wide GIS strategy co-ordinated by the Merseyside Information Service (MIS). This followed a lengthy evaluation carried out on behalf of MIS in which the performance of ARC/INFO was compared to that of several other leading GIS products. For these, and other technical reasons, the GIS elements of the project focus around the use of ARC/INFO.

The fifth and closely related objective was to assess what information was likely to be made available from the 1991 Census of Population, and to draw to the attention of the Authority any social and demographic indicators which could be used to target Urban Programme expenditure in future years. An associated aim was to derive a set of comparable 1981–91 Census indicators which could be used to identify changes in the Borough's social composition and its areas of deprivation.

The final objective for the St. Helens project (No. 6, above) was to increase awareness of GIS within the authority by providing training for council officers. This was implemented through a programme of seminars and hands-on training sessions using ARC/INFO in the RRL's GIS laboratory. The training element was regarded as a particularly relevant 'deliverable' from the project in the light of the imminent acquisition of ARC/INFO GIS software by St. Helens.

Detailed descriptions of the project and its findings appear elsewhere (see, for example, Hirschfield et al. 1991 a, b, c). In this chapter, discussion will focus on three main issues: the development of an Urban Programme database, an appraisal of the geographical areas identified for priority action in St. Helens and a spatial analysis of Urban Programme expenditure.

## 2. Development of an Urban Programme database

The main operational task of the project was to assemble the necessary data sets on local conditions and the deployment of Urban Programme resources and incorporate them into ARC/INFO. This involved identifying, assembling and integrating data covering a range of topics and at a variety of scales.

The first task was to gain familiarity with the Urban Programme in St. Helens in order to identify, in general terms, the types of information required by the Council to interpret its deployment and use of resources. This involved close scrutiny of all of the relevant documentation (official Urban Programme guidelines, past submissions and other policy documents) and the discussion of information requirements with Urban Programme managers and staff in other key departments, most notably, social services and planning.

Lack of information is often regarded by local authorities as the major obstacle to implementing effective monitoring procedures. In most authorities, however, and St. Helens is no exception, the problem is not so much that there is a lack of information, but the fact that much of it is stored manually in filing cabinets and is spread across a large number of departments and external organisations.

An abundance of useful information about St. Helens was uncovered following the data survey and discussions held with council staff. Several strategically important data sets were held by external organisations, including Merseyside's county-wide research and information unit (the Merseyside Information Service), the District Health Authority, the Office of Population Censuses and Surveys and the Regional Computing Centre based at the University of Manchester. A substantial proportion of the information required was available within the Council, particularly from the Chief Executive's Department and the Planning Department. Fortunately, much of the information was spatially referenced.

Several different types of information were identified. They included:

- *Statistics for Small Areas,*
- *Site locations of community facilities and service delivery outlets,*
- *Information about individuals and households containing a residential address and postcode,*
- *Information on land use,*
- *Information about street networks and transportation routes,*
- *Base maps showing administrative and other boundaries,*
- *Information about individual Urban Programme projects.*

One of the main challenges of the project was to consolidate these disparate data sets by implementing a programme of co-ordinated data assembly.

Confronted with an abundance of data, decisions then had to be taken about how much information to collect, how to capture it, how to organise it, and how to use it for evaluation and monitoring purposes. The main priority was to compile a short list of data sets which could be captured within the relatively short time scale of the project. This was achieved by targeting information of direct relevance to the Urban Programme, namely, background indicators on social, economic, and housing conditions in St. Helens and information about the deployment of Urban Programme resources. This was implemented through an intensive programme of data capture which involved transfering information between computers, importing data via floppy disk, extracting information from documents and reports and inputting it into the computer using coding sheets and digitising from maps.

Much of the information to be captured was already in machine readable form (e.g. Census indicators, unemployment rates for small areas, mid year population estimates and vital statistics). Most of the information about current and historic investment through the Urban Programme, however, was not computerised and had to be captured from Project Information Forms (PIFs) and other documents. This necessitated designing an appropriate coding sheet, an exercise carried out in collaboration with relevant members of staff within the authority.

Once captured, the information was organised and stored as a series of topic-based data modules. Two types of data module were defined:

- Background Data Modules, containing statistics for census tracts within St. Helens (referred to as enumeration districts and wards) on basic demography, social conditions and economic status.

- Operational Data Modules, containing information on Urban Programme expenditure, community facilities, basic infrastructure and administrative boundaries.

The content and structure of these data modules was as follows:

Background Data Modules:

- Demographic (e.g. age, sex, migration, births, deaths, migration, population estimates)
- Social (e.g. social, deprivation and health status indicators, geodemographics)
- Economic (e.g. employment, unemployment, occupations, economic activity)
- Cross-sectional (i.e. a multi-topic file containing selected demographic, social, economic and Urban Programme variables).

Operational Data Modules:

- Urban Programme (e.g. objectives, expenditure, beneficiaries, output measures);
  (a) Individual Projects Database; (b) Statistics for wards.
- Infrastructure and Services (e.g. street network, schools, clinics, residential care establishments, community facilities, area offices)
- Boundaries (e.g. enumeration districts, wards, priority areas, personal services areas).

Considerable effort was devoted to the creation of the background data modules on urban conditions in St. Helens. A wide range of small area statistics information was assembled from a variety of sources for the 18 wards and 365 enumeration districts (EDs) used in the 1981 Census of Population in St. Helens. Information at ward level (see Fig. 2) included demographic, social, and economic indicators from the 1981 Census, the latest mid-year population estimates by broad age groups and household estimates, unemployment counts by age and sex (including some estimated rates), estimates of car ownership for 1988, vital statistics and health status indicators. In addition, ways were identified in which it would be possible to supplement this information with data drawn from other sources within the Authority. For example, data relating to support service workloads and take-up rates could be extracted from the Authority's recently installed social services client information system CRISSP. Finally, a complete set of geodemographic codes from the Super Profile area classification was also assembled for the study.

The term geodemographics has come to be adopted in recent years to refer to the development and application of small area typologies that can be used to gain a better understanding of spatial variations in consumer behaviour, social problems, lifestyles, medical conditions, the delivery of services and deployment of resources. It has emerged from the use of more effective means of manipulating census data, together with significant improvements in classification methodology, which have led to the production of multi-dimensional classifications of small areas using cluster analysis (see Brown 1990 and 1991 for a review of recent developments). Geodemographic analysis has been used in this pilot study pri-

**Fig. 2.** St.Helens 1981 Census wards: ward location map

marily as a means of identifying different types of residential area or neighbourhood in St. Helens in order to produce a social profile of the existing Urban Programme priority area.

Much of the operational information, particularly site-specific data on community facilities and administrative boundaries, was stored as separate coverages
in ARC/INFO. Information about each of the 116 Urban Programme projects
was captured using the coding sheet designed by the RRL in conjunction with the
Council. For each Urban Programme project, details were recorded about the
nature of the scheme, the capital and revenue expenditure involved, the ward or

wards in which the project was located, the intended target beneficiaries from the project, expected outputs and the policy objectives which the project was expected to serve.

Two data files were created from this information:

- an individual project-level database which was subsequently supplied to St. Helens MBC in dBase IV format; and
- a set of small area statistics recording the average annual Urban Programme expenditure rates per household by ward.

The statistics in each file were broken down by project categories and target beneficiary groups. The resultant database on current and historic investment that has been funded through the Urban Programme has opened up a range of possibilities for the analysis of resource deployment on a borough-wide basis, at ward level and over time. For example, the level of resources directed at specific beneficiaries or committed to broad policy areas (such as social conditions, local economy or the environment) can be readily identified. This form of analysis can also be produced as a time-series to reveal any changes in emphasis which might have taken place in recent years.

A variety of spatial analyses were also possible, such as, for example:

- identifying levels of expenditure by ward for different categories of project;
- identifying spatial variations in expenditure on intended beneficiaries from projects (e.g. small firms, unemployed, elderly, children) by ward;
- identifying annual variations in the spatial distribution of expenditure;
- identifying relationships between Urban Programme expenditure and the incidence of deprivation at ward level;
- revealing the distribution of expected outcomes (e.g. jobs created, buildings improved) by ward; and
- identifying the spatial implications (e.g. expenditure by ward) of meeting the various Urban Programme policy objectives.

The simultaneous on-line storage of selected data modules in ARC/INFO (such as demographic information and boundary data) enabled a number of cross-referencing operations to be performed. Several of these were undertaken in order to address some of the research questions defined at the beginning of the project. Here we shall restrict discussion to two of these issues. The first of these is the appraisal of the definition of priority areas in St. Helens; the second relates to the targeting of Urban Programme projects to specific areas and beneficiary groups.

## 3. Priority area appraisal

The priority areas employed in targeting Urban Programme expenditure in St. Helens were defined in 1983 in accordance with guidelines set out by the Department of the Environment (1983) for identifying areas experiencing acute problems of deprivation and urban stress which are thus likely to require some form of priority action. The approach was based upon the specification of a series of indicators derived from the 1981 Census and a method for combining them to produce a composite Deprivation Index for each of the 365 EDs or small areas. We note that the selection of variables employed, and the arbitrary weightings applied to them, have been the subject of critical comment elsewhere (Brown 1988). Nevertheless, the index served as a basis for ranking the small areas and a starting point for the identification of priority areas. The EDs for which a critical threshold index value was exceeded were plotted on a map and the effective boundary defined after taking into account local factors, such as, for example, the location of main roads, open space, barriers to movement, and so on. This resulted in the splitting of some 13 EDs, with 133 whole EDs falling inside and 217 entirely outside the priority area boundary. In 1981, the population living within the priority area (69 080) accounted for 36.5 per cent of the St. Helens total. It is important to bear in mind this proportion when interpreting the analysis reported below.

An initial review of Urban Programme priority areas was undertaken using a selection of small area statistics and geodemographic variables drawn from the demographic and social data modules. Tests were carried out to establish the extent to which the Urban Programme priority area contained different sections of the population, including the most deprived residents in St. Helens. The basis of the approach was to calculate the proportion of the Borough's 'deprived' population in a particular target group that was 'captured by' or found in the priority area, based upon an analysis of enumeration district level data.

Table 1 shows that 53.3 per cent of males experiencing unemployment were located inside the priority area, an area within which the rate of incidence of male unemployment was 46 per cent greater than the mean rate, compared with 73 per cent of the mean rate outside (index values of 146 and 73). Other census derived measures featured in Table 1 show, in varying degrees, that the mean values of the indicators relating to the population living inside the priority area boundary are significantly greater than those outside the boundary. It is suggested that, as part of the assessment of the continuing appropriateness of the priority area boundary, contemporary statistics relating to Urban Programme target groups should be analysed in a similar manner to that illustrated in Table 1.

As part of this exercise the boundary of the priority area was digitised as an ARC/INFO coverage. This was used in carrying out another stage of the appraisal of the priority area definition in conjunction with the geodemographic data held in the social data module. The latter exercise will not be described here in detail. However, some of the graphical output will be used to illustrate one of the benefits of the link between the Urban Programme database and Arc/Info GIS software.

An effective total of ten area types are distinguished at the Super Profile Lifestyle level of aggregation of the Super Profile national classification of the 130 000 enumeration districts used in the British 1981 Census. The labels attached to each of these area types are based upon examination of a table which compares the Lifestyle cluster mean values of the 65 variables used in deriving the classification with the corresponding national mean values. A verbal 'pen picture' can be painted by drawing attention to relatively high or low values in comparison with the national average (see, for example, Brown 1991).

Table 1. St. Helens Urban Programme priority needs area appraisal: breakdown of selected deprivation indicator populations between 1981 Census enumeration districts inside (In) and outside (Out) the priority needs area boundary. (Including proportion of St. Helens population/household total for indicator that is 'captured' by the area, and comparison of rates of incidence inside and outside the boundary)

| Deprivation indicator | Number of EDs with non-zero indicator total | | Total indicator numerator in St. Helens | Proportion of indicator pop/Hhld total | | Rate per 1000 denominator pop/Hhlds | | Index (mean rate set to 100) | |
|---|---|---|---|---|---|---|---|---|---|
| | In | Out | | In | Out | In | Out | In | Out |
| Resident population | 148 | 230 | 189250 | 35.6 | 63.5 | 107.2 | 54.0 | 146 | 73 |
| 1. Male unemployment rate | 143 | 223 | 6946 | 53.3 | 46.7 | 48.2 | 19.6 | 158 | 64 |
| 2. Households in accommodation lacking a bath or inside WC | 119 | 121 | 2012 | 60.3 | 39.7 | | | | |
| 3. Overcrowded households living >1.5 persons per room | 83 | 61 | 241 | 62.8 | 37.2 | 6.0 | 2.2 | 164 | 60 |
| 4. Households with no car | 148 | 225 | 27888 | 50.9 | 49.1 | 563.7 | 335.5 | 133 | 79 |
| 5. Pensioners living alone (as a proportion of all households) | 148 | 213 | 8853 | 45.7 | 54.3 | 160.7 | 117.8 | 119 | 87 |

Note: enumeration districts = 365 (13 split); 1981 population: inside = 69080 outside = 120170

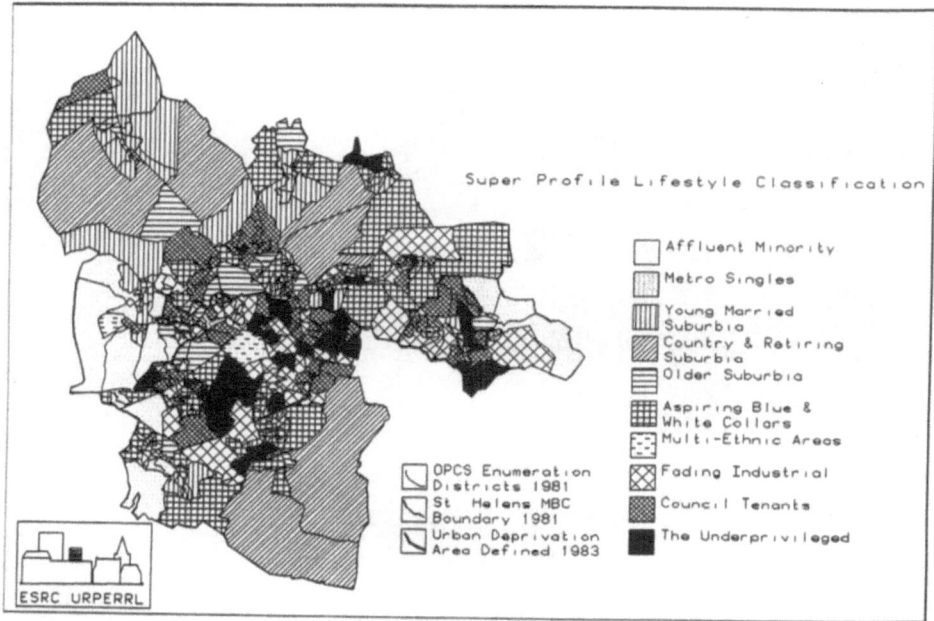

**Fig. 3.** Assignment of 1981 Census enumeration districts to Super Profile Lifestyles: conventional choropleth map – with priority needs area boundary superimposed

Fig. 3 illustrates the assignment of the 365 EDs in St. Helens to the ten Lifestyles and the superimposition of the priority area boundary on this ED base. The monochrome choropleth map is less easy to interpret than the coloured original. One feature of the map is that shading is applied to the entire area of each ED. This means that the entire area of a relatively *large* and *thinly populated* ED is given the same treatment as a relatively *small* and *densely populated* ED which is more likely to be found in the inner or more central parts of St. Helens.

In Fig. 4, a variant on the original map is presented in which the shading is restricted to the parts of the ED which are known to be devoted to residential development. This practice represents an effort to overcome some of the problems and scope for misinterpretation which are encountered when using more traditional methods of analysis (an issue discussed further in Brown et al., forthcoming). This is made possible by digitising the boundary of the built-up/residential area and using the overlay feature in ARC/INFO to achieve the desired effect. In Fig. 4, it is apparent that much of the visual impact of the original map is attributable to the dominant visual influence of a small number of relatively large, rural EDs within which only a small proportion of the area contains population.

Although the restriction of shading to the built-up areas has helped to overcome much of the above effect, it has also helped to highlight another more substantive outcome of this exercise. This relates to the inevitable failure of the area-based targeting of expenditure to reach all of those who can be identified as being in need of assistance, or to whom additional resources are directed. This can be illustrated in Fig. 4 by noting that most of the EDs falling in the 'Under-

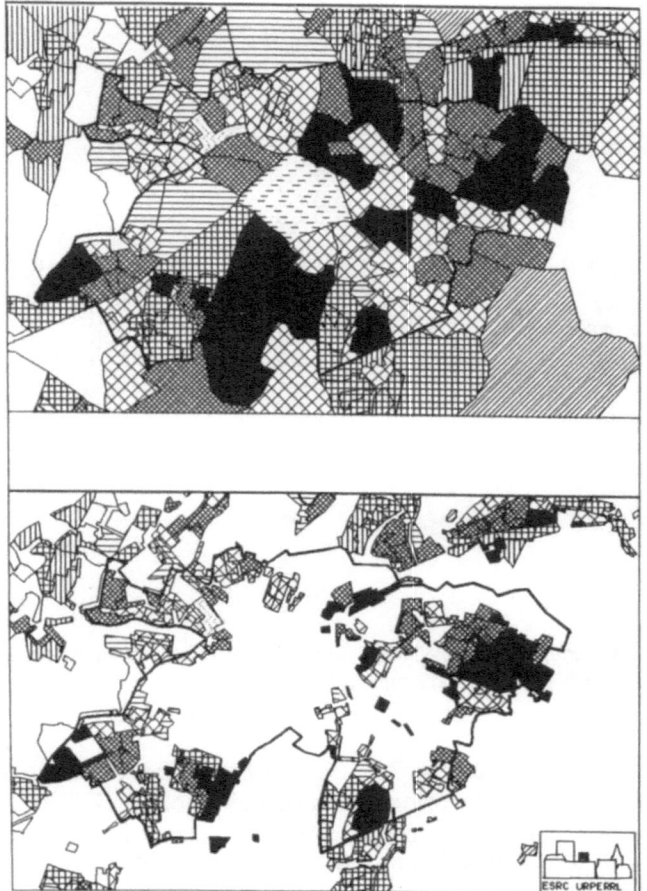

**Fig. 4.** Windowed plot of Super Profile Lifestyle composition of 1981 Census EDs in the western priority needs area: conventional choropleth map and map with shading restricted to residential areas

privileged Britain' Lifestyle cluster (a likely target for discriminatory expenditure) are found inside the priority area boundary. However, there are several EDs in this Lifestyle outside the boundary (to the north east of the windowed area) which, it can be argued, are likely to contain population who are equally in need of or deserving of priority attention.

## 4. Spatial analysis of Urban Programme expenditure

The second major application of the database included a spatial analysis of the impact of Urban Programme funded projects. By bringing together data from the various modules at ward level, it was possible to carry out a series of correlation analyses between Urban Programme expenditure rates and selected social, eco-

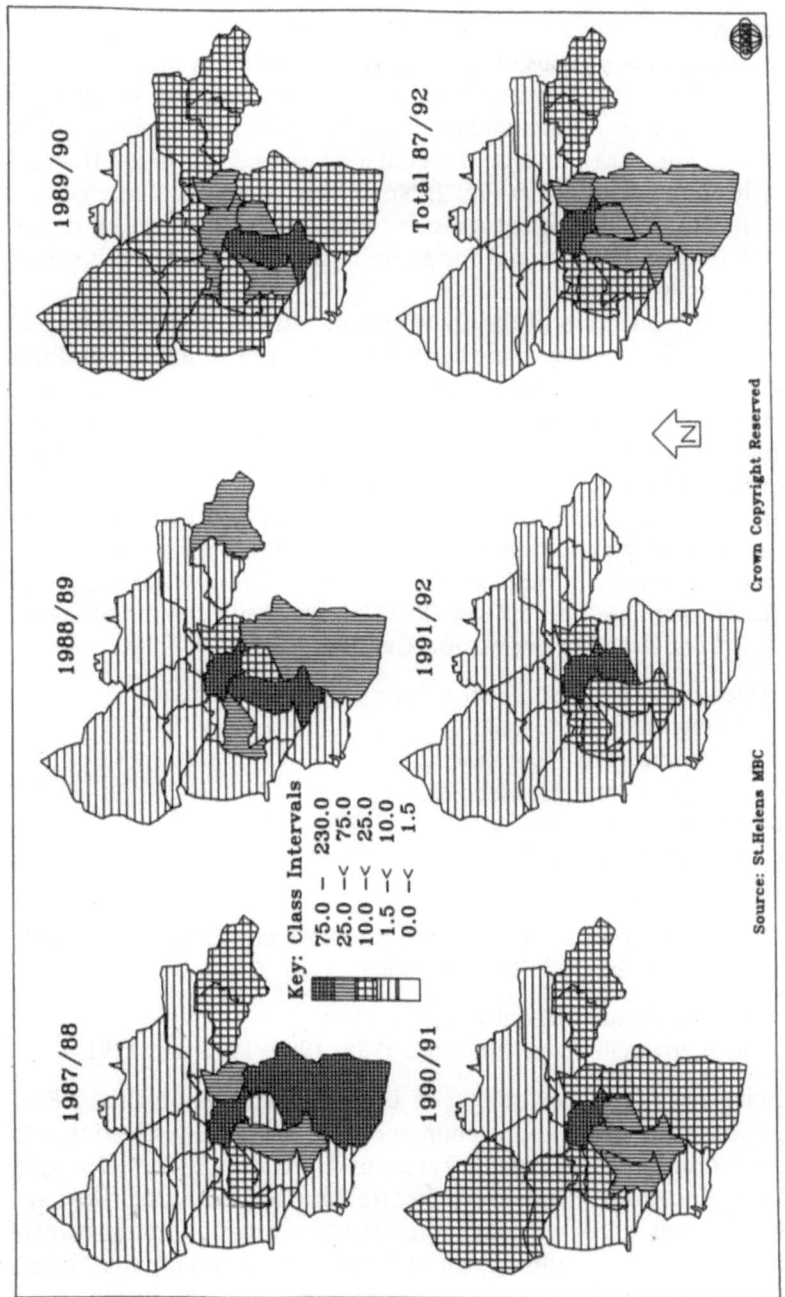

**Fig. 5.** St. Helens Urban Programme: annual expenditure per head 1987—92 for 1981 census wards

nomic and deprivation indicators. Correlating the total expenditure on Urban
Programme projects in each year with the Department of the Environment's composite Deprivation Index produced further interesting results.

Before discussion focuses on the outcome of these correlation analyses, we can
note how the spatial pattern of resource deployment over time can be observed
by mapping the annual expenditure rates per head (in £s) for the individual wards.
This is illustrated in Fig. 5 with respect to overall total expenditure in the 18 wards
in St. Helens which are identified in Fig. 2. No further comment is offered here
beyond noting how useful this format appears to be in drawing attention to patterns in both the temporal and spatial variation in the incidence of both overall
and different categories of expenditure.

In order to aid interpretation of the above spatial patterns, a series of tests of
association or correlation analyses were carried out between various expenditure
rates and selected social and economic indicators. The Pearson Correlation Coefficient can be used to measure both the strength and the direction of the association between pairs of variables at ward level and has been used extensively in area-
based studies to identify relationships of this type.

In the pilot study, correlation coefficients were calculated between various
social, economic and Urban Programme indicators for St. Helens wards drawn
from the multi-topic cross-section data module (see above). Two sets of social and
economic indicators were used in this way. The first set comprised the following
measures derived from the 1981 population Census:

● The Department of the Environment's Basic Deprivation Index,
● Percentage of Unemployed Persons,
● Percentage of Overcrowded Households,
● Percentage of Single Parent Households,
● Percentage of Pensioners Living Alone,
● Percentage of Households Lacking Exclusive Use of Basic Amenities (e.g.
  bath and inside WC).

The second set was made up of more recent indicators which represent conditions
in St. Helens in 1987 and beyond. They included:

● Percentage of Households without a Car in 1988
● Estimated Unemployment Rates for 1987, 1988, 1989, 1990 and 1991.

Prior to the main analysis, which sought to compare patterns of Urban Programme expenditure with social and economic conditions, an investigation was
carried out to determine whether or not key areas of deprivation in St. Helens had
changed markedly since 1981. This involved correlating the 1981 DoE Deprivation Index for St. Helens wards with the more recent information on car ownership and unemployment. The results appear in Table 2 and are highly significant.
They show particularly strong positive correlations between the incidence of
deprivation at ward level in 1981 (the DoE Index) and measures of economic
deprivation (i.e. unemployment and the low income proxy 'households without a
car') in the late 1980's and early 1990's.

The conclusion that we can draw from this is that there has been remarkable
consistency over time in the deprivation rank of St. Helens wards. Refinements

**Table 2.** Relationship between ward level 1981 Department of the Environment deprivation index and recent social indicators

| | |
|---|---|
| Households without a car 1988 | 0.96 |
| Unemployment rate: | |
|   January 1987 | 0.95 |
|   January 1988 | 0.93 |
|   January 1989 | 0.94 |
|   January 1990 | 0.97 |
|   January 1991 | 0.96 |
| Population density 1988 | 0.45* |
| Standardised mortality ratios (SMRs) 1983–1985 | |
|   All causes of death | 0.77 |
|   Ischaemic heart disease | 0.67 |
|   Lung cancer | 0.56 |
|   Respiratory diseases | 0.45* |
| Infant mortality rate | |
| 1981–1989 | 0.27* |

* Not significant

**Table 3.** Relationships between expenditure by category of project and deprivation

Deprivation indicators derived from 1981 Census

| Project category | DoE index | Unemploy-ment rate | Over-crowding | Single parents | Lone pensioners | Lacking basic amenities |
|---|---|---|---|---|---|---|
| All projects | 0.59* | 0.69* | 0.65* | 0.53* | 0.17 | 0.28 |
| Economic | 0.47* | 0.55* | 0.44 | 0.24 | 0.28 | 0.35 |
| Social | 0.62* | 0.69* | 0.73* | 0.61* | 0.15 | 0.24 |
| Environmental | 0.62* | 0.74* | 0.67* | 0.70* | 0.09 | 0.25 |
| Housing | 0.23 | 0.31 | 0.46* | 0.59* | −0.26 | −0.12 |

Deprivation and health status indicators post 1981

| | Households without a car 1988 | Unemployment rate 1991 | SMR all deaths | SMR ischaemic heart disease |
|---|---|---|---|---|
| All projects | 0.64* | 0.62* | 0.38 | 0.69 |
| Economic | 0.53* | 0.53* | 0.43 | 0.68* |
| Social | 0.70* | 0.68* | 0.41 | 0.52* |
| Environmental | 0.62* | 0.61* | 0.28 | 0.67* |
| Housing | 0.19 | 0.13 | −0.14 | 0.13 |

* Significant at the 5 per cent level

to the analysis must await publication of the 1991 Census Small Area Statistics. However, the results suggest that the DoE's 1981 Deprivation Index broadly reflects present day differences in deprivation levels between the 18 wards.

Results from the main correlation analyses between Urban Programme expenditure and the deprivation indicators appear in Table 3. The first set of relationships to be examined was those between the various categories of Urban Programme expenditure by ward and deprivation levels. The key results from the analysis revealed the following patterns or relationships:

**Table 4.** St. Helens wards ranked in descending order of deprivation

| Ward | National Z-score | Annual average expenditure per household (£) | Deprivation rank | Sextile |
|---|---|---|---|---|
| Parr & Hardshaw | 11.50 | 142.02 | 1 | 1 |
| Marshalls Cross | 8.73 | 34.60 | 2 | 1 |
| Queens Park | 8.44 | 14.11 | 3 | 1 |
| Grange Park | 7.72 | 24.82 | 4 | 2 |
| West Sutton | 7.67 | 23.33 | 5 | 2 |
| Broad Oak | 7.42 | 25.10 | 6 | 2 |
| Newton West | 6.94 | 11.69 | 7 | 3 |
| Newton East | 4.73 | 19.06 | 8 | 3 |
| Thatto Heath | 2.98 | 23.04 | 9 | 3 |
| Moss Bank | 1.48 | 7.15 | 10 | 4 |
| Haydock | 0.87 | 8.57 | 11 | 4 |
| Windle | 0.62 | 9.41 | 12 | 4 |
| Sutton & Bold | − 1.09 | 39.67 | 13 | 5 |
| Blackbrook | − 2.14 | 7.82 | 14 | 5 |
| Eccleston | − 4.52 | 7.10 | 15 | 5 |
| Rainford | − 4.71 | 8.35 | 16 | 6 |
| Rainhill | − 4.93 | 6.56 | 17 | 6 |
| Billinge & Seneley Green | − 5.83 | 5.67 | 18 | 6 |

Deprivation rank: 1 = most deprived, 18 = least deprived

- A significant positive correlation (+ 0.59) between total Urban Programme expenditure by ward and the DoE Deprivation Index. This suggests that expenditure in general has tended to be concentrated in the most deprived areas.
- A greater tendency for spending on social and environmental schemes (correlation coefficient 0.62) to be concentrated in the most deprived areas than expenditure on economic projects (correlation coefficient 0.47).
- The absence of any significant relationship between spending on housing projects and variations in levels of deprivation across the 18 wards (correlation coefficient 0.23). This reflects the confinement of expenditure on housing projects to only one of the wards.

Substituting households without a car in 1988 for the DoE Deprivation Index produced an even stronger positive correlation between Urban Programme expenditure and deprivation, particularly for social projects for which the correlation coefficient reaches 0.70.

The relationship between expenditure levels and deprivation becomes clearer when each ward is ranked in descending order of deprivation level (based on the DoE Index) and the corresponding annual average expenditure rate per household is displayed. This is shown in Table 4. Although there is clear evidence of a positive relationship (i.e. the higher the deprivation the greater the expenditure) there are also some anomalies. For example, expenditure in Queens Park, and in Newton West was considerably *lower* than one would expect on the basis of their deprivation rank. Conversely, expenditure in Sutton and Bold, and to a lesser extent in Thatto Heath, was higher than expected. The level of expenditure in Parr

**Table 5.** Average annual expenditure per household by category of project and deprivation sextile

| Deprivation sextile | All projects £ | Economic £ | Social £ | Environ-mental £ | Housing £ |
|---|---|---|---|---|---|
| 1 | 63.04 | 42.12 | 14.91 | 5.92 | 0.07 |
| 2 | 40.29 | 10.88 | 11.68 | 4.64 | 13.08 |
| 3 | 17.92 | 3.82 | 12.13 | 1.88 | 0.07 |
| 4 | 8.30 | 3.02 | 3.68 | 1.51 | 0.07 |
| 5 | 18.16 | 11.53 | 4.86 | 1.68 | 0.07 |
| 6 | 6.68 | 2.69 | 2.40 | 1.50 | 0.07 |

Note: Sextile 1 is most deprived; Sextile 6 is least deprived
Composition of deprivation sextiles:
1 Parr & Hardshaw/Marshalls Cross/Queens Park
2 Grange Park/West Sutton/Broad Oak
3 Newton West/Newton East/Thatto Heath
4 Moss Bank/Haydock/Windle
5 Sutton & Bold/Blackbrook/Eccleston
6 Rainford/Rainhill/Billinge and Seneley Green

and Hardshaw was certainly considerably higher than would seem on first sight justified, given the observed levels of deprivation in that ward. Undoubtedly, this has more to do with its town centre location than with its deprivation status.

The expenditure levels can be further summarised by collapsing the 18 wards into six divisions of three wards each (i.e. into sextile groups). Expenditure levels by sextile and project category appear in Table 5. Generally, expenditure levels fall consistently with deprivation sextile until the fifth sextile when they rise once again. The inflated expenditure level in the fifth sextile is largely attributable to disproportionate resource allocations in Sutton and Bold. However, the significance of the relationship between expenditure and deprivation is most evident when the lowest sextile (6) is compared with the highest (1).

The relationship between the deployment of resources and deprivation over time is shown in Table 6. A number of points emerge from this part of the analysis. In terms of total expenditure (i.e. on all projects), there appears to have been a steady improvement in the extent to which resources were targeted into areas of greatest need during the first three years of the Urban Programme. This is evident from the correlation between the annual expenditure rates and the DoE Deprivation Index (column 1). This was as low as 0.29 in the first year (1987/88) but rose to 0.60 by the third year (1989/90). However, during the last two years, the targeting of resources in general appears to have been less sensitive to variations in deprivation, with the correlation coefficient falling in the range 0.52−0.53. In contrast to this general picture, the correlation has strengthened with the census indicators relating to lone pensioners and people living in accommodation lacking basic amenities.

In addition to the analysis of expenditure by project category, statistics were also produced on resources directed towards the intended beneficiaries from the Urban Programme in St. Helens. This was achieved by capturing, directly from the Project Information Forms, details of expenditure intended for 'client' groups

**Table 6.** Relationships between annual expenditure on all projects and indicators of deprivation at ward level

| Deprivation indicators derived from the 1981 Census | | | | | | |
|---|---|---|---|---|---|---|
| Year of expenditure | DoE index | Unem-ployment | Over-crowding | Single parents | Lone pensioners | Lacking basic amenitis |
| 1987/88 | 0.29 | 0.43 | 0.60 | 0.29 | −0.10 | 0.04 |
| 1988/89 | 0.58 | 0.67 | 0.68 | 0.62 | 0.10 | 0.25 |
| 1989/90 | 0.60 | 0.64 | 0.73 | 0.73 | 0.07 | 0.21 |
| 1990/91 | 0.52 | 0.60 | 0.51 | 0.41 | 0.23 | 0.29 |
| 1991/92 | 0.53 | 0.61 | 0.42 | 0.37 | 0.28 | 0.35 |

| Deprivation indicators post 1981 | | | | | | |
|---|---|---|---|---|---|---|
| | Households without a car 1988 | Unemployment rate for year | | | | |
| | | 1987 | 1988 | 1989 | 1990 | 1991 |
| 1987/88 | 0.40 | 0.47 | | | | |
| 1988/89 | 0.58 | | 0.67 | | | |
| 1989/90 | 0.62 | | | 0.69 | | |
| 1990/91 | 0.57 | | | | 0.58 | |
| 1991/92 | 0.55 | | | | | 0.58 |

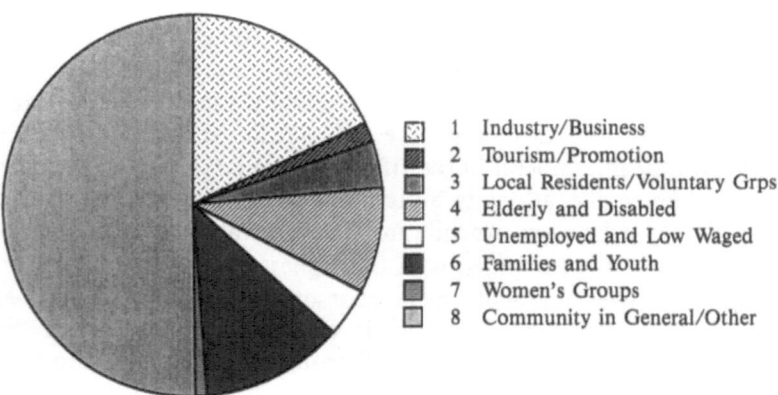

1  Industry/Business
2  Tourism/Promotion
3  Local Residents/Voluntary Grps
4  Elderly and Disabled
5  Unemployed and Low Waged
6  Families and Youth
7  Women's Groups
8  Community in General/Other

**Fig. 6.** St. Helens Urban Programme: average annual expenditure project by beneficiary group 1987/92

within the local community (e.g. young people, elderly and disabled people, etc) and for others (e.g. local industry/commerce) which were targeted through specific types of Urban Programme project.

Approximately half of the expenditure was intended to benefit the community in general. About one fifth of the total was earmarked for the industrial and business sectors and a further fifth was directed at families, youths, elderly and disabled people (Fig. 6). Expenditure patterns by beneficiary group also revealed some marked spatial differences, as evidenced by comparing expenditure by ward

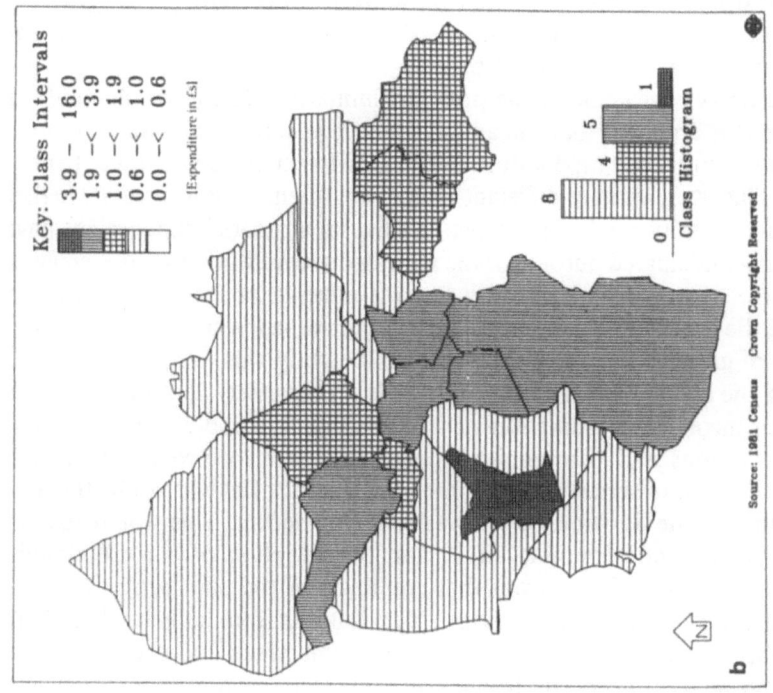

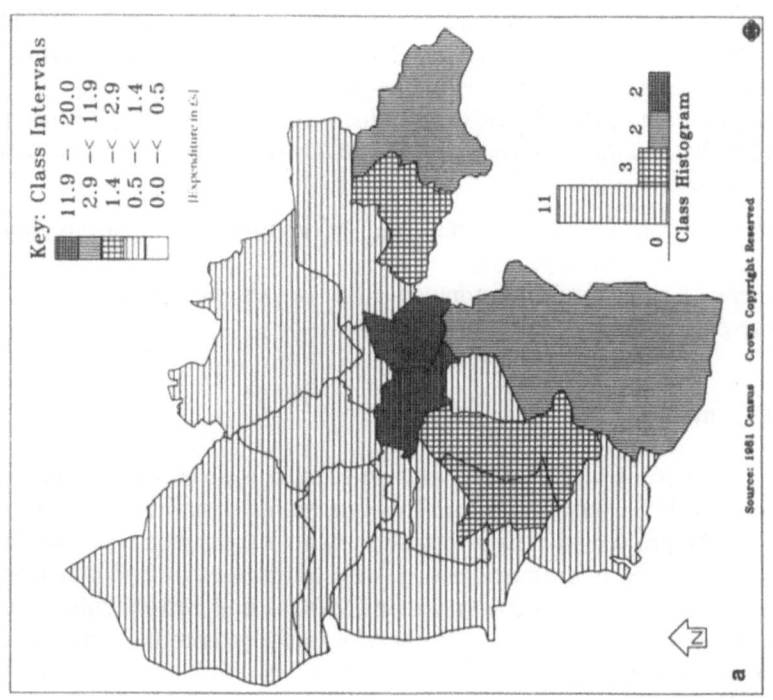

**Fig. 7a,b.** St. Helens Urban Programme: average annual expenditure by household 1987/92 on families and young people (a) and elderly and disabled people (b)

on elderly and disabled people with that on families and young people (Fig. 7a and 7b). The extent to which the intended beneficiaries actually received the resources is, of course, another question.

We must acknowledge some of the principal limitations of the spatial analyses presented here. It is apparent that the analysis has been concerned with examining the distribution of proposed expenditure between different areas and target beneficiary groups. This is an entirely different issue from attempting to assess whether that expenditure actually reached its target or what its ultimate impact might have been. Identifying anticipated outputs of the allocated expenditure (e.g. jobs created, buildings improved) would go some way towards meeting objectives which relate to establishing the extent to which Urban Programme expenditure is being concentrated in areas of greatest need. However, in the pilot study, this was not attempted.

Identifying the actual effects of expenditure is an extremely complex task. For example, it can involve the estimation of likely benefits to individuals, families, businesses, institutions and community groups which are located in or operate from individual wards. In some cases, the intended beneficiaries will not be the resident population but the day-time or non-resident population. This would require some form of social and economic audit, and the adoption of a methodology which is fundamentally different from that employed in the pilot project.

One way of gathering intelligence about actual outputs from individual projects would be to implement a programme of structured interviews and site visits. The interviews would be used primarily to identify social and economic achievements (e.g. number of jobs created, number of visits to special advice centres, etc). Site visits or ground surveys could be employed to ascertain the extent to which environmental improvements had been carried out (such as, length of road improved, buildings cleaned) and housing schemes implemented.

The primary purpose of the interviews and site inspections would be to generate a series of qualitative and quantitative performance measures which would be used to address the following issues:

● the extent to which social, economic and environmental conditions have changed as a results of the project;
● an assessment of what would have happened in the absence of the project;
● how far the project has been cost-effective.

## 5. Concluding comments

The St. Helens pilot study was in many ways an ambitious project which sought to achieve a great deal in a very short time. This chapter has described some of the main features of the project and the information handling requirements that the Urban Programme places upon participating local authorities. It has illustrated how, as a consequence of the project, St. Helens MBC is now better equipped to meet these requirements and has gained a number of useful end products, including a core of resource managers trained in the concepts, operation and application of GIS methods.

The project concluded with a final report which contained over twenty recommendations. These included:

- Recommendations on how the authority should improve its recording, handling and presentation of information;
- Recommendations on how the 1991 Census might be used to review the definition of priority areas;
- Recommendations on how the Council's Urban Programme submission might be improved; and
- A special recommendation on how the Council might improve its information on project outputs and performance by instituting a rolling programme of site visits and targeted interviews to assess local impacts of individual schemes.

Other deliverables have included the data modules and digitised boundaries, computer drawn maps and statistical tables produced during the course of the project. The assembly and capture of information in ARC/INFO format should pave the way for Urban Programme managers and others to take full advantage of the benefits arising from the imminent introduction of GIS to St. Helens.

The authority has been given a further incentive to take advantage of these opportunities by the latest guidance notes issued by the Department of the Environment (1991) for the preparation of the next round of Urban Programme expenditure bids. These place greater stress on the mapping of priority areas, spatial targeting and the analysis and presentation of spatial data. This suggests that there will be growth in demand from other authorities for the methods that have been employed, on an experimental basis, in the pilot study.

The major limitation of the study was its short duration. This meant that it was not possible to capture as much spatially disaggregated information as had been envisaged. For example, it would have been useful to have been able to map the spatial location of individual Urban Programme schemes and to relate them to other policies and initiatives, or to produce population profiles for likely catchment areas. It would have been particularly useful to have been able to acquire address referenced information about benefits of individual schemes, together with up-to-date social indicators from the Social Services Department.

We are confident that the methodology developed for the pilot, especially the procedures for capturing operational data from the Urban Programme, can be successfully applied elsewhere.

*Acknowledgement.* The authors gratefully acknowledge the support and assistance provided by the Chief Executive's Department of St. Helens Metropolitan Borough Council. Funding for the pilot project was itself provided through the British Government's Urban Programme and the assistance of the Merseyside Task Force in making this financial support available also is gratefully acknowledged. Some of this research has been made possible by the grant (No. A504 28 5003) awarded by the Economic and Social Research Council (ESRC), which led to the establishment of URPERRL, as one of eight Regional Research Laboratories (RRLs) that were established in the main phase of the ESRC's £2m RRL Initiative.

# References

Brown PJB (1988) A Super Profile based affluence ranking of OPCS Urban Areas. Built Environment 14:118–134

Brown PJB (1990) Geodemographics: A review of recent developments and emerging issues – Towards an RRL Research Agenda, Regional Research Laboratory Initiative Discussion Paper 5. Economic and Social Research Council, London

Brown PJB (1991) Exploring geodemographics. In: Masser I, Blakemore M (eds) Handling geographical information: Methodology and potential applications. Longman, London, pp 221–258

Brown PJB, Hirschfield AFG, Batey PWJ (1991) Applications of geodemographic methods in the analysis of health condition incidence data. Papers in Regional Science: J Reg Sci Assoc Int 70:329–344

Brown PJB, Hirschfield AFG, Marsden J (forthcoming) Analyzing spatial patterns of disease: Some issues in the mapping of incidence data for relatively rare conditions. In: de Lepper MJC, Scholten HJ, Stern RM (eds) The added value of geographical information systems in public and environmental health. Kluwer Academic Publishers, Dordrecht and World Health Organisation European Office, Copenhagen

Department of the Environment (1983) Urban deprivation information note 2. Inner Cities Directorate, DoE, London

Department of the Environment (1991) Action for cities: Urban Programme user guide, April 1991, DoE, London

Hirschfield AFG (1989) The study of poverty and deprivation in developed countries: A selective review of data sources. Approaches and analytical techniques, URPERRL Working Paper 2. Department of Civic Design, University of Liverpool

Hirschfield AFG, Barr R, Batey PWJ, Brown PJB (1989) Urban Research and Policy Evaluation Regional Research Laboratory. Mapping Awareness, 3:34–37

Hirschfield AFC, Brown PJB, Batey PWJ (1992) Data assembly, information systems and policy relevant research: Urban research and policy evaluation regional research laboratory two years on. Mapping Awareness 6:46–50

Hirschfield AFG, Brown PJB, Marsden J (1991 a) The development of a prototype geographical information system for Urban programme impact assessment and planning in St. Helens. Report submitted to St. Helens Metropolitan Borough Council

Hirschfield AFG, Brown PJB, Marsden J (1991 b) Database development for decision support and policy evaluation. In: Worrall L (ed) Spatial analysis and spatial policy using geographic information systems. Belhaven, London, pp 152–187

Hirschfield AFG, Brown PJB, Marsden J (1991 c) Database development for monitoring the Urban programme: A St. Helens prototype, URPERRL Working Paper 24. Department of Civic Design, University of Liverpool

Rao L (1991) The structural design of the Telford Urban Policy Information System, URPERRL Working Paper 22. Department of Civic Design, University of Liverpool

Worrall L (1990a) Issues in the application of GIS in urban and regional policy making, URPERRL Working Paper 17. Department of Civic Design, University of Liverpool

Worrall L (ed) (1990b) Geographic information systems: Developments and applications. Belhaven, London

Worrall L (ed) (1991) Spatial analysis and spatial policy using Geographic Information Systems. Belhaven, London

Worrall L, Rao L (1991) The Telford Urban Policy Information Systems project. In: Worrall L (ed) Spatial analysis and spatial policy using geographic information systems. Belhaven, London, pp 127–151

# Dynamic GIS models for regional sustainable development

**Vassilios K. Despotakis[1], Maria Giaoutzi[1], and Peter Nijkamp[2]**

[1] Department of Geography, National Technical University, Zographon Campus, GR-Athens, Greece
[2] Department of Economics, Free University, De Boelelaan 1105, NL-1081 HV Amsterdam, The Netherlands

**Abstract.** This chapter gives a new conceptual framework for GIS modelling and offers an empirical modelling application for the Greek Sporades islands. Various spatial and non-spatial simulation experiments focusing on sustainable development of these islands are presented.

## 1. Introduction

There is an increasing need for improving and extending existing Geographical Information Systems (GIS), leading to integrated policy-oriented systems which may assist planners in complex evaluation problems (Fischer and Nijkamp 1992). Such systems should provide analysts with several relevant and feasible development scenarios (options). In this context there is also an urgent need for operational models aiming at reaching ecologically sustainable economic development (SD). On the basis of a compound evaluation, optimum scenarios which meet specific sustainability criteria (Van den Bergh and Nijkamp 1990; Van den Bergh 1991; Opschoor and Reijnders 1991) may then be selected by the user. Moreover, there are several barriers for developing such a system (Nijkamp and Scholten 1991). The main difficulty arises from the fact that we have to link traditionally complex and data-driven procedures (i.e., GIS procedures; see for example Burrough 1983; Scholten and Stillwell 1990; Openshaw 1990) with procedures that monitor economic-ecological interactions (World Commission on Environment and Development 1987; Opschoor 1991 a, b, c). Once we have established the above linkage, we may arrive at a user-friendly and operational system for evaluating SD within the region to which the system is applied. Simplicity and clarity of the system is a necessary characteristic if any strategic or policy decision has to be supported by this system.

In this chapter we present some preliminary results for such a hybrid GIS-SD system which was developed in the GIS Research Laboratory of Free University, Amsterdam. The system was also numerically tested in a 90 km (North-South direction) by 100 km (East-West direction) area of the Greek Sporades Islands in the Aegean Sea (Giaoutzi and Nijkamp 1989). To further evaluate the scenario results, we linked the GIS-SD system to a Decision Support System (DSS), that was originally designed to operate on non-spatial data (Janssen 1991; van Herwijnen and Janssen 1989). In this introductory section we will first start with some background remarks on spatial modelling.

One of the most efficient, computer compatible and elegant scientific approaches of understanding and deepening the processes of real-world phenomena is *modelling*. Models can be defined as idealized, presupposed processes and perceptions that are used to *approximate* the phenomena in the real world. In de Vries (1989) models are accurately and intuitively defined to be "mental maps" of reality. The idealization or completeness of the phenomenon we study is achieved through logical deduction, where the attention is oriented from the whole to its parts, seeking to find simplified relationships, interdependencies and influences (Rokos 1989) it has with respect to the other parts. Then, by induction, the whole is again regenerated by its (now explored) parts to its original compound parts (observed by analysis instruments and senses) and is next again compared to reality. The more the deviations this regenerated system (i.e., model) has from reality, the more new deduction-induction processes (i.e., new theory) are required, together with observations, to form an acceptable system (model). "Acceptable" system means here that the deviations of this system from reality cannot be precisely observed or measured. Fig. 1 (Mueller 1987) gives an illustrative view of what a model is in connection with the real world.

In Fig. 1 the concept of approaching the "real world" which has a non-regular geometric shape (meaning a shape that we cannot easily understand and model) with a model (a circle, which is an object of regular shape) is presented. The "remaining" of the real world that has not been modelled is shown as the "subtraction" between the real world's irregular shape and the circle. This residual area is always unknown and if its largest distance from the model cannot be measured or observed by the most advanced technological tools, then our model is ideal for the current spatial and time dimensions referred to. However, the evolution of technology, but also the expansion of human knowledge about social and economic phenomena, arms the researchers with measuring instruments of every kind which can detect reality's deviations from old existing models that have not been taken into account. Then the old models may be improved so that they contain the (systematic) residuals from reality that have been measured. This process continues permanently at both the micro and macro scale of the phenomena, improving our understanding of them.

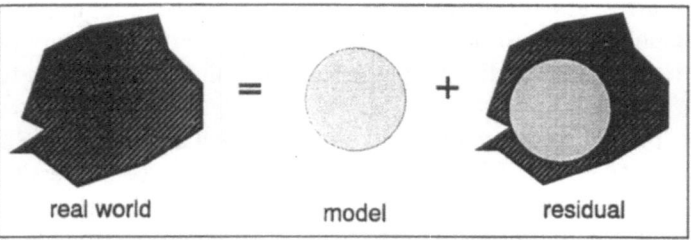

**Fig. 1.** The concept of modelling

We will subdivide here the phenomena under investigation into two different broad categories, both dynamic over time, for which a link is searched, viz:

- *geographical phenomena* and
- *economic, ecological and social phenomena.*

The basic tool that we will use to analyse the geographical phenomena is GIS dynamic spatial modelling, while the corresponding planning tool for analysing the economic, ecological and social phenomena is the notion of ecologically sustainable economic development (shortly: SD) of a system. Since the publication of the Brundtland report (1987) by the World Commission on Environment and Development, the notion of sustainability has become a major orientation for planning at local or regional levels. These two approaches will be finally linked together for a more general methodology with the aim to link sustainability to GIS.

Spatial non-dynamic dispersion and spatial diffusion models have been already constructed by various researchers in the past. Some examples are the Hägerstrand diffusion model (Morrill et al. 1988), gravity models (Haynes and Fotheringham 1988), and transportation models (Werner 1988; Hagishima et al. 1987) etc. In all these models the development process is regarded "frozen" on time, and the object propagation in space due to this motion is deterministically calculated by the models at any specific time point. Spatial flow models that used only the distance as a spatial parameter indicated strong spatial correlation of the model residuals (Baxter 1987), and thus model misspecification may occur when not all spatial registrations (e.g. a 3-D local or national reference coordinate system) are properly taken into account. On the other hand, several studies that aimed at using GIS for monitoring (mainly urban) development have been also carried out in the past (Méaille and Wald 1990; Lo and Shipman 1990). These approaches, although giving very useful results for monitoring urban motion, do not incorporate scenario generation techniques, so that the regional sustainability criteria can hardly be applied, not only in an "external event" scenario mode but also in a "policy" and "behavioural" scenario mode. Finally, pioneering studies in applying GIS for "conservation databases generation" (a concept which is close to sustainability considerations) have also been conducted in the past (see e.g., Ahearn et al. 1990), but again the spatial dynamics was not considered.

It is thus evident from the above discussion that there is a missing node between the field of GIS modelling and non-spatial modelling which would be necessary to integrate most of the benefits from both fields in a dynamic sense. In this contribution, we aim at providing this missing link, in both a theoretical and operational framework.

In section 2 non-spatial modelling concepts are discussed, while in section 3 the discussion is focussed on spatial modelling theory, where both are examined in a dynamic framework.

By creating in this way the desired background for the necessary methodologies and tools that are needed, the concept of *generalized stocks-layers* is introduced in section 4. This concept aims at providing the theoretical modelling link between GIS and SD modelling as already discussed above. Error analysis considerations are provided in Section 5, which aim to determine the error sources that contribute to the total error budget of the above hybrid models. Finally, preliminary numerical results concerning the application of the GIS-SD system to the Sporades Islands in the Aegean Sea, Greece, will be outlined in section 6.

In the next section the general concepts of modelling (non-spatial) economic-ecological-social phenomena are given first, so that the main characteristics of existing models of this type are briefly explained.

## 2. Non-spatial models

The concepts and definitions of SD can be summarized in the following statement of the World Commission on Environment and Development (1987) in the so-called Brundtland report: "Sustainable development is a pattern of development that meets the needs of present generations without jeopardizing the ability of future generations to meet their own needs". See also Nijkamp and Soeteman (1990), van Pelt et al. (1990), Opschoor and Reijnders (1991), Van den Bergh (1991), and Archibugi and Nijkamp (1989).

Traditionally, these economic and ecological SD studies, being complicated in themselves, are dynamically modelled using mainly *time* as an independent variable. For example, the general form of a dynamic economic or ecological model can be written as a set of ordinary simultaneous differential equations as follows (see also de Vries, 1989)

$$\frac{dX_i(t)}{dt} = F_i(X_1(t) , \quad X_2(t) , \quad \ldots, X_n(t); \ \Phi; \ t) \tag{1}$$

where $X_1(t)$, $X_2(t)$, $\ldots$, $X_n(t)$ are components of the $n$-th dimensional vector space $V^n$ (state vectors) which are fully describing the behaviour of the system under investigation; $\Phi$ is the vector of the parameters that characterise the system's structure and context; and $t$ is the independent (continuous) time variable.

The initial state vector of the system is assumed to be known: $X_1(t_0)$, $X_2(t_0)$, $\ldots$, $X_n(t_0)$ are given at the initial time $t_0$ and the system (1) is computed for $t \geq t_0$. The main reasons that make a system of differential equations suitable for dynamic modelling are the following:

a) In nature, we usually know the theoretical velocities (flows) or accelerations (rates of flows) of some quantities more accurately than the quantities themselves. This stems from the fact that we have already models available which can describe the motion of an object, a stock etc. very accurately if the *forces* that produced this motion are (assumed to be) known. A traditional example from a mechanical analogue is the equation of motion of a body under the influence of a (constant in time) force $F$:

$$m\frac{d^2r}{dt^2} = F \tag{2}$$

where $r$ is the position vector from the origin of the coordinate system to the body of mass $m$ which accepts a force of magnitude $F$. Any complications in (2) can easily be formulated in terms of additional forces (which will be included in the right hand side of the equation) and additional displacements (which will be included in the left hand side of the equation), always according to Newton's fundamental law. The simulation of dynamic phenomena in social sciences can also be successfully approached by introducing (2) as a basic "motion" or "development" model. In such type of social models any spatio-temporal movements of objects (urbanization, tourism, migration etc.) are inserted into the left hand side of (2) and are to be interpreted as movements that were caused by external forces of any type (attractiveness, job opportunities, building efficiency etc.) that reside in the right hand side of (2). In our work we will use this concept to link geographical (on the one hand) with socio-economic and ecological (on the other hand) models.

b) There is usually little or no knowledge at all about the analytical solutions for systems of the form (1). This is mainly due to the fact that system (1) contains in the right hand side all (or part of the other) state vectors that describe the system; thus an analytical solution may be very difficult, or sometimes impossible. The physical meaning of this is that the change in time (or motion) of each state vector is influenced by all the other system's state vectors and parameters, thus making the model(s) complicated and impossible to have an analytical (and thus deterministic) solution.

c) The *initial* conditions of a system at a specified time $t_0$ can be measured or determined with theoretically infinite accuracy. This is due to the fact that, without loss of generality, we can regard the system under investigation "frozen" in time $t_0$ and either physically measure its state vector at that time, or determine it through easier (deterministic) models. The mechanical analogue from mechanics (equation (2)) is that the initial positional vector of the body $r = r(x_0, y_0, z_0)$ can be specified very accurately either by physically measuring these coordinates, or by specifying them through easier models (e.g., from a "force-free" motion of the body which results in constant velocities).

d) Models of the form (1) can be numerically integrated using standard numerical integration techniques (Press et al., 1986), which are suitable for implementing in today's computers, having a high performance in speed, accuracy of the results and formulation efficiency, no matter how complex systems (1) can be. In general we can also eliminate the round-off errors of the numerical solutions well beyond the desired observational accuracies of our model. More details on numerical integration will be given later in this section.

e) There is full *control* for the results of each time step $dt$ that is used to integrate (1). Thus any *constraints* regarding the "external events" that result from the solution of (1) alone can be added at any stage of the numerical solution, allowing thus for efficient *scenario generation*.

It should be noted that complex second- or higher-order differential systems can always be reduced to first-order systems by proper substitutions and thus turn out to be of the form (1). In a reverse way, formulation (1) does not necessarily mean that our system is a system of first-order differential equations, since the structure of this system's formulation can equivalently result in a system of second- or higher-order differential equations.

Continuing the analogue concept of mechanics when studying economic-ecological models we can view the original state vectors $X_i$ as a system of *stocks* $S_i$ which dominate the operation of the system. In this case the derivatives $dS_i/dt$ are the *flows* of the stocks through time, and the right hand side of (1) can be a function of stocks, flows, constants, and, of course, time (see also Hordijk and Nijkamp 1977). Other functions of stocks, flows or constants that can enter the right hand side of (1) are called *converters*. A specific modelling tool, STELLA, has been developed in (Richmond et al. 1987), from where the majority of the above terminology has been adopted, and which we will briefly discuss here.

The generation of a model in which the modeller assigns the stocks, flows, converters and constants is done in an interactive graphical way, where a menu of these main modelling "objects" is available (Fig. 2). The above modelling steps result in the following system of differential equations (equations (3)):

$$\frac{dS_i}{dt} = F_i(S_1, S_2, \ldots, S_n; C_1, C_2, \ldots, C_k; t) \tag{3}$$

with the initial conditions (equations (4))

$$S_1(t_0) = \text{INIT}(S_1) = S_{10}$$

$$S_2(t_0) = \text{INIT}(S_1) = S_{20} \tag{4}$$

$$S_n(t_0) = \text{INIT}(S_n) = S_{n0}$$

An example from mechanics can again be given. Consider the spring-mass system (Figure 3).

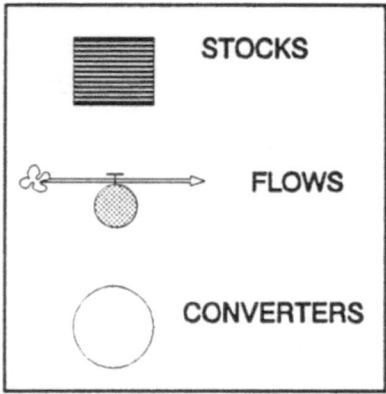

**Fig. 2.** The basic menu entities used for modelling in STELLA software

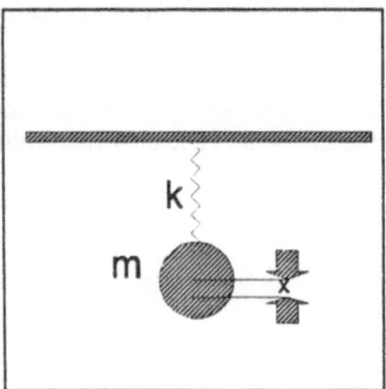

**Fig. 3.** The mechanical example of spring-mass system

The differential equation of motion for this system is a second order differential equation:

$$m\,d^2x/dt^2 = -kx \tag{5}$$

where
$m$ is the mass of the spring,
$k$ is Hook's spring constant, and
$x$ is the displacement.

The above system can be reduced to a system of two first-order differential equations, as follows:

$$dx/dt = v \tag{6}$$

$$dv/dt = -(k/m)*x$$

Thus the system under investigation is fully represented with two stocks, namely the stock of velocities $v$ and the stock of displacements $x$. The flow of $x$, $dx/dt$, is the velocity and the flow of $v$, $dv/dt$, is the acceleration expressed as in (6). The

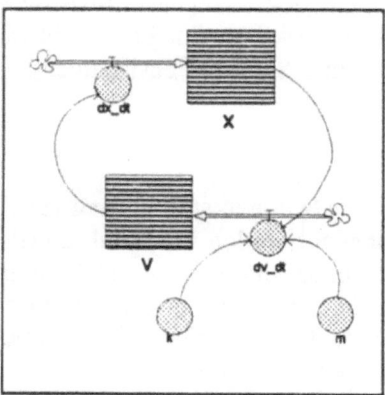

**Fig. 4.** The spring-mass model as a STELLA diagram

converters are in our case the constants $k$ and $m$. The corresponding diagram using the notations of Figure 1 is shown in Figure 4.

Assuming that the initial numerical values for the stocks and the constants given by the user are e.g. $m = 1$, $k = 1$, INIT $(x) = 1$, INIT $(v) = 0$, then as soon as the user finishes up "sketching" the model, the following system of first order differential equations has been created (equations (7)):

$$x = x + dt * (dx/dt)$$
$$\text{INIT}(x) = 1$$

$$v = v + dt * (dv/dt)$$
$$\text{INIT}(v) = 0 \tag{7}$$

$$dx/dt = v$$
$$dv/dt = -(k/m) * x$$

$$k = 1$$
$$m = 1$$

When we run the model for e.g. $t_0 = 0$ till $t = 24$ (hours) with a time step $(dt)$ of 0.125 (hours), then the sinus and cosinus graphs for stocks $x$ and $v$ appear correspondingly. This is evident from the analytical solutions of (7), which are of the form $x = a \sin(wt)$, $y = a \cos(wt)$.

The modelling and simulation language on which the above tool is based is DYNAMO, which is one continuous simulation language. Other simulation languages are: *Continuous*: Continuous System Modelling Program (CSMP); *Discrete*: GPSS, SIMSCRIPT, GASP and SIMULA (see Wyman 1970).

From the above the need to properly utilize this valuable tool which has been designed for *non-spatial* dynamic modelling for modelling *spatial* phenomena is evident, *within* a geographical information system. The proper integration involves several steps which will make possible the efficient data input-output, modelling and scenario generation. A first, very important step, that has to be taken is the creation of an interface between the existing modelling tool (written in *Apple Operating System*) and the majority of the existing geographic modelling packages (written in *IBM* or *Microsoft Diskette Operating System*). For the creation of this interface we selected an efficient, tested and "certified" language among the scientist's computer languages, the *FORTRAN* language (Microsoft, 1989). The logic in forming the above interface followed similar lines as in the procedure of forming the corresponding models in a STELLA environment.

## 3. Spatial modelling

Now we will discuss the field of spatial models (see Scholten and Stillwell 1990), by expanding the discussion presented in the previous section in spatial models.

In writing the systems of equations (1) we assumed that there is only *one* independent variable, namely time $t$. We could then investigate the interaction of various different quantities by relating them to stocks and assigning the proper relations for their flows, intermediate converters and constants.

On the other hand if we forget for the present time this dynamic concept of modelling and assume that we examine an object (e.g. a region) "frozen" in space and time by using geographical spatial modelling, the following concepts can be defined:

A specified region on the surface of the earth (or either above or below it, in a generalized context) presents, in general, a variety of phenomena that need to be analyzed (Fig. 5). The complexity of the region under investigation is exponentially increasing with:

a) the scale of the region, starting from the large scales of $1 : 500$ till the small scales of $1 : 1\,000\,000$,
b) the types of land uses of the region, starting from areas with a single land use (e.g., agriculture only), and proceeding to areas with multiple land uses and urban areas,
c) the geographic location of the area (geographic latitude, longitude and elevation), and
d) the frequency of the terrain changes within the specified region, starting with smooth areas and proceeding to rough and mountainous areas.

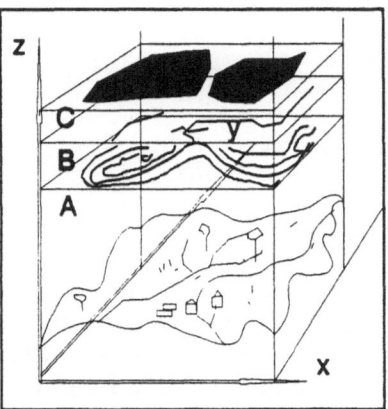

**Fig. 5.** The concept of overlaying a region with layers A, B, C, etc

In Fig. 5 we can see a "sample" region on the earth's surface presenting a landscape with some houses, rivers, roads etc. The region is analysed in terms of *layers* (Burrough, 1983) of information, each layer assuming to be as independent as possible from the others (avoiding redundancy of information) and as complete as possible in the attributes it is representing (avoiding loss of information). The origin of the spatial three-dimensional $(x, y, z)$ reference system is located "somewhere" on the earth and is usually defined by surveying and geodetic methods. All layers are assumed to refer to the same coordinate system, i.e. they are geometrically corrected. We thus form a *spatial* model of the real world which is analytical and tries to decompose the real world into somewhat simpler components, which can be mathematically described and measured. We can then write the general equation for a spatial static process $R(x, y, z)$ (in analogy with equation (1)):

$$R(x, y, z) = F(A, B, C, \ldots, Q) \text{ with}$$

$$A = A(x, y, z)$$

$$B = B(x, y, z) \tag{8}$$

$$\vdots$$

$$Q = Q(x, y, z)$$

The layers $A, B, C, \ldots$ can be one of the following attributes, or aspects, of reality:

- Topography (Digital Elevation Model, DEM),
- Land Uses,
- Hydrology,
- Transportation,
- Thematic information,
- Demographic data,
- Population distribution,
- Watersheds,
- Slopes, aspects gradients of the terrain,
- Cadastral information,
- Public utilities (sewing, water etc. networks),
- Remotely sensed satellite images,
- Digitized air-photographs,
- Meteorological data (rain, heights, temperatures etc.),
- Pollution data,
- Ecological species distributions,
- Historical data (documents), etc.

It is therefore obvious that in $A, B, C, \ldots Q$ point, line areal and volumetric types of information can appear. Each one of these layers can then be regarded as a map of the same region and properly geometrically registered with the other maps. To digitize all above layers so that they can be included in a fully computerized geographical information system, the following should be considered.

Depending on the *resolution* that each of these layers requires (or superimposes depending on the data collection method that was a priori followed), the *representation* of these layers may be done in any of the "traditional" computer data structures, namely in a *raster* or in a *vector* format. Thematic-type of depicted information, land-uses, raw and classified satellite data, digitized air photos etc., are usually stored in a raster format; cadastral information, roads and public transports, watersheds etc., that normally require high internal representation accuracy are stored in vector format; elevation models, slopes, gradients, aspects etc. can be stored in either raster or vector format.

Let us assume that we have stored layers $A, B, C, \ldots$, etc. in a geographic system, forming a database. Each layer represents a specific type of spatially referenced information of the form:

$$K = K(x, y), \quad K = A, B, C, \ldots, Q \tag{9}$$

To be in a position to properly access the database we created we need first to *classify* the stored information into meaningful and efficient numeric intervals.

For example, if $Z = Z(x, y)$ in (9) represents a digital elevation model, then most probably the range of the elevation function $Z$ will be between zero and some positive number, say 1000 m. All the intermediate numbers in (0, 1000) can appear in this data base. To be able to access this information in a more "user-friendly" way, we have to assign *classes* of information, e.g.:

Class 1: [0 m, 300 m]; low elevations
Class 2: [300 m, 700 m]; medium elevations
Class 3: [700 m, 1000 m]; high elevations.

In this respect the original file is "transformed" through the use of look-up tables to a file in which alphanumeric descriptions for each classification scheme become possible. This zoom-out of the layers is very important for the analyst since he can reduce the resolution level of the original data base for analysis purposes.

After we decide on the classification schemes, we obtain the classified layers (equations (10))

$$A: A_1(x, y) , \quad A_2(x, y), \ldots, A_l(x, y)$$
$$B: B_1(x, y) , \quad B_2(x, y), \ldots, B_m(x, y) \tag{10}$$
$$\vdots$$
$$Q: Q_1(x, y) , \quad Q_2(x, y), \ldots, Q_p(x, y)$$

Since $l, m, \ldots, p$ are not large numbers (usually in the order of 10 for land uses, 16 for elevations, 5–6 for slopes and aspects etc.) we can assume that, in this thematic-type of data representation, there is a generalization of information and thus the classified data bases can then be stored in raster format. This of course does not mean that any higher accuracy of the original databases should not be present any time we need it.

In doing so we arrive now at the next step of spatial modelling, which is a design of a tool that will enable us to simultaneously *access* a part or the whole of the created databases of layers, i.e. will allow us to:

- *Edit* the created layers either in a graphical or in a text (manual) form;
- *Update* the layers with new input data information that may become available, by digitizing or rasterizing other sources of information;
- Perform any *algebraic operations* among them and create synthetic (new) layers;
- Perform *statistics* (e.g. multivariate analysis) and *correlations* (e.g. cross-correlation coefficients) between the different layers;
- Perform *simultaneous queries* of logical and numerical form;
- Display *graphically* the results to maps, reports or tables for further processing and decision making.

In other words we now seek for a geographic system which will have inherent capabilities of a *relational data base management system,* together with extra statistical analysis capacities and graphical representation of the results.

A number of GIS spatial analysis tools exist that partially or completely attempt to fulfil the above requirements. The choice always depends on the specific application. The tool that was used in this study for multiple-layers creating and accessing is *SPANS* (SPANS 1988). In this tool the user can enter geographic data from digitizers, manually, or by transforming existing types of data. The various layers of information can be created from point, line, or areal data in vector or raster form with a user-specified classification scheme. There is a further internal compacting of the original data to *quadtrees*, a data compression method. All the created layers can be stored in a quadtree format in a user-specified resolution which depends on the a priori established scale of the region and the data accuracy, transformed in spatial dimensions. The highest resolution that can be recovered from a quadtree map is the highest raster resolution of the original data. The simultaneous processing of the layers (up to fifteen at the same time) can be done while the layers have been transformed in quadtrees and thus we can erase the original (raster or vector) files. There is also an internal programming language which allows for spatial modelling in a batch mode. The results (tables, reports, maps) can be saved on disk in the form of "slides" for fast displaying and presentations.

For a modified run-length encoding compression form that gives comparable results with the quadtree form, see Despotakis (1990).

## 4. GIS spatial models for sustainable development

We have now reached a crucial point where a *link* between the non-spatial models of type (3) and spatial models of type (10) has to be found. From the discussion of the present and the previous section it can be easily seen that the two ends of the linking thread between SD and GIS have been simultaneously developed: The non-spatial concepts of the models of type (3) are first given to enter the ideas of stocks and flows and their dynamic interrelations over time, but "frozen" in space. Then, a summary of the layering processing, and their dynamic interrelations in space, in geographic modelling is given, considered "frozen" in time. The representative tools from both sides were the STELLA non-spatial modelling from the one side and the SPANS spatial modelling on the other side. Thus in the modelling process in the framework of sustainable development constraints, the next natural question that follows is: "how can we model dynamic phenomena in time, considering also their spatial dimensions, within the framework of sustainable development?" It is evident that an answer to such a question will have to *integrate* both given modelling approaches in one, together with the sustainability constraint.

We begin to investigate this problem by regarding the conceptual equivalence of *stocks* with *layers* (in Murthy et al. (1990) they are also referred to as 'reservoirs' or 'levels' interconnected by flow paths): We could write e.g. for layer $A(x, y)$ in (10) and stock $S_i$ in (3) the following relationships:

$$A_1(x, y) \leftrightarrow S_1(t)$$

$$A_2(x, y) \leftrightarrow S_2(t) \tag{11}$$

$$\vdots$$

$$A_l(x, y) \leftrightarrow S_l(t) \ ,$$

i.e. relate the idea of stocks with the idea of layers by realizing that the spatial contents of a specific classified layer (classes of layer $A$ in the example of (11)) can be regarded to be stocks which dynamically change in time. In this manner we can generalize the concept of dynamic modelling of a specific phenomenon to include all the necessary stocks $S_i$ which are needed to describe the available spatial layers of information for a region. From the above it is evident that if there are $t$ spatial layers available for a region, each with $l, m, \ldots, q$ classes (see equation (3)), then we need $r = l + m + \ldots + q$ stocks (and flows) definitions of the form (10). Thus our generalized spatial modelling of dynamic phenomena takes the form:

$$dS_i/dt = F_i(t, x, y, z; S_i; F_i; C_i) , \quad i = 1, 2, \ldots, r \tag{12}$$

with the initial conditions

$$S_i(t_0, x_0, y_0, z_0) = \text{INIT}(S_i) , \quad i = 1, 2, \ldots, r \tag{13}$$

Then the sustainable development considerations can be embedded in (12) and in each integration step $dt$ by:

1) imposing the necessary *conditions* that stocks $S_i$ should fulfil
2) defining and examining the values of the *indicators* (but also the stocks, flows and converters themselves) as functions of the stocks, flows and converters, and
3) defining and examining alternative *scenarios* which, by the proposing system, can be of any kind: No-policy (external events) scenarios (natural progress of the eco-system), external policies (influence totally or partially one or more functions of the eco-system) and behavioural scenarios (creating and monitoring spatial dynamic behavioural patterns such as migration, urban growth etc.).

The important stage for the above link is to decide on the rules of three types of *stock motion*: Expansion, Shrinkage and Migration.

These rules have to govern the propagation of the stock changes that will result from the dynamics on time solution to the spatial geographic system. Such rules can be built in a raster (or quadtree format) and will reflect the spatial reality of the region. For example, if the dynamic solution gives an increase in "urban stock" of the land-use layer of $10000 \, \text{m}^2$, then the spatial model can equally (this can be an assumption, and unequal changes in $x$ and $y$ can be as well treated) increase the $x$ and $y$ dimensions of this class by the square root of $10000 = 100 \, \text{m}$. The "candidate" squares in the geographic layer that can accommodate this expansion will be determined by overlaying the land-use layer with other layers that can influence this expansion: the layer of slopes, aspects elevation, transportation etc. (see also Méaille and Wald 1990). Of course the interaction of each class within the same layer will also play an important role.

After all the "motion" assignments will have been completed to the geographic system, in a hybrid user-computer manner, then the results (maps, reports etc.) for this time stage will be written on disk, and the numerical integration will proceed to the next step.

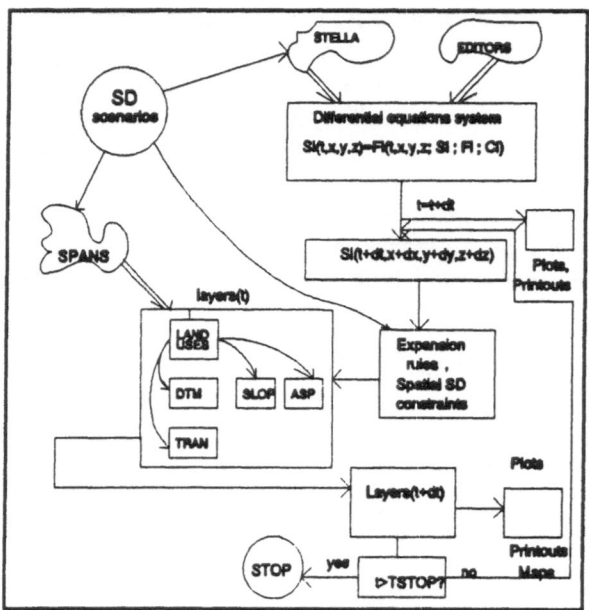

**Fig. 6.** The link between non-spatial and spatial dynamic modelling

The proposed system is shown in Fig. 6: Sustainable development constraints can be entered at the "expansion rules formation" stage, after each integration circle has been completed. These constraints can be of the form (14) (equality constraints) or of the form (15) (inequality constraints):

$$G_{SD}(S_1, S_2, \ldots, S_r) = 0 \tag{14}$$

$$G_{SD}(S_1, S_2, \ldots, S_r) \geq 0 \tag{15}$$

$$G_{SD}(S_1, S_2, \ldots, S_r) \leq 0$$

The user can enter the modelling procedure for *scenario generation* at three points: in the differential equation forming and editing process, in the geographical analysis stage at the "expansion rules formation" after each integration step (Fig. 6). The cooperation of the non-spatial (STELLA) with the spatial (SPANS) system is achieved through the generalized stocks that form the function of layers. The transition to the space domain is achieved by defining the rules of stock motion.

It is evident that the outputs of this system can be further analyzed and evaluated by a decision support system linked into it.

The proposed system in essence employs concepts from the combined fields of economics, geographic information systems, animated cartography (see also Koussoulakou 1990), and surveying.

## 5. Error analysis

We continue the chapter by focusing on the errors that are entered into the model from the various error sources (see also Heuvelink et al. 1989). These error sources can be subdivided into three categories:

- *Numerical integration errors,*
- *Locational and attribute errors in overlaying procedures,*
- *Errors due to mathematical models.*

Each error source is next separately discussed, with the ultimate goal to develop the theoretical error studies frame for the proposed hybrid system. Numerical results will be presented in the future, based on the theoretical discussions that follow:

### Numerical integration errors

These error sources contribute to the general error of the computed quantities when this technique is used in combination with any other modelling (spatial or non-spatial) analysis.

The numerical integration method selected in this study is the Runge-Kutta 4th order method. In terms of accuracy it stands somewhere "between" the Euler method and the advanced *predictor-corrector* methods. In increasing degree of accuracy these methods are as follows (only one variable is given here for simplicity: integrate $y' = f(t, y)$ with $y(t_0) = y_0$.):

*Euler* method:

$$y(t+dt) \approx y(t) + dt * f(t, y) \tag{16}$$

This method is also employed by STELLA for integrating models with limited accuracy demands. Since convergence problems may occur only because of the use of this specific method (i.e. our modelling may otherwise be perfect), it is suggested that this method is rarely used, even if increase in computational time has to be a sacrifice.

*Runge-Kutta* method: $y(t+dt) \approx y(t) + (1/6) * (k_1 + 2k_2 + 2k_3 + k_4)$ with:

$$
\begin{aligned}
k_1 &= dt * f(t, y) \\
k_2 &= dt * f(t+dt/2,\ y+k_1/2) \\
k_3 &= dt * f(t+dt/2,\ y+k_2/2) \\
k_4 &= dt * f(t+dt,\ y+k_3)
\end{aligned}
\tag{17}
$$

The advantage of this method is that it avoids the computations of higher order derivatives, and still approximates the solution using an equivalent of fourth-order Taylor expansion. In place of these derivatives extra values of the given functions $f(t, y)$ are used.

*Predictor-Corrector* methods: They are based in a formula to make a first prediction of the next $y(t+dt)$ value (predictor). This prediction is then followed by an application of a more accurate formula (corrector). This procedure, although complex, has the advantage that with each approximation an estimate of the error can be made and thus we have better control of the propagated error in each integration step.

We give here the most famous predictor-corrector methods: The Milne method and the Adams method (Scheid 1968):

*Milne* method: Uses the predictor-corrector pair:

$$y(t+dt) \approx y(t-3dt) + (4dt/3)(2y'(t-2dt) - y'(t-dt) + 2y'(dt))$$

$$y(t+dt) \approx y(t-dt) + (dt/3)(y'(t+dt) + 4y'(t) + y'(t-dt)) \tag{18}$$

*Adam's* method: Uses the predictor-corrector pair:

$$y(t+dt) \approx y(t) + (dt/24)(55y'(t) - 59y'(t-dt) + 37y'(t-2dt) - 9y'(t-3dt))$$

$$y(t+dt) \approx y(t) + (dt/24)(9y'(t+dt) + 19y'(t) - 5y'(t-dt) + y'(t-2dt)) \tag{19}$$

In both the above methods the computation of the four previous values: $y(t)$, $y(t+dt)$, $y(t+2dt)$ and $y(t+3dt)$ are required before the numerical integration begins.

The *sources of error* for the Runge-Kutta procedure that we utilize, are the following (see also Scheid 1968):

- First, we note that in writing equation (17) the "approximately equal" sign was used. This means that the series is *truncated* at a specified order of expansion and this error source is called *truncation error* $e_T$. It depends on the fifth order derivatives of $y$, $y^{(5)}$, and on the fifth power of the time step used ($dt^5$). The *local truncation error* (i.e. the error made at a specific integration step) and in some cases the total truncation error can be reduced by using the method of *adaptive stepsize*.

- A second error source is the *propagation error* $e_P$. It results from the propagation of error in the initial value(s) $y_0$ throughout the numerical integration procedure, and depends also on the magnitudes of the first, second, third and fourth derivatives of $y$. It can be (although in an elaborate way) computed by applying the law of propagation of errors in (17). To reduce this error source we have to reduce as much as possible the errors in the initial values.

- Implementing the procedure in a computer, the *round-off error* $e_R$ is always present. It also propagates through each step of integration, and sometimes it can be the dominant error in the whole procedure. The treatment of this error is achieved by increasing the internal numerical accuracy of the computations (and, in turn limiting the available memory). In FORTRAN this is done by transferring the real numbers byte domain from 4 (single precision REAL*4) to 8 (double precision REAL*8).

- In approaching a continuous phenomenon in nature by a set of differential equations and solving this set as above by a *discrete* time stepsize $dt$, we introduce an error which we may call *discretion* or *discretization error* $e_D$. Reducing $dt$ as much as possible does reduce this error, but it gives an exponential increase in the round-off error (and, of course, the amount of necessary computations and memory requirements). It is thus suggested that for the treatment of this error the *adaptive stepsize* control is again used.

From the above four different sources of errors we can form the *total numerical integration error e*, assuming that the four separate errors are *uncorrelated* as:

$$e = (e_T^2 + e_P^2 + e_R^2 + e_D^2)^{1/2} \tag{20}$$

The ratio of the total error to the exact solution (which is sometimes impossible to estimate if the exact solution is not known), called the *relative error* $e_r$, is of high importance since if the exact solution grows larger in absolute value, then also a larger total error can probably be tolerated.

We finally give some remarks about the *stability* and the *convergence* of the solution obtained by using (17). The solution is defined to be *stable* if any single error made in applying the Runge-Kutta (or any numerical integration method) to $y' = Ay$ (see also de Vries 1989) has an effect which imitates the exact solution behaviour. On the other hand the *convergence* of the solution to the exact solution is important and it mainly depends on the existence and behaviour of the higher order derivatives.

## Locational and attribute errors in overlaying procedures

We should notice here that the *error propagation* of the original data into the formed layers should be considered as carefully as possible. The various error sources that enter the overlaying procedures of real-world phenomena (see Fig. 5) could be summarized as follows:

- *Round-off error* $l_R$, which results from the same source as it was previously described,
- *Discretion error* $l_D$, which results from the discretized approximation of continuous phenomena. This error is much larger in raster and in quadtree than in vector representations. This error can be diminished by choosing as small a spatial stepsize as possible ($dx*dy*dy$), always stopping at the point where the $l_R$ error resulting by this procedure becomes larger than the discretion error itself.
- *Propagation error* $l_P$, which results from the propagation of the errors $\sigma_K(x, y)$ in the original data $K(x, y)$, $K = A, B, C, \ldots, Q$ when forming synthetic layers $U(x, y)$ as functions of the original layers: $U(x, y) = f(A, B, \ldots, Q)$. This error depends strongly on the first derivatives of functions $A, B, C, \ldots Q$, their variance-covariance matrices, and their error correlations that may exist between them. If layers $A, B, C, \ldots Q$ are uncorrelated, then their errors are also regarded to be uncorrelated. In real-world applications, however, this is not the case, since the extraction of the original data that form layers $A, B, C, \ldots Q$ from various techniques (maps, air-photographs, satellite sensors) can sometimes be strongly dependent to each other. We can somehow control this error by a proper data editing to detect and disregard *blunders* (robust estimation), and by comparing a data source with another (independent) data source of the same region.
- *Locational error* $l_S$, which results from the uncertainty in position the various data layers imply. Even after the proper geometric registrations there remains an uncertainty in position coordinates $(x, y, z)$ which can be expressed by the variance-covariance matrix of the (computed or measured) coordinates (equation (21)):

$$\Sigma_{xyz} = \begin{pmatrix} \sigma_x & \sigma_{xy} & \sigma_{xz} \\ \sigma_{yx} & \sigma_y & \sigma_{yz} \\ \sigma_{zx} & \sigma_{zy} & \sigma_z \end{pmatrix} \tag{21}$$

One way to study this error is through Monte-Carlo methods, i.e. changing randomly the positional coordinates $x, y, z$, perform geographic operations (i.e. compute $U = f(A, B, C, \ldots, Q)$) and compare the results with an expected value of this geographic operation (i.e. a mean value of $U$).

- *Attribute error* $l_A$, which results from the errors in labelling the various classes $A_1, A_2, \ldots, A_l$ in (11). This error can be regarded as a blunder type of error, and can be removed with field control.
- The *total error* $l$ of the overlaying procedure can then be estimated, assuming that all its error components are uncorrelated through:

$$l = (l_R^2 + l_D^2 + l_P^2 + l_S^2 + l_A^2)^{1/2} \tag{22}$$

Concluding this section on spatial modelling, the theoretical concepts of layers, their geometric registrations, and their incorporation into a relational data base management system were given, together with some error analysis considerations, considered to be essential a-priori knowledge for the desired link between SD and GIS modelling.

*Errors due to mathematical models*

The mathematical spatial and non-spatial GIS-SD models are only approximations to the corresponding real world's models. This means that even if the numerical integration errors and the overlaying errors were zero, there would still exist an error source due to the imperfection of the mathematical models themselves. This error source has usually a systematic pattern, and this is how it can be detected by instruments, provided its magnitude is higher than the instrument's sensitivity. It is only through this type of error research that we can improve our knowledge about the real phenomena. By detecting them, we usually arrive at more precise and refined mathematical expressions of reality. However, for the detection of such errors, high external observations precision and minimization of the error propagation is assumed.

## 6. Empirical results

The GIS-SD system developed as described above was numerically applied to the Greek Sporades islands located at the central and western part of Aegean Sea (Fig. 7).

There are two main conflicting objectives that appear in the Sporades area and, consequently, are dealt with in this study: (1) regional economic development and (2) environmental protection. The economic activities of the approximately 20,000 inhabitants of the region (1990) are mainly based on tourism and fishery. The dramatic increase of tourism in the past 30 years in Greece has also influenced in descending order the islands: Skiathos, Skopelos, Alonnisos and Gioura. Pilion is expected to receive spill-over effects from tourism and agriculture. Dur-

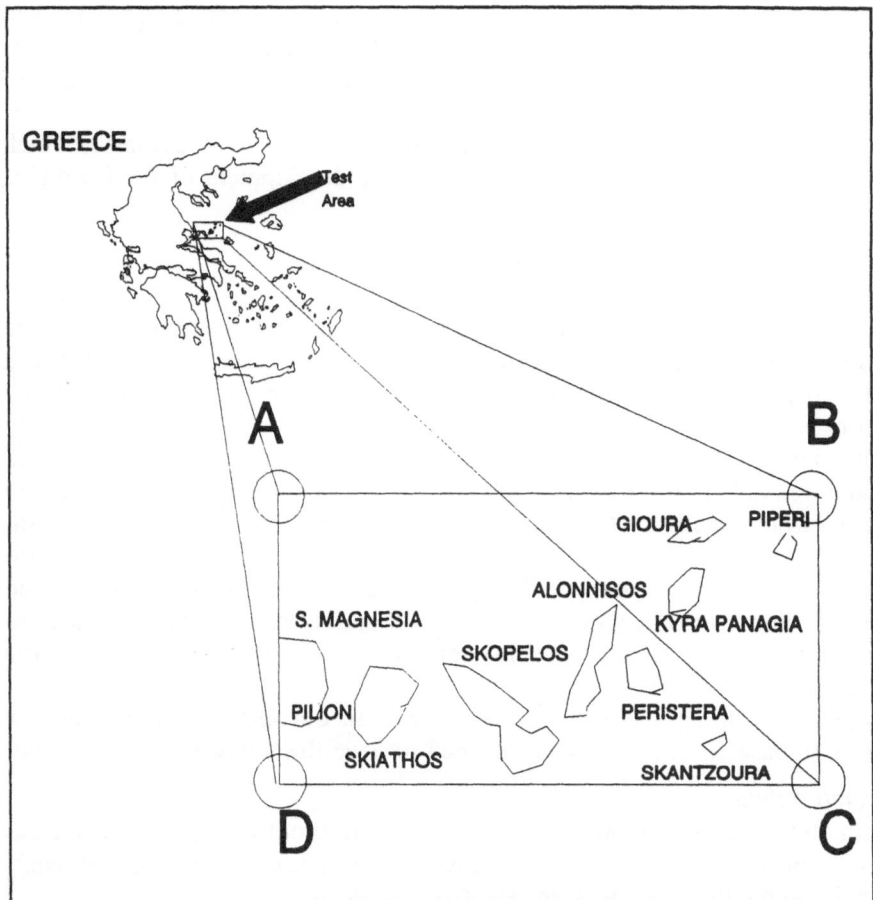

**Fig. 7.** The location of the Northern Sporades test area

ing the summer there is a strong increase of population on the islands due to tourism (domestic and foreign). This population increase often exceeds the population carrying capacity of the islands, and may thus result in abrupt high resource demands. This in turn may cause irreversible processes in the resource stocks. Similar effects caused by the abrupt changes in economic activities due to the tourism have been studied in Despotakis (1988 and 1991).

From the above discussion it becomes apparent that the SD conflicts on the study area are dynamic phenomena. Some of these conflicts can also be expressed directly in spatial dimensions (e.g., the expansion of urban land at the expense of forest areas). But it is only through the use of a spatio-temporal hybrid GIS-SD dynamic model, that we can satisfactorily approach and simulate the mechanics the various SD conflicts. For such a purpose, a meaningful policy model for this area should have the following features:

● it should focus on the dynamic simulation in space and time rather than being an optimization model (see Despotakis and Giaoutzi 1990);

- it could assist researchers in examining the dynamic economic and ecological conflicts of our study area in their physical dimensions (three dimensions for spatial reference together with one dimension for dynamic reference over time);
- it would provide alternative strategic solutions by means of scenarios which should be based upon SD constraints for the development of the region.

### Data requirements

The data requirements for such a GIS-SD hybrid model are determined by (1) the spatio-temporal resolutions selected (i.e., 100 m × 100 m ground pixel size, and a simulation period of 1 year), and (2) the data needs for a successful GIS-SD model calibration.

The data supply may not necessarily be in a one-to-one correspondence with the data demand. This means that (1) digital data may have been collected and processed which finally are not used as an input into the GIS-SD model, but rather form an integrated digital data base, and (2) necessary digital data for the GIS-SD model may not have been available for the period the model was developed and tested. The data collection process focused mainly on obtaining as much information as possible about the socio-economic and natural environment of the study area.

In general, the necessary data input into both our GIS-SD model and the digital data base constructed, was designed to have the following data features:

### Non-spatial data
- data on the socio-economic reality of the region (productivity and income per economic sector, tourism, fishery, houses, energy used per household, etc.),
- demographic data (population, age pyramid etc.),
- ecological data (ground water, sea water quality, forest fires, wild life data etc.).

### Spatial data
- terrain elevation and sea depth data,
- land use data,
- distance data from important land uses such as urban land, forests etc.,
- road transportation network data etc.

The combined SD-GIS model was mainly applied to the island of Alonnisos. After the necessary spatial and non-spatial data requirements were set, the relevant data were collected from the various (mainly Greek) sources: National Statistical Service, Greek Military Geographic Service, Ministry of Environment, Planning and Technical Works etc.

Based on the selected data, we run our GIS-SD model as follows (see for more details Despotakis, 1991).

Two scenario runs – generated by successively excluding and including road transportation on Alonnisos island – were carried out to demonstrate the efficient use of our GIS-SD system for monitoring urban development under different urban attraction conditions. The attraction layers generated by our system-

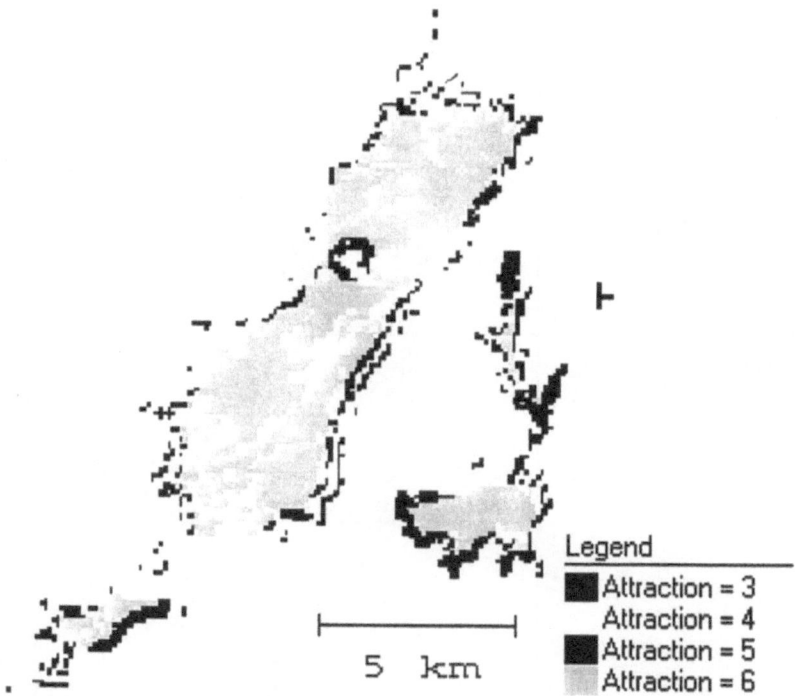

**Fig. 8.** Suitability analysis excluding transportation in behavioural scenarios

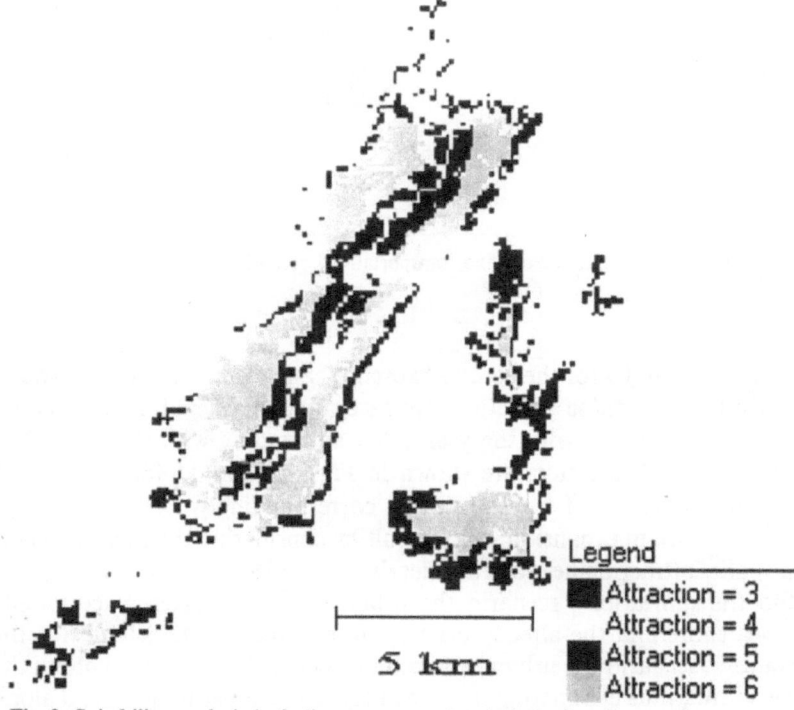

**Fig. 9.** Suitability analysis including transportation in behavioural scenarios

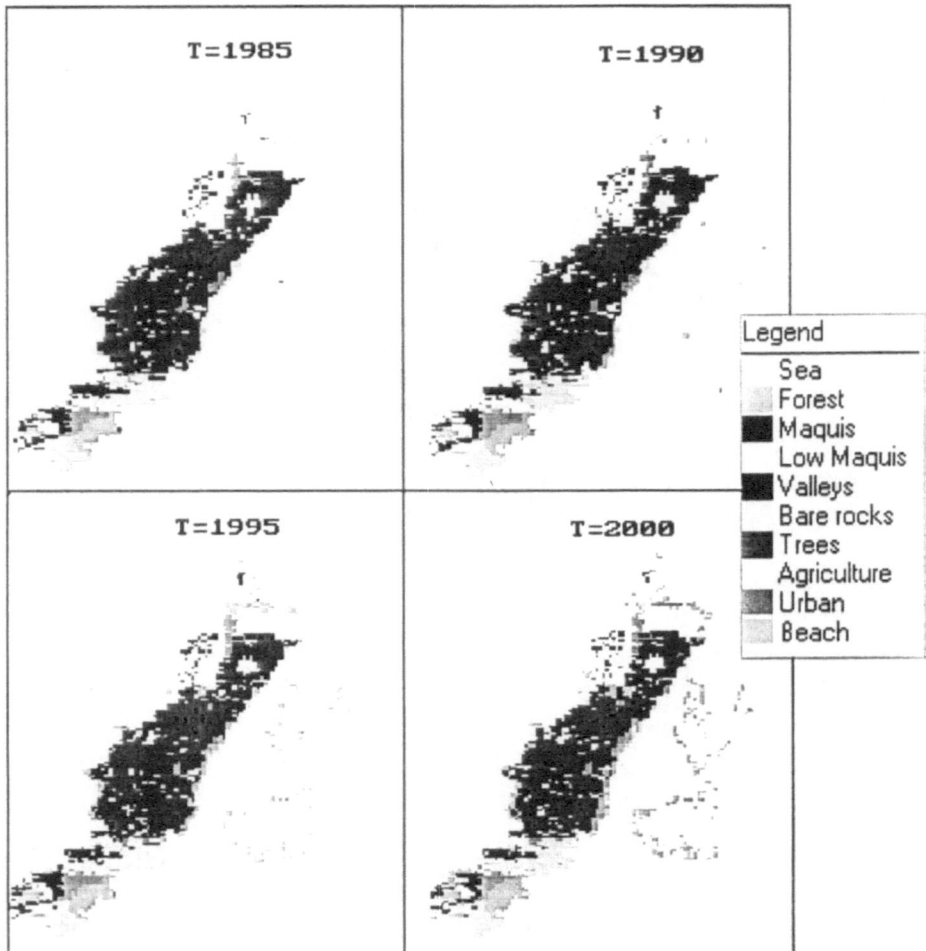

**Fig. 10.** Dynamic land use changes for the "no transportation" scenario

are shown in Figs. 8 and 9 for the case of "absence" and "presence" of transporation, respectively. Using these attraction layers we run our system for a simulation period of 15 years, starting with the year 1985, and ending with the year 2000, for every five years. The results are shown in Figs. 10 and 11 for the case of "absence" and "presence" of transportation, correspondingly.

The effects on urban expansion which result by considering the road transportation as a spatial attraction network are clearly depicted in the above two figures: for the "no transportation" scenario the urban expansion takes place mainly across the sea shore and the already existing urban areas of the island; for the "transportation" scenario the urban expansion presents clusters spread along the roads of the island, thus eliminating the size of the urban areas to be spread along the sea shore. Our GIS-SD system provides these results in the form of raster

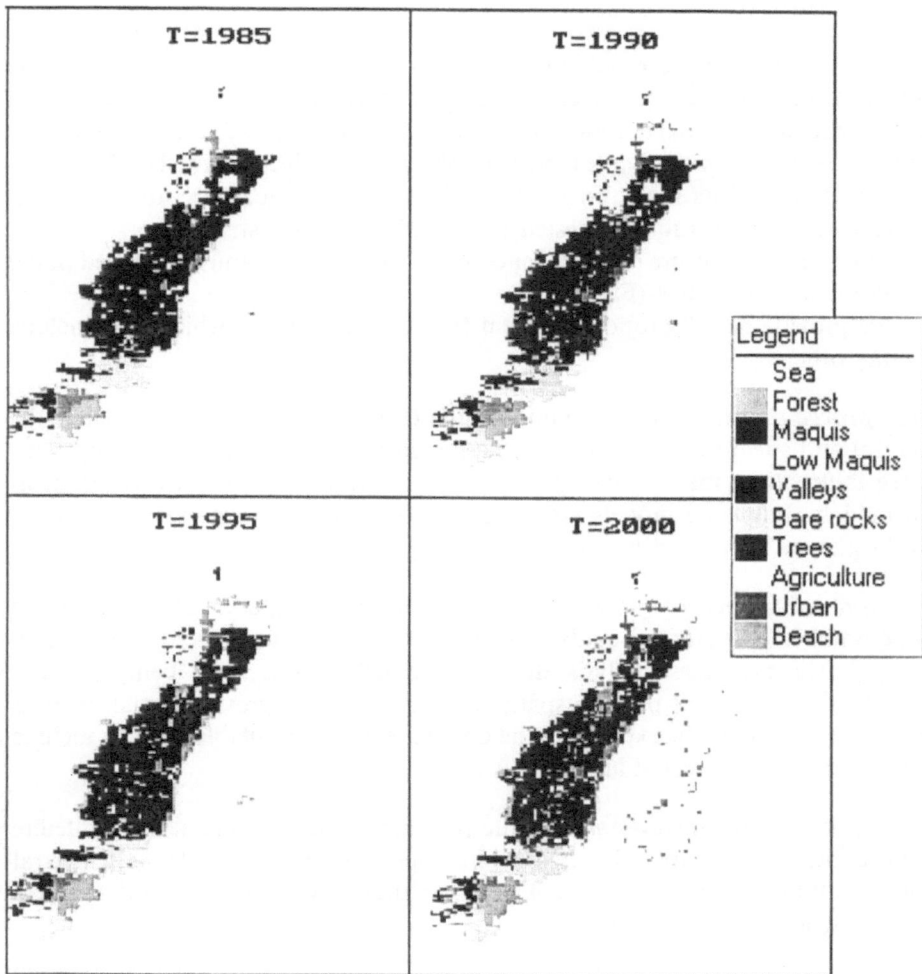

**Fig. 11.** Dynamic land use changes for the "transportation" scenario

images which may also be used for animation applications, so that more intuitive information may be extracted from the simulation results. These two scenario results can be inserted into a DSS as follows.

For each criterion (e.g., nature, tourism, etc.) an evaluation map is generated which corresponds to the selected alternative for a specific year. A weighted summation for each evaluation map may result in an unambiguous performance indicator. This number may then be inserted into its corresponding position in a plan effect table. Multi-criteria analysis (see, e.g., Janssen 1991) may then be used to rank the selected alternatives. Using the above approach a transformation from a two-dimensional space to the one-dimensional space is carried out. Alternative methodologies may result in a rigorous two-dimensional DSS using multi-criteria analysis (see, e.g., Guariso and Werthner 1989).

## 7. Outlook

The GIS-SD system described in this study mainly operates on raster data; at the present time, where economic-ecological procedures are spatially simulated by a GIS, the reduced spatial accuracy of rasters does not seem to create any problems. Since the topology is easily preserved and the spatial objects are well-defined with rasters, these factors make up for their limited spatial accuracy. In the future, other data structures may be tested such as the quadtree structures.

In general, the future developments of our system are mainly identified in the operationalization stage (Fig. 12).

We provide here 7 proposed system future developments which may include among others:

*a) Improved models:* spatial and non-spatial models that are used to simulate the natural and socio-economic reality of a region are often based on assumptions; these assumptions may not be true, especially in the case of complex environmental models. Improved models are needed in order to overcome assumptions or "wild guesses" for specific processes.

*b) Improved spatial analytical tools:* although the non-spatial analysis offers a large number of potential analytical tools (time series analysis, stochastic and deterministic processes analysis etc.), the spatial analysis field, being a newly developed field, is still poor in spatial analysis tools. Improved spatial analysis tools may be derived by expanding the one-dimensional available tools to include two- or three-dimensional analysis.

*c) Integration with remote sensing:* the potential of the remote sensing satellite data resides in their capability of providing the researchers with (1) multispectral and (2) multitemporal raster data. These data may serve as additional data layers for the region under investigation.

*d) Improved error analysis modules:* an improved error analysis module which may operate in parallel with the whole GIS-SD system is necessary. Such a module would provide detailed information on the expected quality of the data and the model results. Then, in case we would like to increase the output quality, we would either collect more data or to have a better model specification.

*e) Use of quadtree structures for spatial data:* the structures used in our study were mixed raster, vector and quadtree structures; the main layer-stock link within the dynamic simulation was carried out using raster layers. If all the layer operations between the original and derived spatial data are to be carried out in a quadtree mode, this will result in faster and more efficient computations throughout our system. Thus the raster structure (typical spatial structure for analyzing economic-ecological thematic data) may be substituted by the quadtree structure.

*f) User-friendliness:* making a complex system user-friendly clearly narrows its range of applications and makes the system more rigid. However, user-support tools is necessary to be provided together with the GIS-SD system.

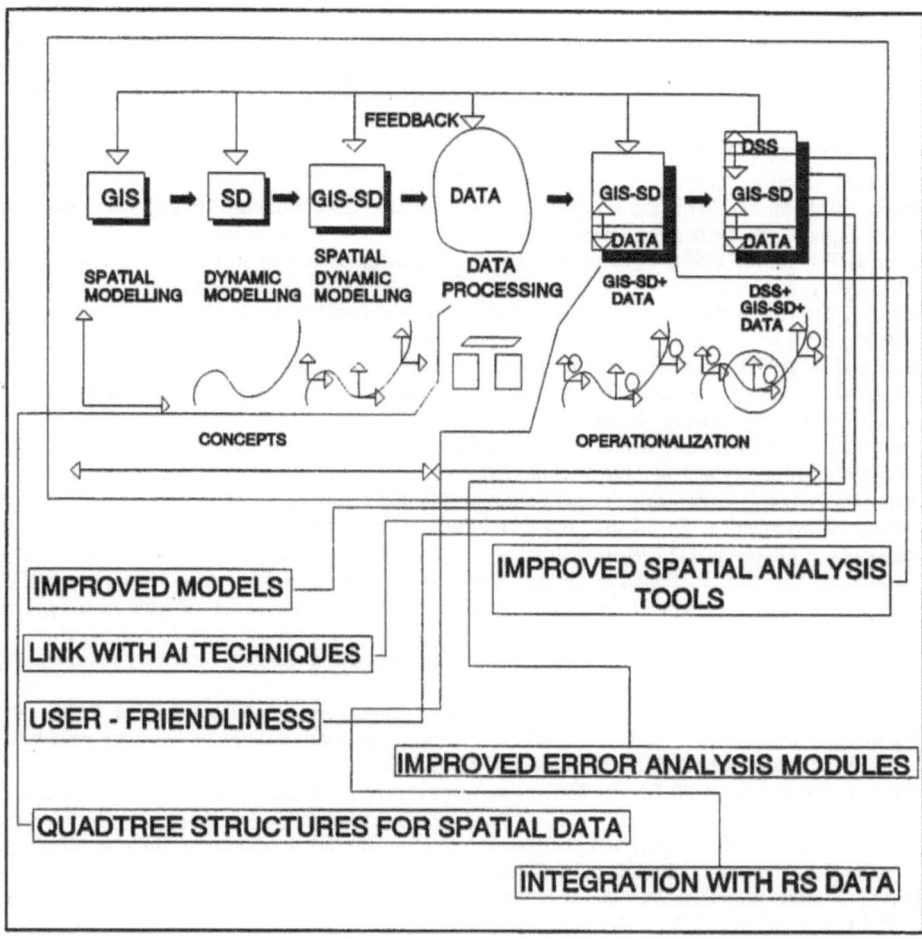

**Fig. 12.** Plausible future developments for our GIS-SD-DSS system; improve the models; in case the predicted quality was sufficient, we would proceed with the model simulation runs

*g) Link with AI techniques:* the recent advances in the field of AI technology, offer a new potential to be explored by a GIS. Especially in the field of spatial analysis, the new AI tools offer tremendous possibilities to be utilized. These include amongst others the utilization of neural networks, pattern recognition methods, group method data handling methods and genetic algorithms for analyzing GIS-SD fuzzy (input or output) information.

# References

Ahearn CS, Smith JLD, Wee C (1990) Framework for a geographically referenced conservation database: case study nepal. Photogram Eng Remote Sensing 56:1477–1481

Archibugi F, Nijkamp F (1989) Economy and ecology: towards sustainable development. Kluwer, Dordrecht

Baxter MJ (1987) Testing for misspecification in models of spatial flows. Environ Plann A 19:1153–1160

Bergh JCJM van den, Nijkamp P (1990) Ecologically sustainable economic development: concepts and implications. Serie research memoranda, Free University, Amsterdam

Bergh JCJM van den (1991) Dynamic models for sustainable development. Ph D Dissertation, Free University, Amsterdam

Burrough PA (1983) Principles of geographical information systems for land resources assessment. Monographs on soil and survey no 12, Oxford Science Publications

Despotakis V (1988) Diachronic monitoring of the land use changes in a subarea of the town Agios Nikolaos Crete for the environmental impact assessment. Presented The Int Conference in Advances in Remote Sensing, Thessaloniki

Despotakis V (1990) A new method for raster files processing. Presented The 1st Greek Conference in Photogrammetry and Remote Sensing, Athens

Despotakis V (1991) Sustainable development planning using geographic information systems. Ph D Thesis, Department of Economics and Econometrics, Free University Amsterdam

Despotakis V, Giaoutzi M (1990) Towards a new approach for digital data bases management in environmental planning. Presented at the 30th Congress of Regional Science Association, Istanbul

Fischer MM, Nijkamp P (1992) Geographic information systems and spatial analysis. Ann Reg Sci 26:3–17

Giaoutzi M, Nijkamp P (1989) A strategic information system and planning model for marine park management. EC report G11

Guariso G, Werthner H (1989) Environmental decision support systems. Ellis Horwood, Chichester

Hagishima S, Mitsuyoshi K, Kurose S (1987) Estimation of pedestrian shopping trips in a neighborhood by using a spatial interaction model. Environ Plann A 19:1139–1152

Haynes K, Stewart A (1988) Gravity and spatial interaction models. Scientific geographic series, vol 2. Sage, Beverly Hills

Herwijnen van M, Janssen R (1989) Definite: a system to support decision on a finite set of alternatives. Institute for Environmental Studies, Free University, Amsterdam

Heuvelink GBM, Burrough PA, Stein A (1989) Propagation of errors in spatial modelling with GIS. Int J Geogr Inf Syst 3:303–322

Hordijk L, Nijkamp P (1977) Dynamic models of spatial autocorrelation. Environ Plann A 9:505–519

Janssen R (1991) Multiobjective decision support for environmental problems. Ph D Dissertation, Free University, Amsterdam

Koussoulakou A (1990) Computer assisted cartography for monitoring spatio-temporal aspects of urban air pollution. Ph D Dissertation, Delft University Press

Lo CP, Shipman RL (1990) A GIS approach to land-use change dynamics detection. Photogram Eng Remote Sensing 56:1483–1491

Méaille R, Wald L (1990) Using geographical information system and satellite imagery within a numerical simulation of regional urban growth. Int J Geogr Inf Syst 4:445–456

Microsoft Corporation (ed) (9 89) Microsoft Fortran Reference Manual. Redmond, WA

Morrill R, Gaile GL, Thrall GI (1988) Spatial diffusion. Scientific geographic series, vol 10. Sage, Beverly Hills

Mueller II (1987) Advanced satellite geodesy, class notes. The Ohio State University, Columbus, Ohio

Nijkamp P, Soeteman F (1990) Ecologically sustainable economic development: Key issues for strategic environmental management. Serie research memoranda. Free University, Amsterdam

Nijkamp P, Scholten HJ (1991) Information systems: Caveats in design and use. In: Harts J, Ottens HFL, Scholten HJ (eds) EGIS '91, Proceedings of the Second European Conference on GIS, EGIS Foundation, pp 735–746

Openshaw S (1990) Spatial analysis and geographical information systems: A review of progress and possibilities. In: Scholten HJ, Stillwell JCH (eds) Geographical information systems for urban and regional planning. Kluwer Academic Publishers, pp 153–163

Openshaw S, Wymer C, Cross A (1991) Using neural nets to solve some hard analysis problems in GIS. In: Harts J, Ottens HFL, Scholten HJ (eds) EGIS '91, Proceedings of the Second European Conference on GIS. EGIS Foundation, pp 797–807

Opschoor JB (1991a) Economic instruments for sustainable development, in proceedings: sustainable development, science and policy. Conference report, Bergen, 8–12 May 1990, Norwegian Research Council for Science and the Humanities, Oslo 1990, pp 249–269

Opschoor JB (1991b) GNP and sustainable frame measures. In: Kuik O, Verbruggen M (eds) In search of sustainable development indicators. Kluwer Academic Publishers, Dordrecht, pp 39–45

Opschoor JB (1991c) Economic modelling and sustainable development. In: Gilbert AJ, Braat LC (eds) Modelling for population and sustainable development. Routledge and the International Social Science Council, New York, pp 191–210

Opschoor JB, Reijnders L (1991) Towards sustainable development indicators. In: Kuik O, Verbruggen M (eds) In search of sustainable development indicators. Kluwer Academic Publishers, Dordrecht, pp 7–29

Pelt van M, Kuyvenhoven A, Nijkamp P (1990) Project appraisal and sustainability. Project Appraisal 5:139–158

Press WH, Flannery BP, Teukolsky SA, Vetterling WT (1986) Numerical recipes: The art of scientific computing. Cambridge University Press, Cambridge

Richmond M, Peterson S, Vescuso P (1987) An academic users guide to STELLA. High Performance Systems Inc, Lyme, New Hampshire

Rokos D (1989) The dialectic character of development. A scientific tool for its approach (revised form of the paper in the A'Congress for the Inter-Scientific Approach of Development, National Technical University of Athens, 1988) Scientific Thought, vol 44, pp 33–41

Scheid F (1968) Theory and problems of numerical analysis. McGraw-Hill

Scholten HJ, Stillwell JCH (1990) Geographical information systems: the emerging requirements. In: Scholten HJ, Stillwell JCH (eds) Geographical information systems for urban and regional planning. Kluwer, Dordrecht, pp 3–14

Spatial Analysis System (SPANS) (1988) User's manual. Tydac Technologies Inc

Vries de HJM (1989) Sustainable resource use: an enquiry into modelling and planning. Ph D Dissertation, Universiteitsdrukkerij Groningen

Werner C (1988) Spatial transportation modeling. Scientific geographic series, vol 4. Sage, Beverly Hills

World Commission on Environment and Development (1987) Our common future. Oxford University Press, Oxford and New York

Wyman FP (1970) Simulation modelling: A guide to using SIMSCRIPT. Wiley, London

# Modelling catchment areas

## Towards the development of spatial decision support systems for facility location problems

**Michel Grothe and Henk J. Scholten**

Department of Economics, Free University, De Boelelaan 1105, NL-1081 HV Amsterdam, The Netherlands

**Abstract.** In this contribution GIS is used as a tool for spatial assignment problems in the field of facility location planning. The integration of spatial data, spatial analysis and modelling capabilities and evaluation models into computer-based Decision Support Systems is considered. Some examples illustrate the potential benefits of GIS technology as a decision support tool.

## 1. Introduction

Recently the service sector has been facing drastic changes. Various (international) developments in demography, socio-economic conditions and technology have had strong impacts on the activities of public and private service facilities. In the financial sector in the Netherlands, for instance, large banking institutions and insurance companies are merging, which has had an enormous impact on the number, characteristics and capacity of consumer-oriented facility outlets. However not only in the private service sector, but also in the public service sector these developments have important consequences for the future.

These dynamic developments in the public and private service sector need a constant monitoring in order to optimise the relationship between demand and supply. In this context the geographic location is of crucial importance, as demand and supply differ in space. Interaction in space is necessary to provide demand with the services needed. Although sometimes supply is moving because of mobile services, most of the time demand is moving. In facility location assessment and optimisation, the interaction between demand and supply is monitored. Facility location optimisation emphasises the search for the optimal location: considering the constraints and determining the 'best available' location. Facility location assessment concerns the evaluation of an existing service facility. Monitoring refers to the performance over a period of time by evaluating the performance of supply in relation to the location of demand, the location of competing suppliers and available transportation. Assessment research can then be seen as an extension of optimisation research.

Facility location assessment and optimisation monitors the performance of a single facility or a network of facilities and optimises the location pattern. For many decades already facility location assessment and optimisation is the subject matter of several disciplines. Not only spatial planners, geographers and regional scientists but also operation researchers and management scientists are handling facility location problems.

Information Technology nowadays offers new possibilities for tackling facility location problems. Increasing availability of data, increasing capabilities of computer systems, the development of computer-based systems for decision support and even the development of Artificial Intelligence in Information Systems Technology offers new perspectives for decision problems to be handled. For spatially related decision problems, Geographical Information Systems (GIS) have been developed. This field of Information Systems Technology handles explicitly spatial data. For facility location planners and decision-makers GIS technology can offer new possibilities and opportunities for assessment and optimisation of facility location problems. By integrating the GIS technology with computer-based Decision Support Systems (DSS) a framework for spatial decision support can be established.

In this contribution an inventory will be made of the role of catchment areas in solving facility location assessment and optimisation problems in urban areas. Methods for modelling catchment areas using a demand-oriented approach will be examined and included in a design for Facility Location Assessment and Optimisation Systems (FLAOS) that support decisions associated with facility location assessment and optimisation problems.

The overall objective is to illustrate that Spatial Decision Support Systems Technology can be used effectively to support decisions on facility location assessment and optimisation problems.

The following research activities and issues have to be addressed to meet this overall objective:

- An overview of facility location theory and the identification of a conceptual framework for catchment area definition.
- Development and implementation of instruments aimed at modelling catchment areas to support facility location assessment and optimisation.
- Application of FLAOSs in facility location planning problems.
- Evaluation of the use of the concept of catchment areas to support decisions in facility location assessment and optimisation.

The above indicated fields of interest are integrated in a research framework (see Fig. 1) and will be explored in the next sections on the basis of a first examination.

## 2. Location theory and the concept of catchment areas

### 2.1 Introduction

The central theme in facility location theory is the search for an optimum pattern of supply locations. Most of the conceptual literature on location theory has fo-

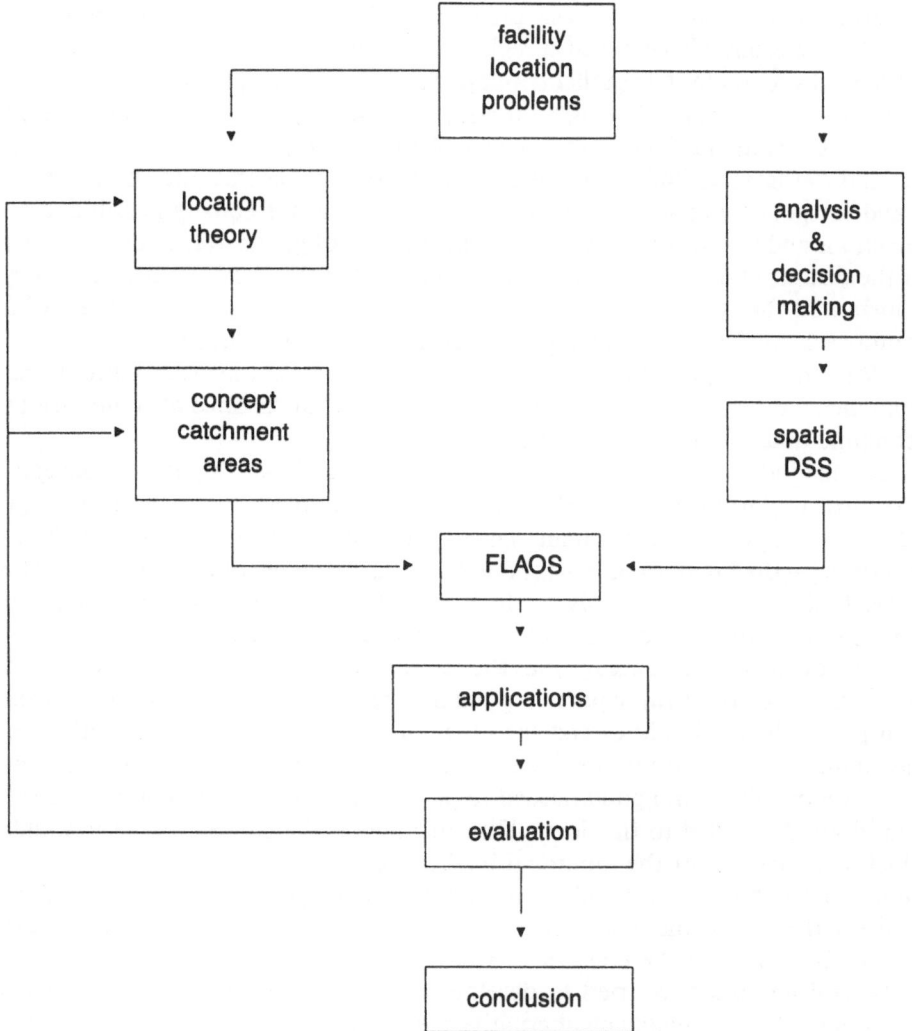

**Fig. 1.** Preliminary outline of research framework

cused on relationships between supply location and the surrounding catchment area. Relationships have been studied at different scales of inquiry – ranging from a full set of activities to a particular activity – and examined both from empirical viewpoints and theoretical abstractions. Theoretical work about catchment areas started with the Central Place Theory of Christaller (1933). Central Place Theory (and Systems) has been offering permanently a framework for an immense amount of literature on theoretical modifications and empirical applications.

## 2.2 Location theory of catchment areas

The most important aspect of Christaller's theory was the idea of a functional interdependence between a town and the surrounding rural area. This was by no means an innovation, but his basic premise that the main characteristic of a town is to be the economic centre of a region, was a new framework for the study in settlement geography. Christaller developed a model of a real world pattern of settlements based on plausible assumptions, the central place model. He assumed first of all a boundless and homogeneous plain. This plain was uniformly inhabited, and the farmers had everywhere the same level of income and the same demand for goods and services. Travel across the plain was equally possible in all directions and the costs of travel and transportation of goods were a function only of the distance traveled. Christaller assumed further that both farmers and consumers and the producers of goods and services were rational individuals who would seek to minimise their costs and to maximise their profits.

With these simplifications, Christaller described a central place pattern that contained a minimum number of urban places and that satisfied also the various economic and behavioural conditions.

For his inductive model he introduced and defined two important concepts. The most important idea was the range of a central place function. This range had both an upper and lower limit. The upper limit was the key concept in Christaller's hexagonal pattern of market areas and the hierarchy of central places. The upper limit was defined simply as the largest distance the dispersed population is willing to travel in order to buy a good offered at a place, a central place.

Christaller did acknowledge that the economics of the supply side also would affect the range. If a firm required a given number of consumers to reach a break-even point, then it would extend the range of business activity sufficiently into the countryside to capture this level of demand: the minimum amount of consumption of this central good needed to pay for the production or supply of the central good. Related to this lower limit of range is the concept of a threshold, which is a measure of the minimum level of demand needed to ensure that the supply of a good or service will be profitable. The range of a good or service establishes the size of the market area necessary to provide the economic support for the firm offering the good or service.

Several authors have tried to develop a more realistic central place theory. Rushton (1971) formulated an alternative theory based on more realistic consumer behaviour. He replaced the distance minimising assumption by the concept of subjective preference behaviour. Subjective preference structure is an ordered collection of spatial alternatives. This order is the result of a subjective valuation of travelling distance on the one hand, and the relative attractiveness of alternatives on the other hand. Rushton illustrated that with these preference structures the characteristics of Christaller's model would change.

Timmermans (1979) demonstrated both Christaller's distance minimising behaviour as Rushton's space preference structure as being not observed in reality. He conceptualised spatial demand behaviour in terms of multi-purpose trips and developed an alternative model in which the impact of multipurpose trips on the size and shape of catchment areas was included. The service area increasing impact of multipurpose trips was then implemented in a more realistic theoretical framework.

The classical and alternative theories have been subjected to empirical research, especially concerning spatial demand behaviour.

Golledge et al. (1966) found that demand travel behaviour was not always based on distance minimising behaviour. They tried to find some regularity in the spatial decision process of the consumer; the consumer would visit the city with the largest attractiveness.

Rushton (1969) regarded, in later empirical extensions, the consumer as a decision-maker, who is faced with the evaluation of a unique combination of spatial alternatives. This spatial decision process was determined by city size and distance, the so called 'revealed space preference'.

### 2.3 Towards a conceptual framework

Christaller introduced a theoretical basis for the conceptualisation of the concept of market area to develop a pattern of settlements. In this theory the size of a market area was determined by the range of a good or service. This relation between demand and supply was based on the distance minimizing and profit maximizing conditions. Later on, adjustments were needed at a theoretical and empirical level. Density and distribution of population in a region − as opposed to the hypothetical world of the theory − would affect the range. In the same way alternative consumers' behaviour, like multi-purpose trips, would influence the range of a service center.

Central place theory provides a major framework to describe and explain aspects of both the system of supply of goods and services and the spatial behaviour of demand. It is primary useful as a reference framework which needs to be expanded and adjusted to the particular situation (see Fig. 2).

The classical models had a single, system-wide objective and assumed a stable environment in which demand visited a single facility. Modern models have multiple objectives and attempt to deal with multiple purpose consumer trips, complex demand and supply behaviour, and uncertainties in the environment.

This contribution emphasises the use of a demand-oriented approach: allocation of demand through the definition of catchment areas can provide effective

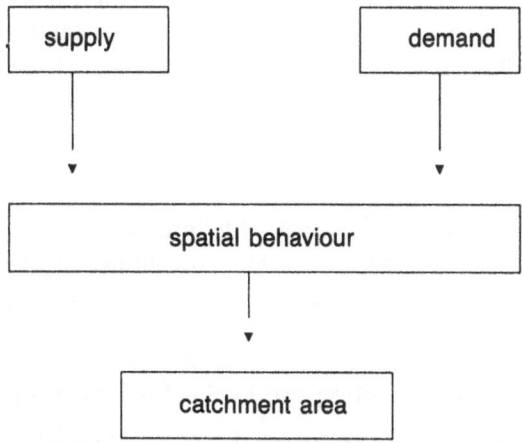

**Fig. 2.** Catchment area framework

possibilities for the spatial assessment and optimisation of public and private service facility location. Demand behaviour in space is modelled on the basis of modifications of and alternatives to the classical central place theory of Christaller.

## 3. Conventional framework for analysing facility location problems

### 3.1 Introduction

Several authors have tried to develop a framework or classification of the facility location problem. Davies and Rogers (1984) made a general distinction between store assessment problems and store location problems. Store assessment research deals with the evaluation and monitoring of existing service facilities at a particular location, e.g. evaluating trading performance or estimating market share. Store location research deals with the search for an optimal location by the analysis of new markets, on sales forecasting of new stores. Both types of location problems rely on the same problem-solving methodologies, methods and techniques, the first one applied retrospectively and the latter one in a future context.

Analysing these two types of facility location problems can simply be seen as matching supply and demand relationships over space. There is a varied demand for different services at different locations and this demand can be satisfied by competing service facilities. Space constraints make that location research is necessary to support matching supply and demand.

A research methodology can support the problem-solving strategy. Such a research methodology should be a systematic process. In Fig. 3 a conventional approach to a solution of facility location problems is illustrated.

First the problem under investigation should be defined and related to the location strategy of the supplier; this will lead to operational concepts. Both strategy and concepts provide input data for analysis and modelling. The alternatives generated will then be evaluated before final decisions are made.

The conventional process is characterised by a serial process in which all components are treated by the analyst who provides the solution to the decision-maker. However, the complexity of many location problems will ask for possibilities for feedback to minimise risk and failure. The various components in this conventional methodology will also be examined.

### 3.2 Location strategy

The objective of a location strategy is to minimise the risk of failure of a location decision. A location strategy essentially is a method for meeting predefined objectives concerning the location of facilities and should be a starting point for facility location assessment and optimisation. A location strategy is influenced by internal and external factors (Mercurio 1984), the nature of the facility itself and the marketplace in which it operates. These factors reflect two kinds of information:

● internal information, e.g. goals, objectives, operating and marketing policies, etc; and

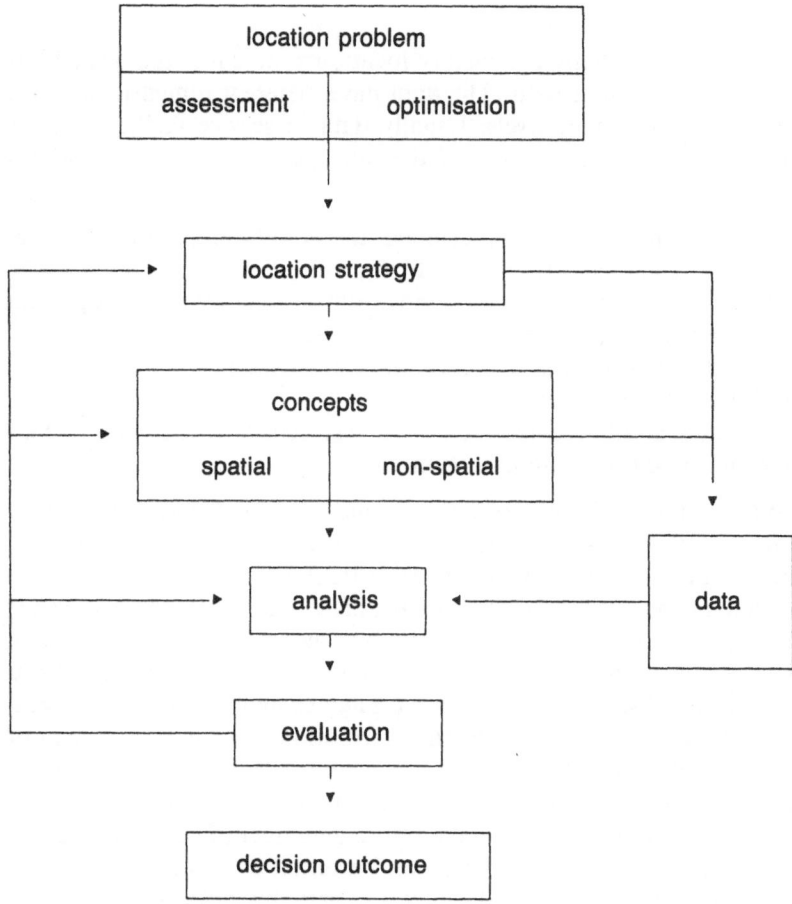

**Fig. 3.** Components in conventional facility location research

● external information about the competitive environment, consumer characteristics, physical environment, etc.

### 3.3 Spatial and non-spatial concepts

For facility location assessment and optimisation it is essential to quantify demand for a range of products and services. A basic requirement for this is information about the characteristics and behaviour of households. Based on these
characteristics the concept of catchment areas can be operationalised. The catchment area provides useful information about the existing socio-economic and demographic characteristics of the service environment. This information is usually
the basis for the assessment of performance by indicating productivity, determining market potential, estimating market shares, etc. It is not always possible and
necessary to develop catchment areas on the basis of information about individual households. Therefore other locational referencing levels are necessary, e.g. census tract, and postal reference.

## 3.4 Data environment

One of the key issues in solving the facility location problem is the availability and quality of necessary input data. The data have different dimensions and is spatially referenced at different levels. Usually data in service facility location planning employed concern the characteristics and location of demand, supply and transportation:

- data at census tract level concerning demand, e.g. population, average income, number of income earners per household;
- market research data, e.g. consumer demography, expenditures and opinions about competing stores (geo-referenced, for instance, by postal code);
- data on competitors, e.g. sizes of stores, parking spaces;
- transportation data, e.g. street networks;
- geographical data about the research area, e.g. census tracts, postal code boundaries and topographic information etc.

As mentioned above the availability of data is crucial. In the Netherlands census data are lacking. However, more or less the same kind of data is available in the form of CBS (Centraal Bureau voor de Statistiek) neighbourhood districts. Market research data collected internally by customer spotting surveys, the family credit card, etc. are also externally available through data service agencies even at a less aggregate level, the 6 digit postal code (ca. 18 households). Data on competitors, especially the retail sector, have recently (in aggregated form) become available through the development of the Dutch Distribution Information System or DIS (Nijkamp and Scholten 1990). Also for the acquisition of geographic information, data service agencies provide the necessary data.

It appears to be necessary to draw on data from a number of different sources. The integration of the existing internal data and external data sources necessary for analysis and modelling will take place by a geographic reference.

## 3.5 Analysis and modelling environment

The methods and techniques used in facility location assessment and optimisation are diverse and should, depending on the nature of the problem and the availability of appropriate data, be combined and related in order to reach an optimal result. Breheny (1988) gives, in relation to a classification of problems, a review of practical methods and techniques in the field of facility location analysis. Two types of techniques have been adopted: store-turnover forecasting techniques and spatial marketing techniques. The forecasting techniques are intended to forecast store performance from knowledge of existing outlets in the same chain, usually called 'analogue' techniques. An analogy is drawn between the performance of an existing and a proposed facility location. All these techniques treat distance explicitly. Multiple regression and spatial interaction models are examples of these spatial forecasting techniques.

Spatial marketing techniques have their origins in the marketing field. The field of geo-demographics is one such an application area in which statistical analysis techniques classify geographical areas according to their dominant socio-economic characteristics.

There is a large variety in techniques which can be applied in facility location assessment and optimisation, each being appropriate in different circumstances, depending on the nature of the problem, the location strategy, the concepts used and the appropriateness of data. A guideline for an optimal approach to apply the different techniques is not available.

### 3.6 Evaluation

Facility location problems are complex and some of their aspects cannot be measured or modelled. Often solutions can only be found by generating a set of alternatives and by selecting from those a solution that appears to be viable. The alternatives generated should be evaluated, while feedback to the primary components in the problem-solving process has to be made. The ultimate objective is to provide the decision-maker with useful and relevant information in order to achieve a effective decision outcome.

### 4. Facility location assessment and optimisation systems

### 4.1 Introduction

The role of the computer in transforming raw data into useful information has become necessary. For the tranformation of geo-referenced data, Geographical Information Systems offer possibilities for storing, linking, analysing and displaying different types of data. By integrating GIS Technology and computer-based Decision Support Systems (DSS), a framework for spatial decision support can be established. In this context recently a shift towards the development of Decision Support Systems for locational planning has been advocated by several authors (Densham and Rushton 1988; Densham and Goodchild 1989; Armstrong and Densham 1990; Beaumont 1991). For the development of Spatial DSSs the experience in developing DSSs for business problems from the fields of Operational Research and Management Science offers a theoretical and operational framework. The development of SDSS however is still in its infancy and many issues are still to be addressed (Densham and Goodchild 1989).

The DSS literature contains sufficient theory and applications for the design, development and implementation of its spatial analogue. Many definitions of DSS emphasise the availability of certain characteristics that also cover the spatial domain. The major characteristics of DSS, and the spatial domain DSS, can be found in the rich literature about DSS (e.g. Spraque and Carlson 1982; House 1983). The DSS approach emphasises semi-structured and unstructured decisions, offers a solution procedure by using models as decision aids to generate and evaluate (multiple objective) alternatives, is easy to use and flexible to adapt to the evolving needs of the user. The problem, analysis and decision environment gives further guidelines for the definition of the necessary characteristics of a DSS.

### 4.2 FLAOS components

The essentials and power of the SDSS approach lies in the integration of different components which where before separately dealt with. In the SDSS architecture of Armstrong and Densham (1990) a SDSS consists of five components; a database management system, an analysis system or modelbase management system, a report and display generator and a graphical user interface (see Fig. 4). An expert system shell can be used to guide the decision-maker. Within this framework also the architecture for FLAOS can be found. Some preliminary comments concerning the components of FLAOS will be made. These comments are based on a previous inventory about modelling demand behaviour for defining catchment areas and designing research agendas on Spatial DSSs for locational analysis (by Densham and Goodchild 1989; Amstrong and Densham 1990 and Beaumont 1991).

### 4.3 Analysis system

For facility assessment, statistical and spatial analysis is applied to estimate productivity, profitability or some other type of performance indicator for (an activity of) a service facility at a certain location.

In this context the model-based approach as illustrated by Clarke (1990) offers tools for modelling performance indicators. Model-based analysis provides a framework for data transformation, integration, updating, forecasting and impact analysis (Clarke 1990).

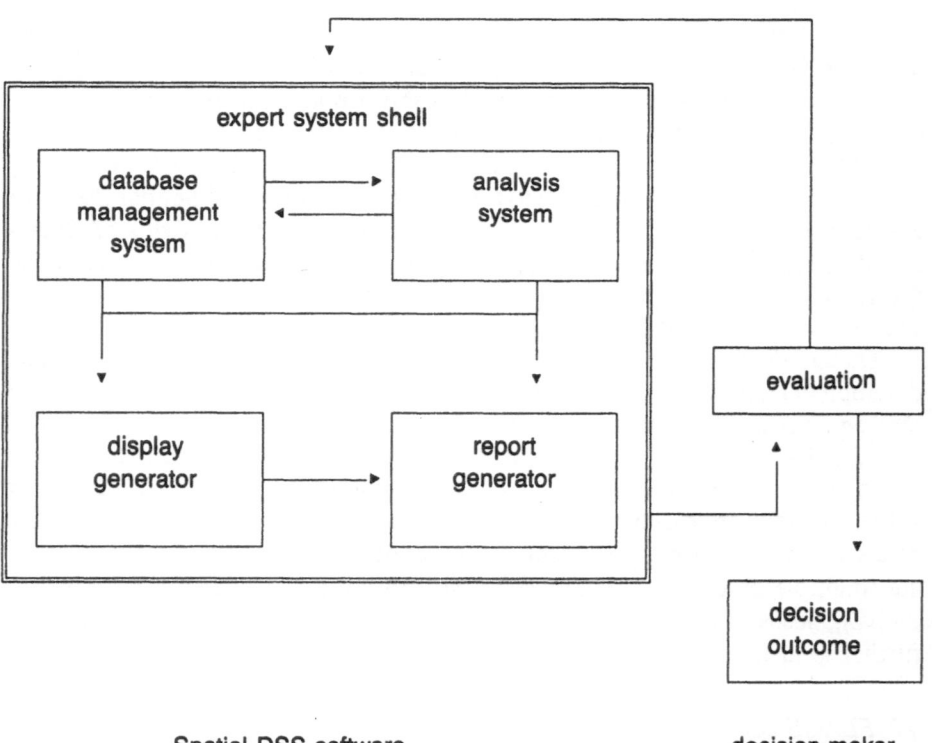

Spatial DSS software                                        decision-maker

**Fig. 4.** Components of spatial decision support systems (Armstrong and Densham 1990)

For service facility location modelling, the definition of catchment areas by modelling spatial demand choice behaviour is inevitable. A support system for modelling spatial choice behaviour offers possibilities for predicting the effects of demand behaviour on service activities (Borgers and Timmermans 1989). Spatial interaction and discrete choice modelling capabilities are examples of this type of models (Golledge and Timmermans 1988).

Techniques for spatial search – evolving from the field of Operations Research and known as location-allocation models – are useful for solving optimisation problems. However, the standard optimisation procedures from Operations Research cannot deal with value judgements in a multiple objective decision-making environment (Densham and Rushton 1989). Facility location problem-solving and decision support is complex and requires the use of spatial analysis models to tackle multiple criteria and objectives involved in locational decision-making. Locational alternatives generated by the analysis and modelling system can be evaluated by the decision-maker or expert using multiple criteria analysis and evaluation techniques (see also Janssen and Rietveld 1990; Carver 1991). The decision-maker generates and evaluates a set of alternative solutions by making value judgements in the analysis process. The expert participation leads to a structuring of the decision-making process and improves its quality. By evaluation the intensity and nature of differences between management criteria can be analysed and alternatives can be ranked according to the weights of location policy priorities. To support this evaluation process Multiobjective Decision Support Systems (MODSS) have been developed (e.g. by Janssen 1991). MODSS offers the decision-maker possibilities for alternatives and criteria generation, criteria weighting procedures, sensitivity analysis and alternative ranking. Graphical presentation of the results provide decision-makers with information in the most concise form: a map. This information is immediately accessible for decision-making. The integration of MODSS in a Spatial DSS environment provides the decision-maker with a more effective tool for multiple objective decision support concerning service facility location problems.

The analysis and modelling system of FLAOS offers possibilities for:

- specification, calibration and goodness-of-fit calculation of behavioural demand models;
- network analysis and spatial interaction models;
- multiple criteria evaluation;
- sensitivity analysis; and
- visual simulation.

The use of expert knowledge in the analysis system can be useful for choosing the appropriate type of models most suitable to the data used and for inferring conclusions from the evaluation (for instance, which is the most suitable alternative and for what uncertainty level will this result change).

## 4.4 Database management system

The core of a FLAOS is the database management system (DBMS). All other components draw upon the database to perform analysis, modelling and display.

In spatial data bases different types of data are represented; geometrical descriptions of spatial entities (represented by point, line and area objects) with topological relationships and attributes of varying spatial and temporal dimensions. The analysis and modelling capabilities of a Spatial DSS ask for a specific database design in order to require and obtain flexible interaction in the decision-making process.

The choice for a topological spatial data model has been discussed by Armstrong and Densham (1990). They examine a methodology for data base design of Spatial DSS and compare available data models on their appropriateness for their use in SDSS. Especially in this field much research still needs to be done.

For the analysis models in FLAOS (e.g., statistical and regression analysis, spatial interaction models and decision models using multiple evaluation techniques) network modelling using arc-node topology and spreadsheet modelling are widely used.

### 4.5 Graphical display and report generator

Spatial data presentation is associated with maps. Cartographic display of maps is undoubtedly increasing the explorative mode of the decision process. Location-based interactive simulation offers possibilities for model specification and calibration, but also for spatial search activities in interaction with the analysis system. For multiple objective decision support models the use of graphics (maps, graphs, bars) is improving the analysis process and can be a decision aid in itself (Janssen 1991). The report generator produces standard format output of the results in graphical, tabular and textual format.

### 4.6 Graphical user interface

One of the main characteristics of Spatial DSS is their ease of use. In this context the development of Graphical User Interfaces is a step forward to more effective interaction between the user and the system. The shift towards Graphical User Interfaces using windows, icons, mouse etc is increasing also in the field of GIS technology. But also new developments in Information Technology for designing user interfaces are dynamic and offer opportunities for a Spatial DSS:

- interactive multi-tasking;
- visual simulation;
- three-dimensional graphics;
- expert knowledge developments for help in selecting appropriate methods; and
- multi-media; graphics, video, voice.

### 4.7 Evaluation

The decision-maker has interaction with the processed data various possibilities to evaluate the outcomes and results by graphical presentation. After evaluation of the results two options exist. First, the decision-maker has gained sufficient information to take a decision; in this case the decision outcome is clear and the decision process is finished. However, often the decision-maker wants to have the possibilities to evaluate different alternatives; then simulation and optimisation procedures are possible in order to optimise the decision-making process and generate a decision. The decision-maker guides the system through a process which is at the same time a problem-solving and learning process.

## 5. Applications

In the past years various research projects in the field of FLAOS have been undertaken at the GIS Research Laboratory of the Free University Amsterdam. Two such application fields will briefly be outlined here.

### 5.1 Spatial interaction analysis for the banking sector

A first example is an application in the Dutch banking sector. At present the GIS Research Laboratory is − in corporation with Geodan (a Dutch GIS research agency) − developing a FLAOS for facility location problems in the Dutch banking sector.

The study area used here as an illustration is the municipality of Arnheim, a middle-large city (130.000 inhabitants) in the Netherlands. Almost 50 banks of 10 different origins are located in a more or less clustered pattern in this city. Data about the bank outlets consist of locational data by x-y coordinates and attribute data about the characteristics of the bank outlet. Furthermore, at a very detailed level information about the population characteristics in the municipality is available; data by 6 position postal code areas (meaning in practice almost 3200 point locations in the city) and 4 position postal code areas with information about incomes, house ownership, family status etc. This information is based on the small area classification data of Geomarktprofiel, a Dutch data service agency. The 6 digit postal code area consists of 12 to 18 households, the smallest available aggregated data level in the Netherlands. Finally, a detailed street network containing address ranges is available for geographical reference and distance calculations (see Fig. 5).

The core of this application framework is the spatial analysis and modelling capability in a GIS environment. As mentioned before, in the field of facility location analysis several models can be applied, e.g. potential models, spatial interaction models and location-allocation models. In this version of the system a choice has been made for the application of spatial interaction modelling techniques. In this environment different models can be run with possibilities to modify interactively the model parameters. An evaluation module, with possibilities for cost benefit and multiple criteria analysis, is available to evaluate alternative locations generated by the model input and the users' choice. With a graphical interface the user can perform interactive modelling by changing, deleting and adding facility locations, run the model and compare the results. A link with standard GIS software is available for several value added functions, like integration of data and final presentation in maps.

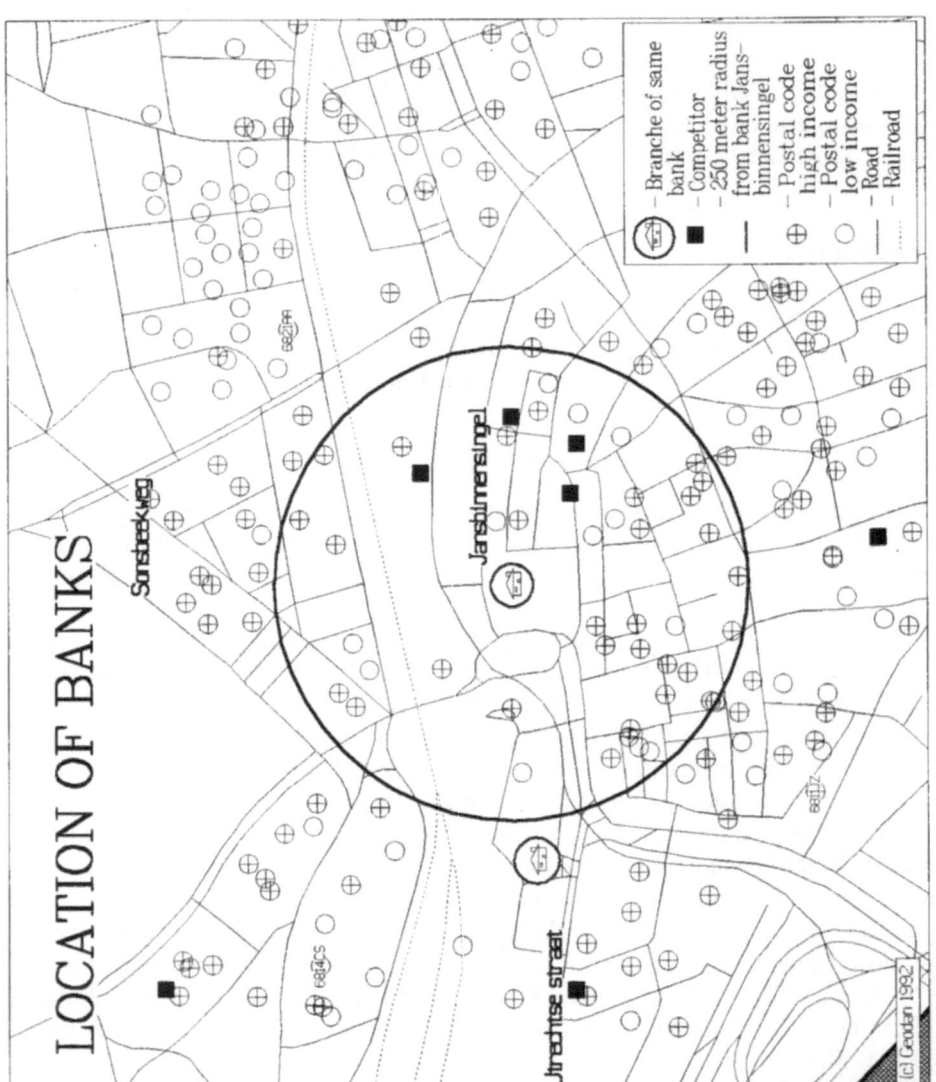

**Fig. 5.** A FLAOS for the Dutch banking sector

## 5.2 Residential zoning for elderly

The second example concerns a pilot study in a residential district in the Amsterdam urban area. In this application the FLAOS concept has been applied in the field of residential zoning for elderly. The study area has approximately 40,000 inhabitants with almost 5,000 elderly, a figure which is in comparison with the Netherlands more than the average. Elderly inhabitants appear to have in general a lower education and gain low incomes. Almost 80% of the house ownership in the study area is private which indicates a housing quality below average.

This study concerned was commissioned by the district council after a preliminary research has highlighted several bottlenecks in housing, service and care facilities for elderly in this area. In three stages of plan development the following goals have been defined:

(1) the identification of residential 'friendly' zones based on objective criteria;
(2) an inventory of actual behaviour and perception of elderly concerning housing, use of facilities, living environment; and
(3) the design of action strategies for an integral housing policy plan.

One of the objectives of this project was an investigation into the participation of individual elderly and local institutions in the planning process. Nowadays urban planning and urban management in the Netherlands is influenced by a new framework for policy development, called 'social renewal'. Social renewal as a policy development approach emphasises the participation of societal institutions and groups, appreciation of individual initiatives, and special attention for the small-scale characteristics of neighbourhoods and special population groups and areas. This innovation in the planning process of urban renewal and urban management is applied in this study to improve the housing situation of elderly; institutions associated with elderly policy as well as individual elderly have been invited to participate in the development of strategies concerning housing policy in their own residential environment.

In our case study we analyse in particular the access of elderly to retail facilities and care service centres. Accessibility of retail facilities and care services centres is measured along a transportation network. Further buffering and overlay operations are used to determine attractive and less attractive areas in the district. Several local participants have drawn their mental map of attractive and less attractive areas in the district. These results have been collected and put into single map layers. By buffering and overlay operations these layers have been combined into attractive and less attractive zones. The use of different types of data makes that a graphical presentation of the results is an important element in the policy preparing process. The communication between decision-makers, policy preparers and local participants in the policy development process has already been improved by the availability of maps (see Fig. 6). Next the map can be interpreted and evaluated, while also possibilities for feedback and feedforward are available. The zoning map gives an indication of potential sites for new developments in the housing, retail and care service sector. For solving the ultimate location problem (which potential site is most suitable regarding the objectives of the various participants) the various objectives of the participants have been translated into a set

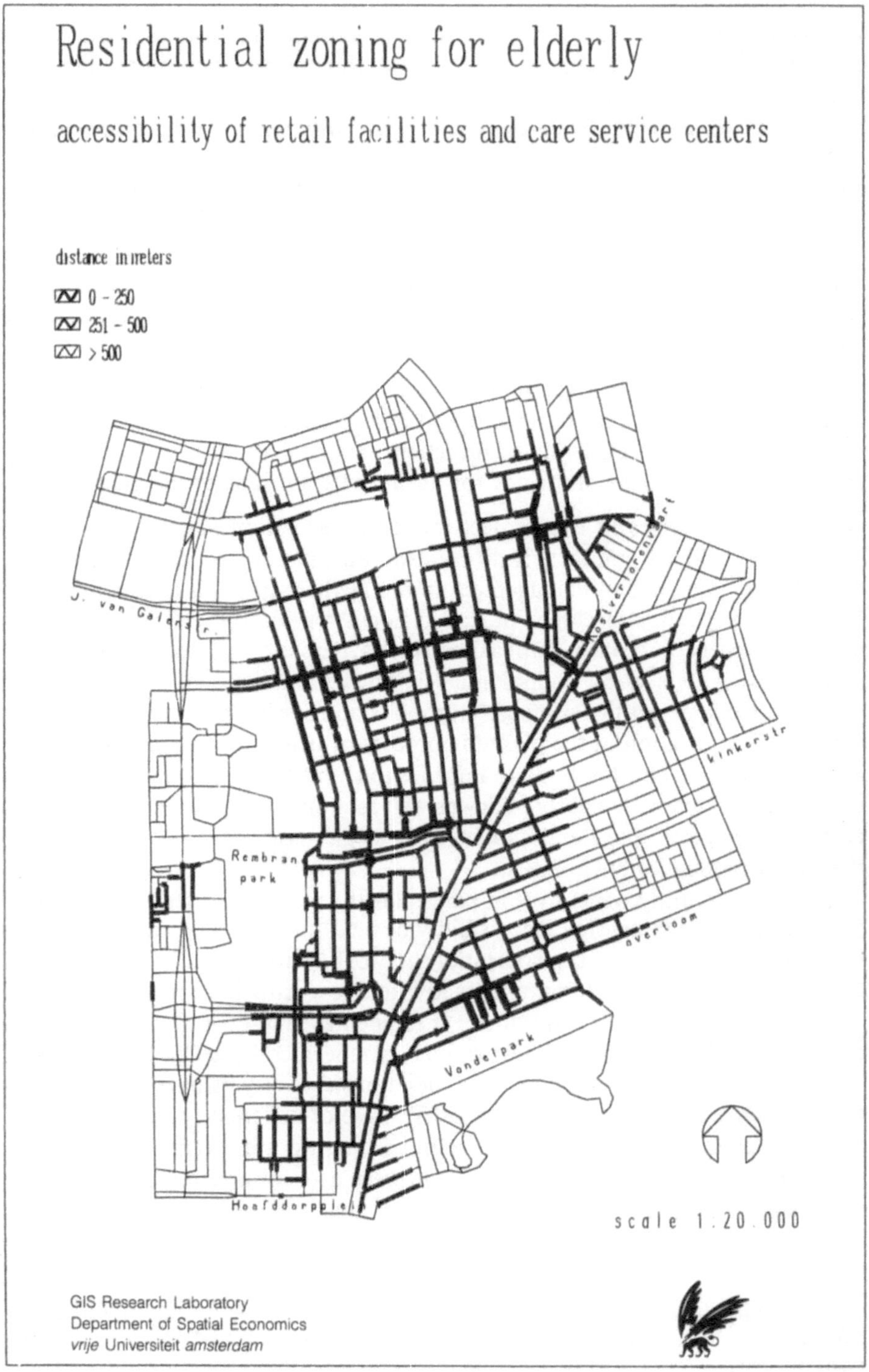

**Fig. 6.** Residential zoning for elderly

of judgement criteria. Then evaluation methods are used to gain insight into the preferences of the different participants. Finally a concensus will be generated about which site is most appropriate for new developments.

## 6. Concluding remarks

The growth of GIS in Europe at the moment exceeds all expectations. According to the analysis of a leading firm in this field, Dataquest Inc this development will within two years exceed the developments in America and in 1995 there will be a market of 1,619 million dollars in Europe alone. These are gigantic sums, but also great challenges. An important proportion of the money will, of course, be used for applications in management systems, such as the AM/FM applications in public utility services. The demand for analytical possibilities will, however, undoubtedly reach the same level, and this perspective offers a great challenge, as mentioned also in the Introduction.

The Facility Location Analysis and Optimisation System is a tool for researchers and analysts who have to provide decision-makers with strategic information. However, the combination of data integration, spatial analysis and modelling capabilities, evaluation methods and interactive graphical analysis makes this approach a step forward to an operational application for the decision-maker.

## References

Armstrong MP, Densham PJ (1990) Database organization strategies for spatial decision support systems. Int J Geogr Inf Syst 4:3–20

Beaumont JR (1991) Spatial decision support systems: some comments with regard to their use in market analysis. Environ Plann A 23:311–318

Borgers A, Timmermans H (1989) A decision support and expert system for retail planning. In: Yeh AGO, Fang S (eds) Proceedings of the International Conference on Computers in Urban Planning and Urban Management, Centre of Urban Studies and Urban Planning, University of Hong Kong, pp 341–352

Breheny MJ (1988) Practical methods of retail location analysis. In: Wrigley N (1988) Store choice, store location and market analysis. Routledge, New York, pp 39–86

Carver SJ (1991) Integrating multi-criteria evaluation with geographical information systems. Int J Geogr Inf Syst 5:321–340

Christaller W (1933) Die zentralen Orte in Süd-Deutschland. Fischer, Jena

Clarke M (1990) Geographical information systems and model based analysis. In: Scholten HJ, Stillwell J (eds) Geographical information systems for urban and regional planning. Kluwer, Dordrecht, pp 165–175

Davies RL, Rogers DS (1984) Store location and store assessment research. Wrigley, London

Densham PJ, Goodchild MF (1989) Spatial decision support systems: a research agenda. Proceedings of GIS/LIS '89, vol 2, Orlando, Florida, pp 707–716

Densham PJ, Rushton G (1988) Decision support systems for locational planning. In: Colledge R, Timmermans H (eds) Behavioural modelling in geography and planning. Croom-Helm, London, pp 65–90

Golledge RG, Timmermans H (eds) (1988) Behavioural modelling in geography and planning. Croom-Helm, London

Golledge RG, Rushton G, Clarke WAV (1966) Some spatial characteristics of Iowa's dispersed farm population and their implictions for the grouping of central place functions. Econ Geogr 42:261–272

Heyden van der RECM (1986) A decision support system for the planning of retail facilities: Theory, Methodology and Application. Dissertation, University of Technology, Eindhoven

House WC (ed) (1983) Decision support systems. Petrocelli, New York

Janssen R (1991) Multiobjective decision support for environmental problems. Dissertation, Free University, Amsterdam

Janssen R, Rietveld P (1990) Multicriteria analysis and GIS: An application to agricultural landuse in the Netherlands. In: Scholten HJ, Stillwell J (eds) Geographical information systems for urban and regional planning. Kluwer, Dordrecht, pp 129–139

Mercurio J (1984) Store location strategies. In: Davies RL, Rogers DS (eds) Store location and store assessment research. Wiley, Chichester, pp 237–262

Nijkamp P, Scholten HJ (1990) New information systems: the use of retail information systems in the Netherlands. Neth J Hous Environ Res 5:209–224

Rushton G (1969) Analysis of spatial behaviour by revealed space preference. Ann Assoc Am Geogr 59:391–400

Rushton G (1971) Postulates of central place theory and the properties of central place systems. Geogr Anal 3:140–157

Spraque RH, Carlson ED (1982) Building effective decision support systems. Prentice Hall, Englewood Cliffs

Timmermans HJP (1979) Centrale plaatsen theorieen en ruimtelijk koopgedrag, een theoretische verkenning en een aanzet voor de formulering en toetsing van een alternatieve theorie. Ergon bedrijven, Eindhoven